Battery Management Algorithm for Electric Vehicles

Rui Xiong

Battery Management Algorithm for Electric Vehicles

机械工业出版社
CHINA MACHINE PRESS

Springer

Rui Xiong
National Engineering Laboratory for Electric
Vehicles, School of Mechanical Engineering
Beijing Institute of Technology
Beijing, China

ISBN 978-981-15-0247-7 ISBN 978-981-15-0248-4 (eBook)
https://doi.org/10.1007/978-981-15-0248-4

Jointly published with China Machine Press
The print edition is not for sale in China. Customers from China please order the print book from: China Machine Press.
ISBN: 978-7-111-60864-6

This Springer imprint is published by the registered company Springer Nature Singapore Pte Ltd.
The registered company address is: 152 Beach Road, #21-01/04 Gateway East, Singapore 189721, Singapore

Foreword

The promotion of new energy vehicles represented by electric vehicles, fuel cell vehicles, and plug-in hybrid vehicles is a worldwide strategy to alleviate climate change and enhance energy security in many countries. In China, vigorously developing the electric vehicle industry has become a strategic development direction for governments, institutes, and automobile enterprises. By the end of 2018, China had been the world's largest producer of electric vehicles in terms of sales volume for four consecutive years. An unprecedented revolution in the automotive industry is happening in China and around the world.

The battery system, which consists of a combination of series and parallel connections of battery cells, can be the ideal power source for the electric vehicle. For the past few years, lithium-ion battery cells have been the most promising energy storage candidates for electric vehicles due to their excellent performances in high energy density, high power density, high energy efficiency, and long cycle life. However, they exhibit strong coupling effects among electric, thermal, and mechanical behaviors in electric vehicle applications, resulting in strong time-varying, ambient temperature- dependent, and nonlinear characteristics. As a complex electrochemical system, there are only a few parameters, including battery terminal voltage, charge or discharge current, and surface temperature that can be measured or monitored from battery cells. In addition, the performance of battery system in an electric vehicle is strongly correlated to a range of factors including ambient temperature, aging status, depth of discharge, dynamic and complicated operation conditions, charge and discharge current rate, and cell inconsistency in the battery pack. These difficulties and challenges have restricted the development of electric vehicles. Therefore, more research into advanced battery technologies is crucial for the safe and efficient operation of electric vehicles.

The book covers the research results of Prof. Rui Xiong in battery management techniques for electric vehicles. This book starts with plans for new energy vehicle development in China as well as around the world and an overview of battery management system (BMS) technologies. It continues by systematically describing methods of battery experiments and basic theories of battery modeling including electrochemical models, equivalent circuit models, and fractional-order models.

After that, it presents detailed algorithm theories and implementation processes of battery states estimation, remaining useful life prediction, low-temperature heating strategies, and optimized charging methods. Finally, the book ends with the research and development process of the battery management system algorithms. Thus, this book covers the necessary background and techniques for the development of core algorithms in the battery management system for electric vehicles.

Professor Xiong has been engaged in extensive research and development on electric vehicles and hybrid electric vehicles, energy storages, and BMS for more than one decade. After receiving his Ph.D. degree in 2014, he started research work as an associate professor in the National Engineering Laboratory for Electric Vehicles, School of Mechanical Engineering, Beijing Institute of Technology, Beijing, China. Since 2019, he was promoted to professor at Beijing Institute of Technology. He has also been an Adjunct Professor at the Faculty of Science, Engineering and Technology, Swinburne University of Technology, Melbourne, Australia since 2017. Over the years, he has been carrying out systematical and in-depth investigations in cutting-edge electric vehicle technologies on battery testing, modeling, active controlling, reliability, durability and safety, all-climate batteries and battery system integration, management and application, yielding outstanding achievements. This book is the concentrated representation of his persistent efforts and experiences in the development of core algorithms in battery management system for electric vehicles.

I highly recommend this book to people who are working in the field of battery management and controlling represented by scholars and researchers, senior undergraduate, and postgraduate students as well as engineers and technicians. First, this book has solid technical content and can provide in-depth theory source and comprehensive reference for those people. Second, it is the quintessence of Prof. Xiong who is the world's leading expert in this field. He has made excellent contributions to the research and development of core algorithms for the battery management system and played a unique role in promoting advanced and intelligent battery management systems for all-climate electric vehicles. In summary, this book is a required material for people who are going to enter the electric vehicle industry and want to acquire core algorithms and relevant technologies of the battery management system for electric vehicles.

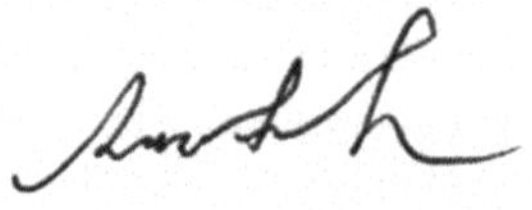

Fengchun Sun
Professor and Academician
Beijing Institute of Technology
Beijing, China

Preface

The development of energy-saving and new energy vehicles is an international consensus and the China's strategic emerging industry, as well as a key development direction for "Made in China 2025". According to the statistics of China Association of Automobile Manufacturers, the annual production and sales volume of new energy vehicles in China reached 1.27 million and 1.256 million in 2018. By the end of 2018, the global sales of new energy vehicles exceeded 5.5 million, and China accounted for more than 53%. China makes significant contributions to reducing energy consumption and carbon emission. Obviously, the new energy vehicle industry has ushered in unprecedented development opportunities, and at the same time led to the rapid development of the lithium-ion battery industry.

The battery system is the technical bottleneck of new energy vehicles, and the battery management technology is the core and key to ensure the high efficiency, safety, and long life of the battery system. Batteries have the characteristics of limited measurable parameters and characteristic coupling, degradation with cycling, strong time-varying, strong ambient temperature dependence, and strong nonlinearity. Therefore, battery state estimation with high precision and strong robustness is the fundamental way to actively manage the battery system. Improving the dynamic modeling accuracy of the battery system, breaking the technical bottleneck of multistate collaborative estimation, strengthening the remaining life prediction and durability management, and ensuring the realization of the expected service life becomes an important part of the battery management system. They are the key to ensure the vitality and sustainable development of new energy vehicle market and also a basic requirement for building confidence to the general public in the new energy vehicle.

This book combines the author's research practice for more than 10 years, and elaborates the technical details of the core algorithm development of the new energy vehicle battery management system. Chapter 1 analyzes the new energy vehicle development plan and the technical indicators of the battery management system in "The 13th Five-Year Plan" of China, and systematically expounds the key points of the design and implementation of the battery management system. Chapter 2 describes the construction of the battery test platform, the experimental method design and the characteristic analysis, and then systematically analyzes the working characteristics of the battery under different conditions of aging, temperature, and charge/discharge current rates, which provides the direction for the development of the core algorithm of the battery management system. Chapters 3–7 systematically discuss the theory basis and construction and implementation details of core algorithms, which include the battery system modeling, state of charge and state of health collaborative estimation, peak power prediction, remaining useful life prediction, low-temperature rapid heating, and optimized charging methods. Finally, the "V" development process of the battery management system algorithm is discussed from the perspective of hardware and software simulation, bench test, and real vehicle verification of the core algorithm. The author strives to make the text accurate and refined, with clear illustrations and make the content systematic, detailed, and advanced, and strives to help readers master the core algorithms of the new energy vehicle battery management system.

This book is under the guidance of my supervisors, Academician Prof. Dr. Fengchun Sun and Prof. Dr. Hongwen He. It is the crystallization of the research and developments of the National Engineering Laboratory for Electric Vehicles in the field of new energy vehicle battery and its management and application. I would like to express my heartfelt thanks to the people who participated in the compilation of this book including Associate Prof. Dr. Quanqing Yu, Postdoc. Yongzhi Zhang, Zeyu Ma, Ph.D. student Cheng Chen, Shanshan Guo, Jiahuan Lu, Jinpeng Tian, Ju Wang, Ruixin Yang, Xinggang Li, Yanzhou Duan, Master student Linlin Li, Zhirun Li, Liang Lv, Kan Wang, and so on.

After years of preparation, we are making efforts to dedicate the latest research progress in the field and the research results and experience in the battery management systems to colleagues and readers, to help innovation and progress in this field, and to promote the technological development. Although the book has been revised many times, it is still not good enough. The main reason is that some work is still not finished and some theories and technologies are still being carried. I hope the readers can understand it. We welcome comments and critical comments to jointly promote the rapid development of research and development of worldwide battery management systems.

Beijing, China Prof. Dr. Rui Xiong

Contents

Abbreviations

AC	Alter current
AEKF	Adaptive extended Kalman filter
AIC	Akaike information criterion
BCT	Box–Cox transformation-based prediction method
BCU	Battery control unit
BEV	Battery electric vehicle
BMC	Battery monitoring circuit
BMS	Battery management system
CC	Constant current
CCCV	Constant-current constant-voltage
CPE	Constant phase element
CTCDC	China typical city driving cycle
CV	Constant voltage
DC	Direct current
DOD	Depth of discharge
DP	Dual polarization
DSP	Digital signal process
DST	Dynamic stress test
DV	Differential voltage
DVA	Differential voltage analysis
e	Electrons
ECM	Equivalent circuit model
EIS	Electrochemical impedance spectroscopy
EKF	Extended Kalman filter
FCEV	Fuel cell electric vehicle
FIT	Fitting-based prediction method
FUDS	Federal urban driving schedule
GA	Genetic algorithm
G-L	Grünwald–Letnikov
HEV	Hybrid electric vehicle

HIF	H_∞ filter
HIL	Hardware in the loop
HPPC	Hybrid pulse power characteristic
IC	Incremental capacity
ICA	Incremental capacity analysis
IDE	Integrated development environment
KF	Kalman filter
LCO	Lithium cobalt oxide battery
LFP	Lithium iron phosphate battery
Li^+	Lithium ions
$Li_4Ti_5O_{12}$	Lithium titanium oxide
$LiCoO_2$	Lithium cobalt oxide
$LiFePO_4$	Lithium iron phosphate
$LiMn_2O_4$	Lithium manganese oxide
LMO	Lithium manganese acid battery
LS	Least square
LSTM RNN	Long short-term memory recurrent neural network
MAAB	Mathworks automotive advisory board
MAE	Mean absolute error
MAEFK	Multi-timescale adaptive extended Kalman filter
MC	Monte carlo
MCD	The multi-constrained dynamic method
MHIF	Multi-timescale H_∞ filter
mPnS	m parallel n series connected battery pack or module
NCA	Nickel cobalt aluminum ternary lithium-ion battery
NEDC	New European driving cycle
NMC	Nickel–cobalt–manganese ternary lithium-ion battery
OCV	Open-circuit voltage
P2D	Pseudo-Two-Dimensional
PHEV	Plug-in hybrid electric vehicle
PNGV	Partnership for a New Generation of Vehicles
PTC	Positive temperature coefficient
RCP	Rapid control prototype
R-L	Riemann–Liouville
RLS	Recursive least squares
RMS	Root mean square
RMSE	Root mean square error
RUL	Remaining useful life
SampEn	Sample entropy
SEI	Solid-electrolyte interphase
SIL	Software in the loop
SimRNN	Simple recurrent neural network
SOC	State of charge
SOE	State of energy
SOH	State of health

SOP	State of power
SPSS	Statistical product and service solutions
TLC	Target language compiler
TLS	Total least squares
UDDS	Urban dynamometer driving schedule
VCU	Vehicle control unit

Chapter 1
Overview of Battery and Its Management

Developing energy-saving and new energy vehicles industry is an international consensus, which is also an emerging industry in China and a key filed established by "Made in China 2025". The battery technology is the bottleneck of the development of electric vehicles (EVs), while the battery management system (BMS) is not only the key to ensure efficiency and safety of the EVs, but also the focus of national core technology competition. In 2017, the medium and long-term development plan for the vehicle industry was jointly issued by the Ministry of Industry and Information Technology, the National Development and Reform Commission, and the Ministry of Science and Technology of the People's Republic of China. They pointed out that the annual production and sales of new energy vehicles in China would reach 2 million, the specific energy of cells would exceed 300 Wh/kg, striving to reach 350 Wh/kg; the specific energy of the system would strive to reach 260 Wh/kg, and the cost would be reduced to less than 1 yuan/Wh by 2020. The annual sales of global new energy passenger vehicles have reached 1.22 million by 2017 with a growth rate of over 55%. According to the statistics of China Association of Automobile Manufacturers, China's annual production and sales of new energy vehicles in 2017 reached 794,000 and 777,000, respectively, ranking first in the world for 3 consecutive years. From January to May 2018, the production and sales of new energy vehicles in China both reached 328,000 with an increased rate of 122.9 and 141.6%, which shows that China is becoming an important role in the development of the international new energy vehicle market.

However, as an alternative for fuel, the battery has the characteristics of low cycle life, slow charging, poor low-temperature discharging capability, and difficult to predict the degradation state, which causes the users' anxiety about driving range and concerns about performance at low temperatures. When the cells used in groups on the EVs, the degradation rate and trajectory of the cells are different and the inconsistency between the cells is particularly obvious due to the differences in the manufacturing and operating environment, which makes it hard to estimate the accurate state of the battery system. When the state of health (SOH) of some batteries degrades sharply, the performance of the whole system will be affected, which will aggravate lifetime decay and even cause safety problems. As a result, it is important

R. Xiong, *Battery Management Algorithm for Electric Vehicles*,
https://doi.org/10.1007/978-981-15-0248-4_1

for BMSs to strengthen the battery energy management, reduce heating time at low temperature, improve remaining useful life (RUL) prediction accuracy, slow performance degradation, and guarantee the expected service lifetime. They are not only the keys to guarantee the vitality and sustainable development of the new energy vehicles market, but also the basic requirements to establish the confidence of the market.

1.1 Developing Plan for the New Energy Vehicle in China

The traditional vehicles driven by internal-combustion engine system, which have dominated the vehicle industry for a long time, are facing a huge transition. In order to seize the opportunity of technology and market, countries all over the world compete to develop technologies of EVs, and actively promote the investment and expansion of the new energy vehicle industry.

As early as the 1960s, China began to trial-produce EVs, due to the limitations of technical conditions and social environment, no breakthroughs were made in technology at that time. It was not until the "overall design of EVs" for National Key Technologies R&D Program of China during the 8th Five-Year Plan Period was approved in 1992 that the R&D of China's EV industry was officially started [1].

During the 10th Five-Year Plan period, China began large-scale organized R&D of EV technology. The layout of "three vertical and three horizontal" was established by the Electric Vehicle Key Project under the 863 Program in 2011. The vehicles including fuel-cell electric vehicles (FCEVs), hybrid electric vehicles (HEVs), and EVs are named as "three vertical", and three key technologies including multi-energy powertrain control system, driven motor, and battery are named as "three horizontal". China launched extensive studies on EV technology, laying a technical foundation for the development of EVs.

During the 11th Five-Year Plan period, China organized the Energy Savings and New Energy Vehicle Key Project, sticking to the layout of "three vertical and three horizontal". China carried out research on key technologies and large-scale industrialization of the electric vehicles, and successfully carried out the demonstration projects such as Beijing Olympics, Shanghai World Expo, Shenzhen Universiade, the ten thousands of new energy vehicles and so on. It is worth mentioning that the Rules on the Production Admission Administration of New Energy Automobiles, published by the National Development and Reform Commission in China, officially defined the new energy vehicles for the first time. The new energy vehicles refer to the vehicles that use the unconventional vehicle fuel as an energy source (or use the conventional vehicle fuel with new vehicle energy device) and integrate the advanced technologies of vehicle energy control and drive. According to the definition, the new energy vehicles can be divided into HEV, battery electric vehicle (BEV), FCEV, hydrogen internal-combustion engine vehicles, and other vehicles with new fuels such as dimethyl ether and high-efficient energy storage capacitors, which indicates that China has begun to implement management and access systems for new energy vehicles and their production enterprises.

The key component technology, the vehicle integration technology, and the public platform technology should be improved, deepened, and upgraded, forming the strategic focus of "three vertical, three horizontal and three platforms". It is pointed out in the 12th Five-Year Plan that breaking through the bottleneck of battery technology is the main task, but the three key technologies including the battery, the electric motor, and the electric control system were also needed to be broken through [2, 3]. In 2015, the new energy vehicles were included in "Made in China 2025", and the target that the battery energy density would reach 350 Wh/kg by 2026 was proposed. In the same year, the emission reduction commitment made by President Xi Jinping at the Climate Conference in Paris was reflected in the 13th Five-Year Plan.

In November 2016, the State Council in China officially issued the National Strategic Emerging Industries Development Plan for the 13th Five-Year Plan, where the strategic position of green and low-carbon industries such as the new energy vehicles, the new energy, and the energy conservation and environmental protection had been clarified once again. As one of the eight major tasks in the 13th Five-Year Plan, the development of the new energy vehicles is the priority of the State Council. The Plan calls for a large-scale application of the new energy vehicles and the overall improvement of the quality and performance of the EVs. Moreover, it requires promoting the R&D of the battery technology, breaking through the technology of battery grouping and system integration, and boosting the cascade utilization of battery. It also emphasizes that the R&D of the battery system should be improved, and the technical targets of the lithium-ion batteries such as high safety, long lifetime, and high-energy density should be realized. It can be seen that China is constantly making efforts in the application and development of the batteries in the EVs, striving to improve the quality of the battery management to promote the successful transformation from a big country to a powerful country in automobile industry.

1.2 Application Requirements for Batteries and BMS

The performance of the EV highly depends on the performance of the battery and BMS. In order to meet the requirements of dynamic performance, safety, economy, and environmental friendliness of the EVs, the battery system should meet the following requirements:

(1) High specific energy. Improving the specific energy of the battery can greatly improve the driving mileage of the EVs, and also reduce the mass and volume of vehicles.
(2) High specific power. Improving the specific power of the battery can effectively improve the vehicle dynamic performance of EVs so that it has excellent acceleration performance.

(3) Long service lifetime. About 50% of the cost of EVs comes from the battery system, so extending the battery lifetime can greatly reduce the use and maintenance cost, thus reducing the cost of the vehicle.
(4) High safety. The high safety of the battery can reduce the probability of dangerous accidents such as fire and explosion of the vehicles which are caused by liquid leakage, short-circuit, collision, and so on.
(5) High reliability. Improving the adaptability of the battery to complex working conditions can effectively prevent the sudden change in battery characteristics caused by drastic changes in the working environment and false operation.
(6) Excellent high/low-temperature performance. The battery is supposed to have strong adaptability to the change of vehicle operating environment and can work normally in a wide temperature range.
(7) Low self-discharge rate. Low self-discharge rate can reduce the capacity degradation rate of the battery and extend its service lifetime.
(8) Low cost. Reducing the cost of the battery can effectively reduce the cost of the vehicles and improve the product competitiveness of the EVs.
(9) Environmental friendliness. The battery will help to establish battery recycling standards and prevent secondary pollution to the environment.

At present, the commercialized batteries used in the EVs can be divided into three categories according to their capacity and output power, namely, the energy-type batteries, the power-type batteries, and the energy/power-type batteries.

(1) The energy-type battery usually has a large capacity and can supply energy continuously, which is often used in BEVs and HEVs. The total energy of this type batteries takes up a large proportion in the vehicle energy configuration, generally exceeding 10 kWh, which is not only beneficial to recover the energy from the vehicle braking feedback, but also can increase the driving range of BEVs and reduce the pollutants emission.
(2) The capacity of the power-type battery is generally small but can meet the instantaneous high power supply, which is often used in light HEVs. This type of battery is mainly used to absorb the energy of the braking feedback, while providing instantaneous extra energy for the starting and accelerating condition of the vehicle.
(3) The energy/power-type battery has high energy density, high power output capability in a low state of charge (SOC) range and high power acceptability in a high SOC range. It gives consideration to both high energy and high power characteristics, which is often used in plug-in hybrid electric vehicles (PHEVs).

To be specific, different EV types have different configurations and working modes, leading to different requirements for the batteries. The following describes the specific requirements for the batteries in the BEVs, HEVs, and PHEVs.

1.2.1 Battery Electric Vehicle

BEV is a new energy vehicle that only uses the battery as the energy storage source. The battery provides the electric energy to the motor through the battery and driving the motor, and thus propel the vehicle. The driving mileage of BEV completely depends on the capacity of the battery. The larger the capacity of the battery is, the longer the driving mileage of the vehicle will be, while the volume and mass of the battery will also increase accordingly. Therefore, the choice of the corresponding battery used in BEV depends on different design goals, road conditions, and driving conditions. The specific requirements can be summarized as follows:

(1) The battery pack should have enough energy and capacity to ensure that the rate of continuous discharge generally does not exceed 1C (C represents the battery charge or discharge current rate. For example, a rate capacity of a cell is 3 A h, and 3 C indicates the value of the charge or discharge current is 3 times of the rated capacity, i.e., $3 \times 3 = 9$ A.), and the peak discharge rate generally does not exceed 3C. For BEV with regenerative braking energy, the battery pack must be able to withstand the pulse charging current of 5C.
(2) The service of the battery should not be affected by deep discharging, and it can support full load discharging when necessary.
(3) BMS is required to feedback and control the real-time states of the battery pack to ensure the safety and efficiency of the battery pack.
(4) The space layout and installation of the battery box require targeted research and design due to the large volume and mass of the battery pack.

1.2.2 Hybrid Electric Vehicle

HEV is an energy-hybrid new energy vehicle that combines the traditional internal-combustion engine system with the electric propulsion system. In general, the battery is the main energy source of the electric propulsion system in HEV. Therefore, during the development process of the HEV, the relevant design of the battery also needs to be considered in detail.

Generally speaking, the battery on HEV does not need a large capacity, but needs to meet the instantaneous high power demand of the vehicle. Different from BEV, HEV has three different structural configurations: series, parallel, and series–parallel leading to different battery design requirements.

(1) The series HEV is powered jointly by the engine, the generator and the battery, and the SOC of the battery system is usually at a high level. Therefore, the requirement of the battery is similar to that of the BEV, but the requirement of the capacity specification is relatively low.
(2) The engine and motor of the parallel HEV can directly provide the driving force, and different power combinations can meet the different power demands

of the vehicle, which means that the vehicle can adapt the battery with a small capacity, but the maximum discharging current of the battery should be more than 20C to meet the instantaneous high power demand of the vehicle when HEV accelerates or climbs.

(3) The series–parallel HEV composed of a series HEV and a parallel HEV, which has all the working modes of the series HEV and parallel HEV. Therefore, it is necessary to comprehensively consider the battery design requirements of series HEV and parallel HEV according to the actual situation.

Although different types of HEV have different requirements for the batteries, there are still some common requirements:

(1) The state of power (SOP) of the battery should be large enough to meet the demand for high power charging and discharging in a short time.
(2) The service lifetime of the battery should be as long as possible to reach about 1000 deep discharge cycles and 400,000 shallow discharge cycles.
(3) The battery SOC should be kept within the range of 50–85% as far as possible.
(4) BMS is required to control and feedback the real-time state of the battery pack to ensure the safety and efficiency of the battery pack.

1.2.3 Plug-in Hybrid Electric Vehicle

PHEV is a kind of HEV that can be charged from the power grid. It has both BEV and HEV operating modes, which can be simply understood as a combination of BEV and HEV. Such vehicles are usually expected to drive dozens of kilometers in the BEV operating mode, and can meet highway driving requirements in the HEV operating mode, while having high-power output capacity in low SOC range. Therefore, the energy density requirement of PHEV batteries is close to that of BEV batteries, and the power density requirement is close to that of HEV batteries. In fact, the price of PHEV is generally higher than BEV and HEV due to the higher requirement of PHEV battery.

1.2.4 Relevant R&D Indicators

The 13th Five-Year Plan of China has clearly defined the relevant indicators for the battery and BMS. In December 2017, the Guidelines for the Application of New Energy Vehicles for Key Special Projects in 2018 issued by the Ministry of Science and Technology clearly stated:

(1) New energy passenger vehicle

The design of the battery and BMS needs to meet the requirements of high safety and high specific energy. The specific energy of the battery system should be more than 210 Wh/kg, while the cycle life should be more than 1200 cycles (80% depth of discharge (DOD), annual temperature distribution). What's more, the absolute value of the estimation error of SOC, SOP, and SOH within the whole life and wide operating temperature range is less than 3%. The maximum temperature difference between cells is less than 2 °C, and the time required for fast charging up to 80% SOC is less than 1 h. The battery system should also meet the requirements of China's standards such as safety and wide temperature range usage, and meet the ISO 26262 ASIL-C functional safety requirements and industry standards. The cost is less than 1.2 yuan/Wh. The disaster analysis and hazard assessment of thermal runaway and thermal diffusion should be completed, and the design, manufacture, and test specifications of the battery system based on vehicle integration should be established.

(2) Electric bus

The design of the battery and BMS needs to meet the requirements of high safety and long lifetime. The specific energy of the battery system should be more than 170 Wh/kg, while the cycle life should be more than 3000 cycles (80% DOD, annual temperature distribution). What's more, the absolute value of the estimated error of SOC, SOP, and SOH within the whole lifetime and wide operating temperature range is less than 3%, the maximum temperature difference between the cells is less than 2 °C, and the time required for fast charging up to 80% SOC is less than 15 min. The battery system should also meet the requirements of China's standards such as safety and wide temperature range usage, meet the ISO 26262 ASIL-C functional safety requirements and industry standards, and ensure that no fire or explosion occurs in the system within 30 min after the cell thermal runaway. The cost is less than 1.2 yuan/Wh. The disaster analysis and hazard assessment of thermal runaway and thermal diffusion should be conducted, and the design, manufacture and test specifications of the battery system based on vehicle integration should be established.

1.3 The Battery

The batteries are mainly responsible for providing energy for the whole vehicle. With the continuous development of battery technology, the battery types are becoming more and more. As the core components of EVs, the performance of the batteries will directly affect the safety, economy and dynamic performance of the vehicle.

1.3.1 Development Background of the Battery

The batteries are essentially rechargeable batteries that can be traced back to the nineteenth century. In 1859, the French scientist Gaston Planté invented the first rechargeable battery, namely, lead–acid battery. It used sulfuric acid as the electrolyte, lead, and lead dioxide as anode and cathode, whose chemical reaction can be reversed by applying an opposite current to recharge the battery [4]. In 1899, the Swedish scientist Waldemar Jungner invented another rechargeable battery, the nickel–cadmium battery [5]. Unlike the lead–acid battery, the nickel–cadmium battery is the first battery using the alkaline electrolyte, which consists of nickel and cadmium electrodes soaking in potassium hydroxide solution. Waldemar Jungner also invented another rechargeable battery, the nickel–iron battery in the same year. However, as the nickel–iron battery generates a large amount of hydrogen during charging, which is inefficient and difficult to seal, Waldemar Jungner did not apply for a patent for this. In 1903, Thomas Edison from the United States patented the design and sold the nickel–iron battery, hoping to lighten and commercialize it, so as to replace the lead–acid battery as the energy source for the early vehicles. However, people quickly discovered that the nickel–iron battery had a short lifetime and was prone to leak, and the performance did not exceed the lead–acid batteries. Though Thomas Edison continued to improve the nickel–iron battery in subsequent years, it was still difficult to reverse the trend of cheap and reliable gasoline engine vehicles all over the world. Until the 1970s, the lithium-ion battery and the nickel–hydrogen battery came out, the development and production cycle of the battery began to shorten greatly [6]. In the following years, the battery has been constantly improved in material structure and technology, and the performance of all aspects of the battery has also been improved rapidly. The battery technology has made great progress and has been gradually applied in the aerospace, defense, military, transportation, and electronic mobile devices.

The battery is the energy source of the EV. In the early days, the EV used to have a short driving mileage and heavy mass and its power performance is difficult to reach the level of internal-combustion engine due to the restriction of the battery technology. The lead–acid battery is the most mature battery in modern times, whose energy density is 30 Wh/kg and cycle lifetime is about 300. It has stable performance, good reliability, and low price. However, the energy density of gasoline reaches more than 10,000 Wh/kg that is about 300 times of that of the lead–acid battery, which means that the EV equipped with lead–acid batteries needs a great space and increases the vehicle mass to obtain the same driving performance [7]. As a result, the lead–acid battery has been gradually replaced by other types of batteries. In addition, the lead–acid battery generates a large number of lead emissions during the process of production, processing, and recycling, causing irreversible environmental contamination, which seriously violates the original intention of the EV development. The nickel–cadmium battery is another early battery. Compared with the lead–acid battery, the energy density and cycle lifetime of the nickel–cadmium battery are increased to 50 Wh/kg and 500, respectively, and it has a higher current density. However, the toxicity of cadmium metal elements is huge, which will result in more serious environmental pollution, so it is difficult to be applied in the EVs. It is worth mentioning that the nickel–cadmium battery also has a strong memory effect [8]. In the case of incomplete charging and discharging, there will be a temporary decrease in capacity, which results in a decrease of driving mileage and makes it difficult to adapt to the working environments of the EV. The Ni–MH battery is a less polluting battery with an energy density of 60–80 Wh/kg. Its overall performance is better than the above two batteries, and it has good characteristics at low temperature. Therefore, the Ni–MH battery was used in batches in hybrid vehicles such as Toyota, Honda, Ford, and Chevrolet in the 1990s. However, the Ni–MH battery has not completely eliminated the memory effect, and there are also a series of new problems such as serious heating during charging and poor charging performance at high current rate. In comparison, the lithium-ion battery has no memory effect and has higher energy density and power density. Moreover, it can reduce the mass of the pack by 40–50% and volume by 20–30%. In terms of cycle life, self-discharge rate and environmental friendliness, lithium-ion batteries are better than the above three batteries. Therefore, the lithium-ion battery has become the preferred battery for EVs [9].

In summary, the advantages of the lithium-ion battery are mainly reflected in the following aspects:

(1) High working voltage. The working voltage of the lithium-ion battery is 3 times than that of the Ni–MH battery and the nickel–cadmium battery, reaching 3.6 V.
(2) High specific energy. The specific energy of the lithium-ion battery is 4 times than that of the nickel–cadmium battery and 2 times than that of the Ni–MH battery, about 200 Wh/kg.
(3) Long cycle lifetime. The cycle life of the lithium-ion battery has reached more than 2000, and can reach tens of thousands of times under low discharging depth, exceeding the performance of the above three batteries.
(4) Low self-discharge rate. The self-discharge rate of the lithium-ion battery is only 6–8%, which is much lower than that of the nickel–cadmium battery (25–30%) and the Ni–MH battery (15–20%).
(5) Memoryless. The lithium-ion battery can be charged and discharged at any time according to demand without reducing performance.
(6) Environmental friendliness. There is no harmful substance in the lithium-ion battery, and it is a non-polluting battery.
(7) Flexible volume. Lithium-ion batteries are easy to meet the layout requirements of battery systems in EVs.

1.3.2 Principles and Classification of the Lithium-Ion Battery

The lithium-ion battery is mainly composed of a positive electrode, a negative electrode, a separator between the positive and negative electrodes, a lithium salt electrolyte, a positive temperature coefficient (PTC) component, and a safety valve.

(1) The positive electrode usually has the space to accommodate lithium ions and the diffusion, which largely determines the overall performance of the battery. At present, the most common positive electrode materials are lithium cobalt

oxide, lithium manganese oxide, lithium iron phosphate, and ternary materials (usually nickel–cobalt–manganese, nickel–cobalt–aluminum) composed of various metal oxides.

(2) The active material of the negative electrode is usually a paste made of carbon material, a binder, and the organic solvent, which is a thin coating on a copper based. In addition, some new types of fast-charging batteries use lithium titanium oxide ($Li_4Ti_5O_{12}$) as a negative electrode material.

(3) The separator is used to isolate the electrons between the positive and negative electrodes, and only lithium ions are allowed to pass through. Generally, it is made of a microporous film of polyethylene or polypropylene materials.

(4) The electrolyte, usually a high-permittivity lithium salt electrolyte mixed with the organic solvents, is responsible for delivering the ions. The electrolyte is chemically stable to the active material and must be able to adapt to the severe redox reaction that occurs during the charging and discharging.

(5) Although the lithium-ion battery generally has an abnormal current cutting device, usually a PTC component, to ensure safety, an abnormal rise in the internal pressure of the battery may occur during use. The safety valve is set to effectively release high-pressure gas and prevent the battery from rupturing.

The basic structure and working principle of the lithium-ion battery are shown in Fig. 1.1, and both the positive and negative electrodes are immersed in the electrolyte. The charge and discharge of the lithium-ion battery is achieved by the process of inserting and releasing lithium ions between the positive and negative electrodes. When the battery is charged, lithium ions move from the positive electrode of the

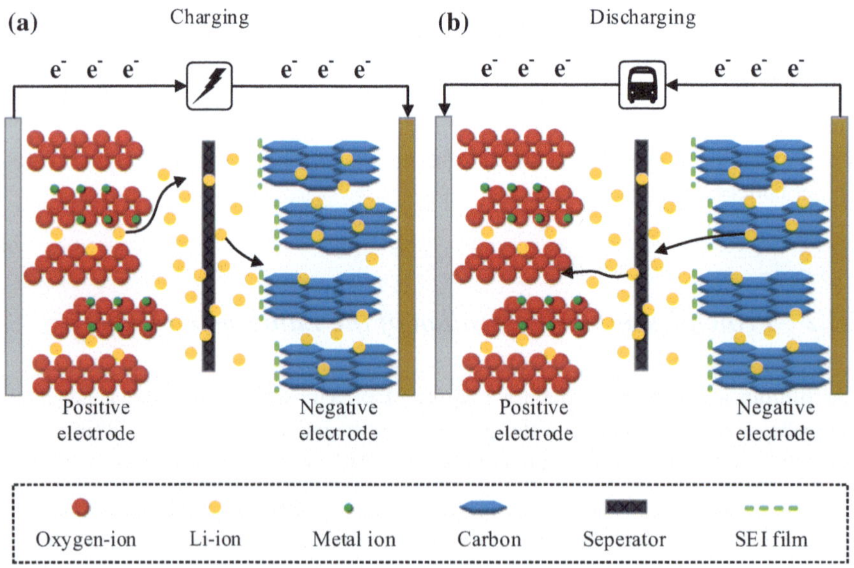

Fig. 1.1 Basic structure and working principle of lithium-ion batteries

battery into the electrolyte. Since the lithium ions in the electrolyte have a large concentration difference near the positive and negative electrodes, they are driven to move toward the negative electrode, and finally embedded the negative electrode through the separator. At the same time, the electrons in the external circuit move from the positive electrode to the negative electrode to form a current, realizing the charging process. So the negative electrode receives electrons to occur a reduction reaction, and the positive electrode loses electrons to undergo an oxidation reaction. The discharging process is the opposite. Lithium ions are deinserted from the negative electrode and embedded in the positive electrode. The electrons in the external circuit also move from the negative electrode to the positive electrode through the load, driving the vehicle to work and realizing the discharging process. So the negative electrode loses electrons to undergo an oxidation reaction, and the positive electrode receives electrons to occur a reduction reaction. When the lithium-ion battery is charged for the first time, a solid interface film called solid electrolyte interphase (SEI) film is formed between the electrolyte and the electrode near the negative electrode to protect the negative electrode from corrosion and prevent the reduction reaction between the negative electrode and the electrolyte. However, the SEI film continuously thickens or dissolves over time, resulting in a decrease in the amount of lithium ions and active materials available for recycling, which has become the main cause of the battery capacity decline [10, 11].

The difference between the positive and negative materials, electrolyte materials and production process of lithium-ion batteries makes the battery exhibit different performance and has different names [12, 13]. At present, the lithium-ion battery on the market is often named according to its positive material. For example, the earliest commercial lithium-ion battery uses lithium cobalt oxide ($LiCoO_2$) as the positive electrode material, so it is called a lithium cobalt oxide battery. What's more, the batteries using lithium manganese oxide ($LiMn_2O_4$) and lithium iron phosphate ($LiFePO_4$) as the positive electrode materials are called lithium manganese acid batteries, and lithium iron phosphate batteries, respectively. In addition, the batteries using the ternary materials lithium nickel–cobalt–aluminate oxide and lithium nickel–cobalt–manganese oxide as the positive electrode materials are, respectively, referred to as nickel–cobalt–aluminum ternary lithium-ion batteries and nickel—cobalt–manganese ternary lithium-ion batteries.

The abbreviation, positive chemical formula and performance characteristics of common lithium-ion batteries are shown in Table 1.1. As can be seen from the table, the lithium cobalt acid battery and lithium manganese acid battery have poor thermal stability and rarely used in EVs. In contrast, the lithium iron phosphate battery and ternary lithium battery have better safety performance, so they are widely used in the EV batteries.

Table 1.1 The abbreviation, positive chemical formula, and characteristics of common lithium-ion batteries

Batteries	Abbreviation	Positive chemical formula	Characteristics
Lithium cobalt oxide battery	LCO	$LiCoO_2$	High-voltage (3.9 V), high specific energy, but potentially fire risk
Lithium manganese acid battery	LMO	$LiMn_2O_4$	The voltage and specific energy are close to that of LCO, the capacity decays quickly, and the thermal stability is poor
Lithium iron phosphate battery	LFP	$LiFePO_4$	Good safety, high power density, low energy density, good thermal stability
Nickel–cobalt–aluminum ternary lithium-ion battery	NCA	$Li_{0.8}Co_{0.15}Al_{0.05}O_2$	The voltage is slightly lower than that of LCO, safety is better than that of LCO, cycle life characteristics are good
Nickel–cobalt–manganese ternary lithium-ion battery	NMC	$LiNi_{1\text{-}x\text{-}y}Co_xMn_yO_2$	Safety is between NCA and LMO, capacity decays faster than NCA

1.3.3 Lithium Iron Phosphate Battery

In 1996, Prof. John B. Goodenough from the University of Texas first revealed in the paper that lithium iron phosphate can be used as a positive electrode material for rechargeable lithium-ion batteries [14]. Since then, the lithium iron phosphate battery has gained extensive attention due to its low cost, nontoxic, natural richness of iron and excellent thermal stability, safety, and electrochemical characteristics. Its internal structure is shown in Fig. 1.2. On the left is an olivine lithium iron phosphate as the positive electrode of the battery, which is connected by an aluminum foil. The middle is a polymer separator that is responsible for separating the positive electrode from the negative electrode, allowing only lithium ions (Li^+) to pass through and not allowing electrons(e^-)to pass through. The right side is a battery negative electrode composed of carbon, which is connected by a copper foil. The electrolyte of the battery is between the upper and lower ends of the battery, and the whole is sealed by a metal casing. The chemical process of charging and discharging of lithium iron phosphate battery is as follows

Charging process:

$$\mathrm{LiFePO_4} - x\mathrm{Li}^+ - xe^- \rightarrow x\mathrm{FePO_4} + (1-x)\mathrm{LiFePO_4}$$

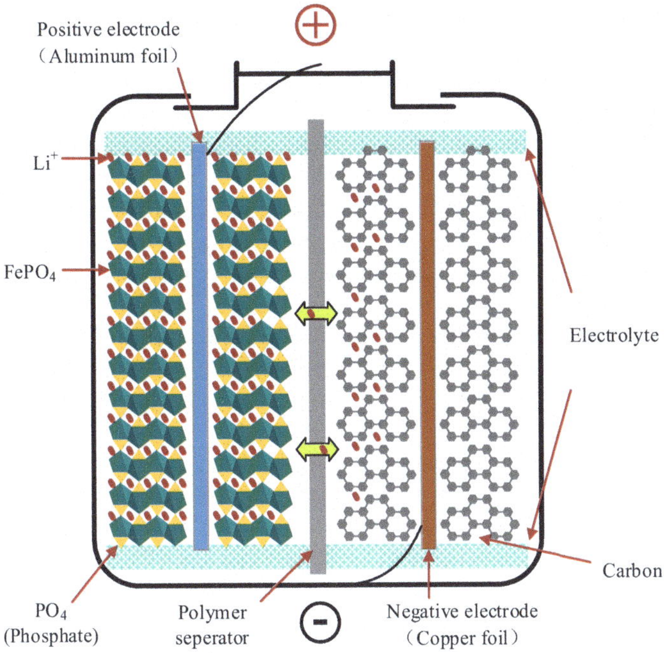

Fig. 1.2 Structural diagram of lithium iron phosphate battery

Discharging process:

$$FePO_4 + xLi^+ + xe^- \rightarrow xLiFePO_4 + (1 - x)FePO_4$$

In general, the lithium iron phosphate battery has a nominal voltage of 3.2 V, a charge cut-off voltage of approximately 3.6 V, and a discharge cut-off voltage of approximately 2.0 V [15]. Due to the different quality and technology of the materials used by various manufacturers, there will be some differences in their performance. The relevant data shows that for standard batteries of the same model (or same package), the difference in capacity can be up to 10–20%. In fact, the capacity of lithium iron phosphate batteries can be divided into three categories: the capacity of small batteries is from a few tenths to a few ampere hours, the capacity of medium batteries is dozens of ampere hours, and the capacity of large batteries is hundreds of ampere hours. There are some differences in the same parameters for different types of batteries.

The advantages of lithium iron phosphate battery are mainly summarized in the following aspects:

(1) High safety. The lithium iron phosphate battery will not burn or explode due to internal or external damage, which significantly reduces the probability of accidents during driving.

(2) Low cost. The chemical raw materials used to make lithium iron phosphate are abundant and easy to obtain.
(3) Excellent cycling performance. The lithium iron phosphate battery can still maintain 95% of the initial capacity after 500 charge and discharge cycles.
(4) High-efficiency output and good charge/discharge characteristic. The standard discharge rate of the lithium iron phosphate battery is 2–5C, the continuous high discharge rate can reach 10C, and the instantaneous pulse discharge (10 s) rate can reach 20C. What's more, it can support fast charge and discharge, does not have memory effect, and maintains normal battery performance even stored in over-discharge conditions.
(5) Environmental friendliness. All raw materials involved in the manufacturing process of the lithium iron phosphate battery are nontoxic and environmentally friendly, so there is no pollution to the environment in the production and utilization.
(6) Good temperature characteristic. The lithium iron phosphate battery not only has good battery characteristics at room temperature, but also can maintain good structure at high temperature. For example, when the external temperature is 65 °C, the internal temperature of the battery can reach 95 °C, and the temperature at the end of discharge is as high as 160 °C, but at this time the battery can still keep the safety and structure intact.
(7) Safe and stable over-discharge performance. Even if the lithium iron phosphate battery is over-discharged to 0 V, there will be no damage or accidents.

Although the lithium iron phosphate battery has obvious advantages, there are still some deficiencies such as the low compaction density that only can reach 1.0–1.4 g/cm^3, which makes the battery have a large specific surface area and volume, and the energy density is low, affecting the lightweight and space utilization of the EVs.

At present, the lithium iron phosphate batteries are widely used in new energy commercial vehicles and some new energy passenger vehicles. Some applications of the lithium iron phosphate batteries in new energy passenger vehicles are shown in Table 1.2.

1.3.4 Ternary Lithium-Ion Battery

The ternary lithium-ion battery refers to a lithium-ion battery using a ternary composite material as a positive electrode material, and is a new type of lithium-ion battery developed in recent years. The most common ternary lithium-ion battery is the nickel–cobalt–manganese battery with $LiNi_{1-x-y}Co_xMn_yO_2$ as the positive electrode material [16]. The composing proportions of the three elements of nickel, cobalt, and manganese can be adjusted within a certain range, which will affect the performance of the battery. According to the different proportions of the three elements, the ternary nickel–cobalt–manganese lithium-ion battery can be subdivided

Table 1.2 Some applications of the lithium–iron–phosphate batteries in new energy passenger vehicles (sorted by manufacturer initials)

Manufacturers	Model name	Battery capacity (kWh)	Driving mileage (km)
BAIC BJEV	EC180	20.3	202
BYD	e5	43	360
	e6	82	400
	Qin EV300	47	300
JAC	iEV40	23	260
	iEVA50 luxury	46.5	390
	iEV6E luxury	29.2	310
	iEV7	24	250
ZOTYE	5008EVL	32	200
	M300EV	32	200

into 333, 523, 811, and other systems. Due to the synergistic effect, the electrochemical properties of nickel–cobalt–manganese composite are better than those of single-component layered oxide, which can compensate for their respective shortcomings. Cobalt can effectively inhibit ions mixing, stabilize the layered structure of the material, and improve the conductivity of the material. Nickel can guarantee the high capacity of the material. Manganese does not participate in electrochemical reaction but is responsible for stabilizing the structure and improving the safety of materials. Such a battery can effectively overcome the problems of the high cost of lithium cobalt oxide, poor stability of lithium manganese oxide, and low capacity of lithium iron phosphate, combining the advantages of lithium cobalt oxide battery and lithium manganese oxide battery, for which gradually received widespread attention from car manufacturers and users. The raw material of the ternary nickel—cobalt–manganese lithium-ion battery contains a noble metal—cobalt, which has a large price fluctuation that will directly affect the price of lithium cobalt oxide. When cobalt is at a high price, the price of ternary nickel–cobalt–manganese material is much lower than that of lithium cobalt oxide, which makes it have strong market competitiveness, but when cobalt is at a low price, the cost advantage of ternary nickel–cobalt–manganese material will be greatly reduced.

The advantages of the ternary nickel–cobalt–manganese lithium-ion battery are summarized as follows.

(1) High energy density. The specific capacity of the ternary nickel–cobalt–manganese lithium-ion battery is high, which can reach more than 145 mA h/g, and the capacity of a 18650 battery can reach more than 3 A h.
(2) Excellent cycling performance. The ternary nickel–cobalt–manganese lithium-ion battery can still maintain 80% of the rated capacity after 500 cycles.

(3) High compaction density. The compaction density of ternary nickel–cobalt–manganese lithium-ion battery can reach more than 3.4 g/cm^3.
(4) High-voltage platform. The nickel–cobalt–manganese battery has a relatively high-voltage platform, the median voltage of the 1C rate discharge can reach 3.66 V, and the median voltage of the discharge at 4C rate is about 3.6 V.

The nickel–cobalt–manganese battery is widely used due to its high energy density. However, there are still some defects such as relatively complex preparation process, poor safety performance, and pollution caused by the nickel–cobalt element.

At present, the nickel–cobalt–manganese batteries are widely used in new energy passenger vehicles. Some applications of the ternary nickel–cobalt–manganese lithium-ion batteries in new energy passenger vehicles are shown in Table 1.3.

Table 1.3 Some applications of the ternary nickel–cobalt–manganese lithium-ion batteries in new energy passenger vehicles (sort by manufacturer initials)

Manufacturers	Model name	Battery capacity (kWh)	Driving mileage (km)
BAIC BJEV	EU5	53.6	570
	EU400	54.4	360
	EX360	48	390
	EC200	20.5	200
Chana Auto	Yi EV300	45	360
	CS15EV	45	360
GAC NE	Trumpchi GE3	47	475
JAC	iEV7S	39	350
JMC	E160	20	200
	E200-JMEV	17.3	170
	E400	41	380
Geely New Energy	Emgrand GSe	52	460
	Emgrand EV	52	450
Chery New Energy	eQ	22.3	200
	eQ1	38	410
	Aryiiza5e	49	420
SAIC Motor Passenger Vehicle Corporation	Roewe ei5	35	401
	Roewe ERX5	48.3	425
DENZA	500	70	635
Tesla	Model S 100D	100	632
	Model X 100D	100	565
NEXTEV	ES8	70	355
ZOTYE	Z500EV	38.7	250

1.4 Battery Management System

The BMS is a product or technology that manages and controls the batteries in some way. Figure 1.3 shows the schematic diagram of a typical BMS consisting of various sensors, actuators, controllers of the algorithms and the signal lines, etc. Its main task is to ensure the safety and reliability of the battery system, provide the state information required for vehicle control and energy management and take appropriate interventions on the battery system in the event of abnormal conditions. Moreover, it is supposed to collect the terminal voltage, current, temperature and other information of the pack and each cell in real time through the sampling circuit, and use the embedded algorithms and strategies to estimate the battery SOC [17], SOH [18], SOP [19] and RUL [20], then output them to the vehicle control unit (VCU), providing the basis for energy management and power distribution control of the EVs.

1.4.1 Basic Functions of the Battery Management System

The main functions of the BMS include data collection, state monitoring, safety protection, charging control, energy management, equalization management, thermal management, and information management.

(1) Data collection
The working environment of the battery in the EV is very complicated. EVs need to adapt to the complex and changing environment, which means that the battery needs to operate in a complex and variable humiture environment all the year-round. In addition, as road conditions and drivers change, the battery needs to be adapted to the rapidly changing load. In order to accurately obtain the working

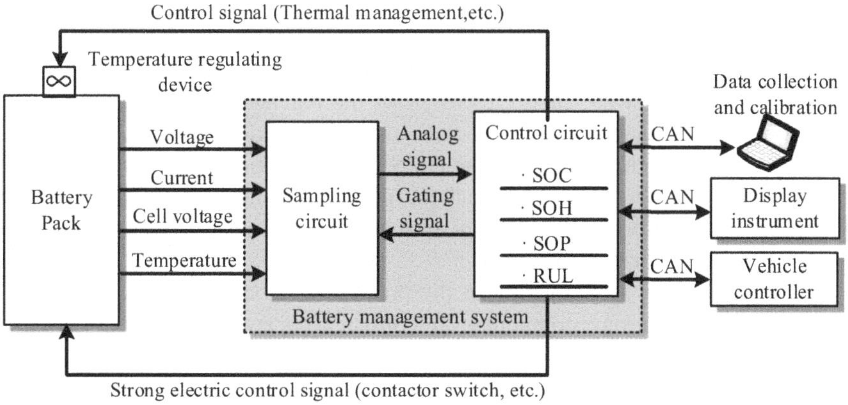

Fig. 1.3 Schematic diagram of a battery management system

condition of the battery and better implement the management countermeasures, the BMS needs to collect the terminal voltage, current, temperature and other information of the pack and each cell in real time through the sampling circuit.

(2) State monitoring
The battery is a complex and nonlinear time-varying system with multiple states that change in real time. Accurate and efficient monitoring of the battery state is the key to the battery cell and its system management, which is also the basis for the energy management and control of the EVs. Therefore, the BMS needs to estimate the states of the pack based on the real-time collected battery data by using the embedded algorithms and strategies, so as to obtain the battery states at each moment, including the SOC, SOH, SOP, and state of energy (SOE), providing support for the real-time states analysis of the battery.

(3) Safety protection
The battery safety protection function mainly refers to the online fault diagnosis and safety control of the battery system. The online fault diagnosis of the battery refers to the fault type diagnosed by the fault diagnosis algorithm based on the collected sensor signals. The faults to be diagnosed in the battery management generally include overvoltage (over-charge), undervoltage (over-discharge), smog, overcurrent, ultrahigh temperature, short-circuit, joints loose, reduced insulation, and electrolyte leakage, as well as the faults of electronic components such as sensors, actuators, and controllers. The BMS needs to make warnings and take corresponding measures to intervene as soon as possible to ensure the safety of EVs after diagnosing the fault types.

(4) Charging control
Since the charging process of the battery directly affects the lifetime and safety of the battery, the BMS usually needs to equip a charging management module to control the charging of the battery according to the real-time characteristics, the temperature of the battery and the power level of the charger.

(5) Energy management
The driving conditions of the EVs is complex. The random driving operations such as rapid acceleration, sudden braking, uphill, and downhill can result in complex and variable dynamic loads. In order to ensure the safe and economical operation of the vehicle, the BMS needs to reasonably control the energy output of the battery and the energy recovery of the regenerative braking based on the collected battery data and real-time status information. For an EV equipped with the hybrid energy system, the BMS needs to optimize the distribution of the respective energy according to the respective state information of the hybrid energy system to ensure the best performance of the composite power supply.

(6) Equalization management
Due to the error accumulation in the production process, transportation and storage, and electronic components, inconsistencies are inevitable between the cells. In order to fully utilize the energy stored in the cells and ensure the safety of the pack, the BMS needs to adopt an active or passive equalization method according to the information of the cells, reducing the inconsistency of the cells.

(7) Thermal management
In normal operation, the battery is not only affected by the ambient temperature but also affected by heat produced in charge and discharge. As a result, the BMS needs to equip the battery thermal management module to determine the intensity of active heating/cooling according to the temperature distribution information, and charge/discharge requirements in the pack. Therefore, batteries can work at a suitable temperature range, and maximize the performance and extend the lifetime.

(8) Information management
The BMS needs to integrate multiple functional modules and coordinate the communication between modules. Due to a large amount of data running, the BMS needs to manage and filter the data of the batteries, store critical data, and communicate with network nodes such as the vehicle controllers. With the development of big data technology, the BMS also needs to interact with the cloud platform in real time to better handle the management problems of the battery and improve management quality.

1.4.2 Topology of the BMS

The design of an EV usually needs to meet certain dynamic performance indicators such as acceleration capability, climbing ability, and maximum speed. Therefore, the cell is usually connected in series or parallel to meet the technical requirements of vehicle design. For example, the Tesla Model S adopts the NCA 18650 ternary nickel–cobalt–aluminum lithium-ion batteries manufactured by Panasonic, with the nominal capacity of the cell being 3,100 mA h. More than 7,000 cells were used in the series–parallel, forming a pack placed on the underframe. The topology of BMS is very important for large-scale battery management.

The topology of the BMS directly affects the cost, reliability, the convenience of installation, and maintenance and measurement accuracy of the system. Generally speaking, the battery monitoring circuit (BMC) and the battery control unit (BCU) form the hardware circuit. According to the structural relationship among the BMC, the BCU, and the cells, the BMS can be divided into centralized topology and distributed topology.

The BMC and the BCU in the centralized BMS topology are integrated on a single printed circuit board (PCB) to realize the functions of data collection, calculation, safety monitoring, switch management, charge and discharge control, and communication with the vehicle controller, etc. The centralized BMS is often used in applications where the battery has low capacity, low total voltage, and small battery systems. The centralized BMS topology is shown in Fig. 1.4. The measurement signals of each cell are transmitted to a single PCB.

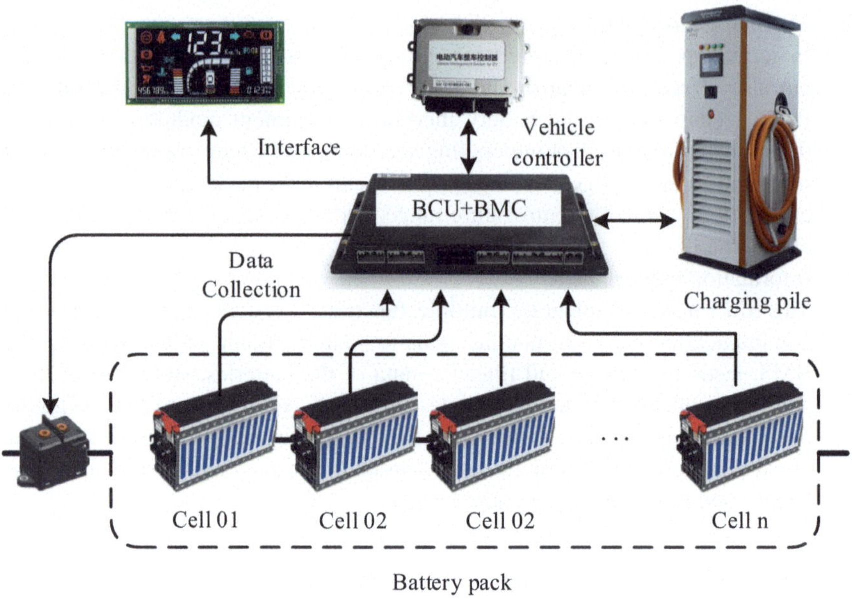

Fig. 1.4 Structural diagram of centralized BMS topology

The centralized BMS topology generally has the following advantages:

(1) High-speed intra-board communication is beneficial to ensure synchronous data collection.
(2) It has compact structure and strong anti-interference ability.
(3) The cost is low, and all the work of the BMS can be completed using only one package.

At the same time, the centralized BMS topology has the following disadvantages:

(1) It produces a large number of complex wires and connectors.
(2) It is difficult to protect the battery system when short-circuits and overcurrent occur in different parts of the system.
(3) Considering the high-voltage safety, sufficient safety clearance must be left between different channels, which eventually leads to the excessive size of the PCB.
(4) All components are integrated into a single PCB, resulting in poor scalability and maintainability.

Different from the centralized topology, the BCU and BMC in the distributed BMS are arranged separately, as shown in Fig. 1.5. The BCU is primarily responsible for fault detection, battery state estimation, switch management, charge and discharge control, and communication with the vehicle controller. The BMC is used to manage the cell voltage, current, and temperature, as well as security and consistency management. The BCU and the BMC are connected by a CAN bus, and any

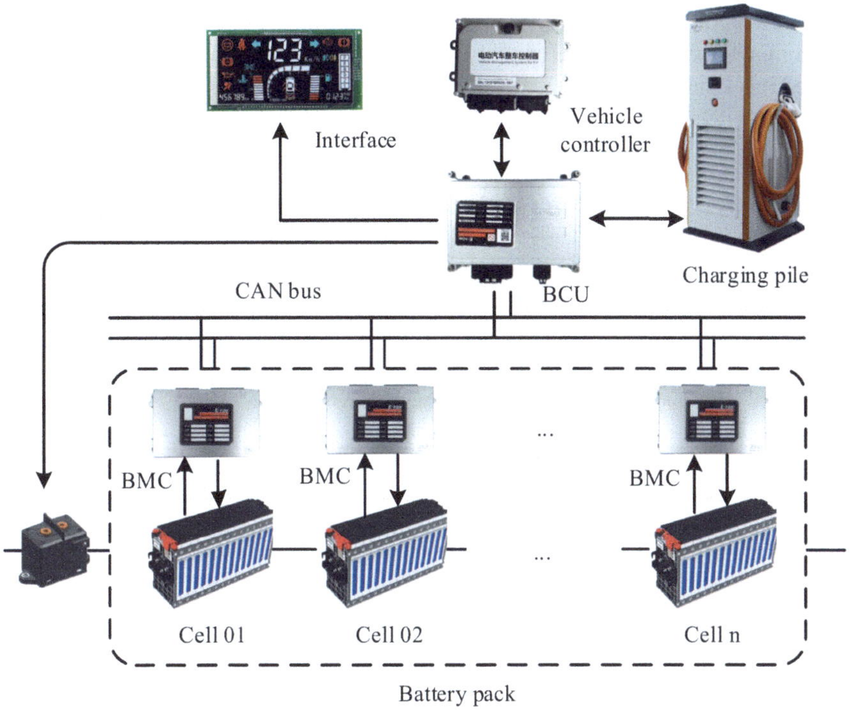

Fig. 1.5 Schematic diagram of distributed BMS topology

BMC can communicate with the BCU. In addition, each PCB of the BMC belongs to a node of the CAN bus and is separately connected to the corresponding cell. Therefore, the communication can also be established among BMCs.

The distributed BMS topology generally has the following advantages:

(1) Since the function of data collection and calculation is separated, the fault is easy to check, and the calculation efficiency is high.
(2) The structure of the system is simplified, and the layout is flexible and applicability is good.
(3) The number of managed batteries can be increased or decreased by placing or removing the BMC PCB near the corresponding battery and connecting it to or unhooking the reserved CAN bus interface.

At the same time, the distributed BMS topology has the following disadvantages:

(1) The large number of components call for more PCBs and steps of installation, debugging, and disassembly.
(2) As it is easy to form network delay and affect the synchronization of the collected data, the design requirements of the communication network are hard to meet.

At present, the distributed BMS topology is most widely used in EVs. For example, commercial EVs such as Tesla Model S, BMW i3, Roewe eRX5, and BYD Qin have adopted this kind of topology.

1.4.3 Development Process of the BMS

The basic development process of BMS is shown in Fig. 1.6. It can be seen that the development of both the battery and the BMS are based on the vehicle requirements

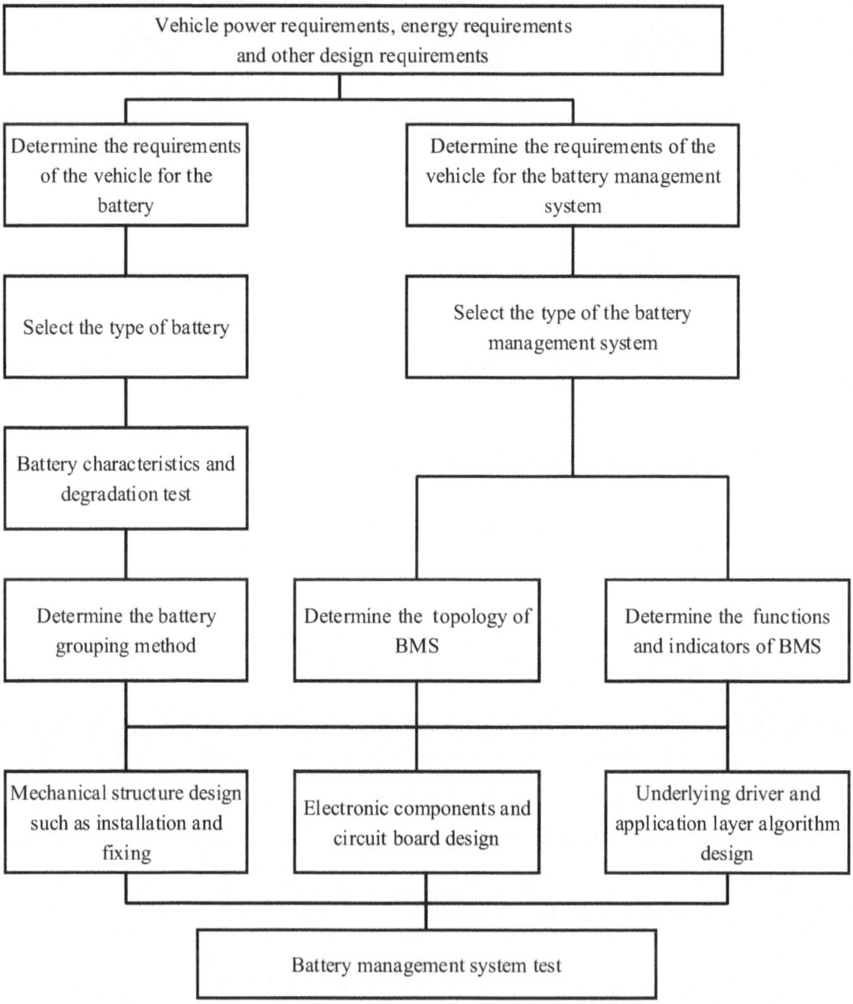

Fig. 1.6 Basic development process of the BMS

of power, energy, and others. In the development of the battery, it is first necessary to select the battery, and carry out a series of characteristics test and cycle life test of the cell to obtain its performance characteristics. The next is to determine the battery grouping method including the number of the cells in series and parallel connection and the specific arrangement. In the development of BMS, it is first necessary to select the type, and then determine the system topology and the basic functions and indicators that need to be met including water-resistant, dust-resistant, seismic, etc. After determining the battery grouping method, it needs to manage the BMS topology and basic functions and indicators, the mechanical structure design, electronic components, and PCB design, and the underlying driver and application layer algorithm design of the system. Different tests of BMS are required before system development is completed to ensure the integrity and security of the system design.

1.5 Conclusion

This chapter focuses on the battery and its management system in new energy vehicles. It outlines the history of development and strategic plan of the new energy vehicles and the battery in China. The operating requirements of the battery and its management system of the BEV, the PEV, and the PHEV are also sorted out. Then, it introduces the R&D indicators of China's 13th Five-Year New energy Special Project on the battery and its management system. After that, this chapter reviews the research background and current status of batteries, and outlines the basic working principles and classification of the lithium-ion battery. Two kinds of mainstream batteries in the EVs, namely, lithium iron phosphate battery and ternary lithium battery, are discussed in detail. Moreover, this chapter introduces BMSs in detail from the perspective of basic functions and topology, and summarizes the basic development process of the BMS, guiding the design and development of the software and hardware systems of the battery.

References

1. Zheng J, Mehndiratta S, Guo JY, Liu Z (2012) Strategic policies and demonstration program of electric vehicle in China. Transp Policy 19:17–25
2. Jia Y (2015) Powering China's sustainable development with renewable energies: current status and future trend. Power Components Syst 43:1193–1204
3. Marquis C, Zhang H, Zhou L (2013) China's quest to adopt electric vehicles. Stanford Soc Inno Rev 11:52–57
4. Smith MJ, Fonseca A, Silva MM (2009) The lead–lead oxide secondary cell as a teaching resource. J Chem Educ. https://doi.org/10.1021/ed086p357
5. Yoshino A (2011) The birth of the lithium-ion battery. Angewandte Chemie-International Edition

6. SR, Fetcenko MA, Ross J (1993) A nickel metal hydride battery for electric vehicles. Science 260(5105):176–181
7. Bode H, Davis J (2012) Lead-acid batteries. J Electrochem Soc 125:187C–187C
8. Sasaki T, Ukyo Y, Novák P (2013) Memory effect in a lithium-ion battery. Nat Mater 12:569–575
9. Gifford P, Adams J, Corrigan D, Venkatesan S (1999) Development of advanced nickel/metal hydride batteries for electric and hybrid vehicles. J Power Sources 80:157–163
10. Xiong R, Li L, Tian J (2018) Towards a smarter battery management system: a critical review on battery state of health monitoring methods. J Power Sources 405:18–29
11. Xiong R, Li L, Li Z, Yu Q, Mu H (2018) An electrochemical model based degradation state identification method of Lithium-ion battery for all-climate electric vehicles application. Appl Energy 219(1):264–275
12. Nitta N, Wu F, Lee JT, Yushin G (2015) Li-ion battery materials: present and future. Mater Today 18(5):252–264
13. Junaidi H (2016) Types of lithium ion batteries. Battery University
14. Gorman J (2002) Bigger, cheaper, safer batteries: new material charges up lithium-ion battery work. Science News
15. Jiang J, Zhang C (2015) Fundamentals and application of lithium-ion batteries in electric drive vehicles. https://doi.org/10.1002/9781118414798
16. Eddahech A, Briat O, Vinassa JM (2015) Performance comparison of four lithium-ion battery technologies under calendar aging. Energy 84(1):542–550
17. Xiong R, Sun F, Chen Z, He H (2014) A data-driven multi-scale extended Kalman filtering based parameter and state estimation approach of lithium-ion polymer battery in electric vehicles. Appl Energy 113:463–476
18. Ma Z, Wang Z, Xiong R, Jiang J (2018) A mechanism identification model based state-of-health diagnosis of lithium-ion batteries for energy storage applications. J Clean Prod 193:379–390
19. Sun F, Xiong R, He H (2014) Estimation of state-of-charge and state-of-power capability of lithium-ion battery considering varying health conditions. J Power Sources 259:166–176
20. Zhang YZ, Xiong R, He HW, Pecht M (2019) Validation and verification of a hybrid method for remaining useful life prediction of lithium-ion batteries. J Clean Prod 212:240–249

Chapter 2
Battery Test

A battery is a typical electrochemical system. The battery test plan established for the battery management system (BMS) studies belongs to the field of experimental science. In order to establish accurate battery models and develop high-performance BMS, it is necessary to design and imply a series of targeted tests to acquire the battery performance under diverse conditions. The quality of the test plan and the experimental data directly affects the rationality and integrity of the battery characteristics analysis, which further affects the accuracy and reliability of the battery model, and ultimately affects the control performance of the BMS. This chapter will focus on the battery system test platform construction, the design of the test methods, the data analysis, and the basic characteristics of lithium-ion batteries [1].

2.1 Battery Test Platform

Battery test platform includes the battery charge and discharge test equipment, the impedance test equipment in the frequency domain, the environmental simulation equipment, and connecting devices.

2.1.1 Charge and Discharge Test Equipment

The charge and discharge test equipment can obtain the voltage, power, capacity, energy, internal resistance/impedance, and temperature of a battery by loading a specific program. Through analyzing and processing the measured data, whether the tested battery meets the requirements of the EV can be examined. Since Arbin Instruments in the United States launched the first computer-controlled battery test system in 1987 [2], various battery charging and discharging equipments have been rapidly developed. In addition to the Arbin Instruments, there are also other famous companies in this field, including MACCOR in the United States, Hioki Ltd. in Japan,

R. Xiong, *Battery Management Algorithm for Electric Vehicles*,
https://doi.org/10.1007/978-981-15-0248-4_2

Digatron in Germany and other enterprises. Major Chinese production enterprises include Wuhan Blue Electronics Ltd., Shenzhen Neware Electronics Ltd., Ningbo Bate Technology Ltd., and Harbin Zimu Technology Ltd. The imported equipments have been continuously updated and improved due to its early development, but the cost is much higher than domestic equipments.

The data in this book are all from the AESA test platform of Beijing Institute of Technology, mainly using the Arbin-BT2000 battery cell and system test equipments, including three battery cell test equipments and two battery pack test equipments. Arbin-BT2000 and its working interface are shown in Figs. 2.1 and 2.2, respectively, and its parameters and characteristics are shown in Tables 2.1 and 2.2, respectively.

On the other hand, a suitable battery clamp is also an important factor in ensuring the smooth running of the battery performance test. Many side reaction products (particularly gas) will accumulate in the battery during charging and discharging process. These side reaction products will lead the battery swelling and bulging, which will affect the electrical performance and safety of the battery. Therefore, a specific clamp should be installed on the battery before testing. Figure 2.3 shows a rectangular battery clamp. Furthermore, since the cylindrical battery is difficult to directly connect to charging and discharging device, it also requires a specific clamp. Figure 2.4 shows a cylindrical battery clamp.

Fig. 2.1 Arbin-BT2000

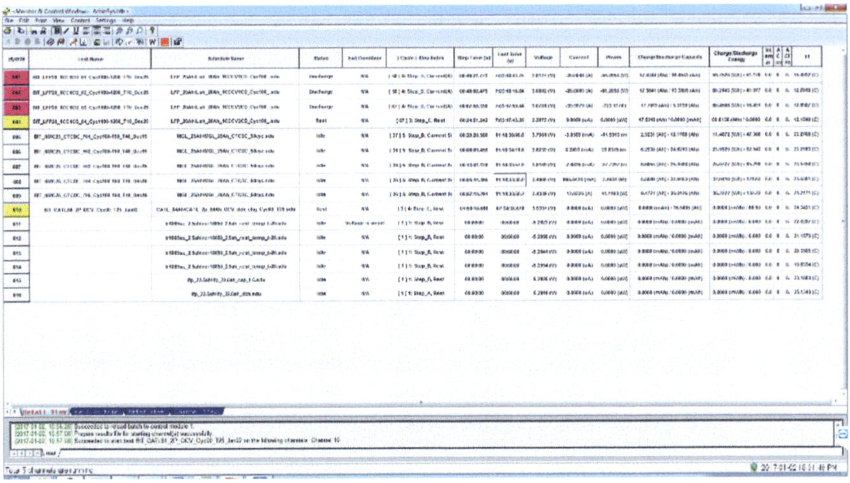

Fig. 2.2 Arbin-BT2000 working interface

Table 2.1 Parameters and characteristics of the Arbin-BT2000 battery cell test device

Equipment specification	Parameters	Equipment characteristics
5 V, 100 A	A total of 16 separate channels, each channel can be used in parallel; the maximum charging and discharging current of each channel is 100 A (divided into 3 ranges, 1 A/10 A/100 A) with the accuracy of 0.05%	Each channel has a voltage range of 0– 5 V and an accuracy of 0.05%; a variety of controllable modes: constant current charging, constant voltage charging, constant-current constant-voltage charging, pulse charging and discharging, ladder, arbitrary programmable control power, and dynamic working condition simulation, etc. The response speed is fast, and the time of the current rise is less than 50 ms The maximum frequency of data storage is 100 Hz; a variety of physical quantities can be stored
5 V, 60 A	A total of 32 separate channels, each channel can be used in parallel; the maximum charging and discharging current of each channel is 60 A (divided into 3 ranges, 0.5 A/5 A/60 A), with the accuracy of 0.05%	
5 V, 30 A	A total of 32 separate channels, each channel can be used in parallel; the maximum charging and discharging current of each channel is 30 A (divided into 3 ranges, 1 A/5A/30 A), with the accuracy of 0.05%	

Table 2.2 Parameters and characteristics of Arbin-BT2000 battery pack test device

Equipment specification	Parameters	Equipment characteristics
60 V, 300 A	One channel, the maximum charging and discharging current of each channel is 300 A (divided into three ranges, 5 A/50 A/300 A), with the accuracy of 0.05%, and the operating voltage range is 2–60 V with the accuracy of 0.05%	A variety of controllable modes: constant current charging, constant voltage charging, constant-current constant-voltage charging, pulse charging and discharging, ladder, arbitrary programmable control power, and dynamic working condition simulation The response speed is fast, and the time of the current rise is less than 50 ms The maximum frequency of data storage is 100 Hz; a variety of physical quantities can be stored
100 V, 300 A	Two channels, the maximum charging and discharging current of each channel is 300 A (divided into three ranges, 10 A/50 A/300 A), with the accuracy of 0.05%, and the operating voltage range is 10 –300 V with the accuracy of 0.05%, supporting CAN bus communication with the BMS	

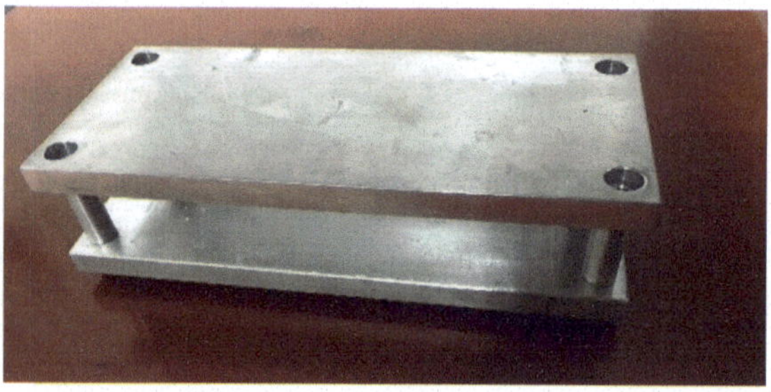

Fig. 2.3 A square battery clamp

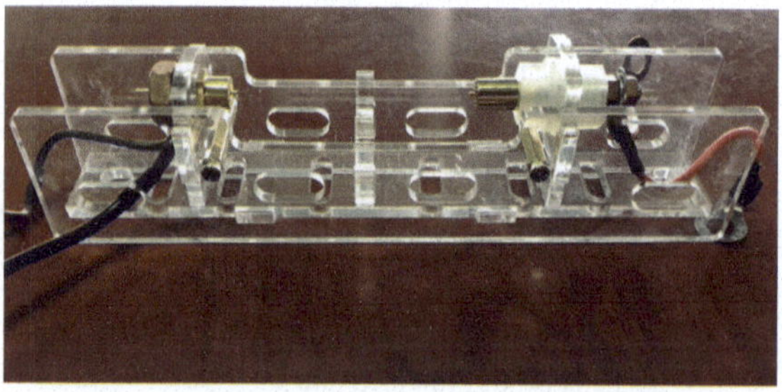

Fig. 2.4 A cylindrical battery clamp

2.1.2 *Impedance Characteristic Test Equipment in Frequency Domain*

Impedance characteristic test equipment in the frequency domain is used for measuring the AC impedance of the battery in a certain frequency range, i.e., EIS. It is usually done by an electrochemical workstation. The Zahner IM6 electrochemical workstation is shown in Fig. 2.5, and its specific parameters and characteristics are shown in Table 2.3. Zahner IM6 electrochemical workstation can acquire the AC impedance of the battery within the frequency range of 10 μHz–8 MHz. It has the functions of high input impedance and high current excitation. The maximum excitation current is 100 A. MITS Pro and Thales software programs installed in the host computer are used to manipulate Arbin-BT2000 tester and Zahner IM6 electrochemical workstation to carry out battery tests. It has the function of designing test programs, loading

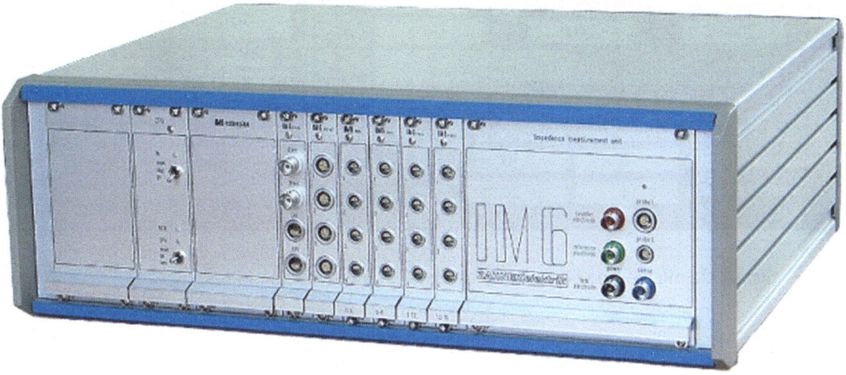

Fig. 2.5 Zahner IM6 electrochemical workstation

Table 2.3 Parameters and characteristics of the Zahner IM6 electrochemical workstation

Serial number	Parameters and characteristics	Values
1	AC impedance spectrum range	10 μHz–8 MHz
2	Maximum linear scan range	±4 V
3	Maximum output current range	±3.0 A
4	With a voltage buffer, when Gain = 0.4, the voltage range is ±10 V; when Gain = 1, the voltage range is 4 V	
5	Unique network application module to support remote start and measurement of instruments	

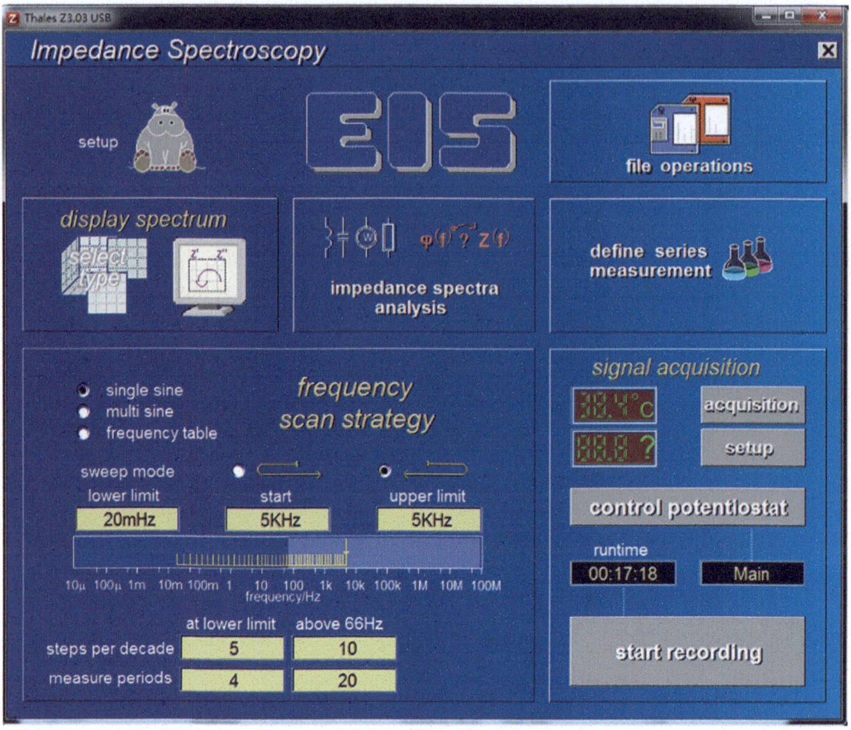

Fig. 2.6 Electrochemical impedance spectroscopy test interface

test conditions, storing test data, etc. Its electrochemical impedance spectroscopy test interface is shown in Fig. 2.6.

2.1.3 Environmental Simulation Equipment

Conditions such as ambient temperature and humidity have a significant impact on the internal resistance, capacity, and charge–discharge characteristics of the battery. In order to explore the effects of temperature and humidity on battery characteristics, we need to control environmental parameters using a temperature and humidity test chamber. In this book, the test platform uses a three-stacked programmable temperature and humidity test chamber, as shown in Fig. 2.7. The thermostat adopts an industrial microcomputer control system, which is suitable for constant high/low-temperature, gradual change, and damp heat environment simulation experiments of various auto parts, electronic parts, and other products. Specific parameters and characteristics are shown in Table 2.4.

Fig. 2.7 Three-stacked programmable temperature and humidity chamber

Table 2.4 Parameters and characteristics of the three-stacked programmable temperature and humidity test chamber

Number	Parameters and characteristics
1	Temperature range: –70 to 180 °C
2	Humidity range: 20–98% RH
3	Volatility: ≤0.5 °C (temperature), ≤2 .5% RH (humidity)
4	Temperature regulation rate: average nonlinearity 3.0 °C/min (heating), average nonlinearity 1.0 °C/min (cooling)
5	Total power: 14.0 kW (380 V three-phase four-wire + ground protection)
6	With constant high and low temperature, gradual temperature change, and damp heat and other functions
7	The refractory grade high strength PU polyurethane foam insulation material is used for the insulation layer
8	With over temperature protection, over current protection, and three color light alarms, and other functions

2.1.4 Battery Test Platform

To complete the comprehensive test of the battery, the AESA built the battery test platform shown in Fig. 2.8 with above battery experimental apparatus. The structure of the battery test platform is shown in Fig. 2.9.

In the battery test platform, the main body adopts a master-slave two-stage control structure, which is composed of a host computer and a lower computer. The host computer uses relevant software to control the lower computer to complete a variety of battery tests. The lower computer comprises a battery charge–discharge test equipment and electrochemical workstation. They are connected with the battery by the connecting line and run separately to test the AC characteristics and the DC characteristics of the battery. At the same time, the battery usually needs to be placed in the temperature chamber to achieve the target temperature and humidity and keep them constant. The collection module is responsible for collecting voltage, current, temperature, impedance and other signals of the battery and transmitting them to the host computer for data collection. The construction of the platform provides a hardware foundation for the test design of the battery, and also provides a powerful guarantee for the development and verification of the core algorithm of the BMS in this book.

It should be noted that this book adopts the data sampling frequency of 1 Hz.

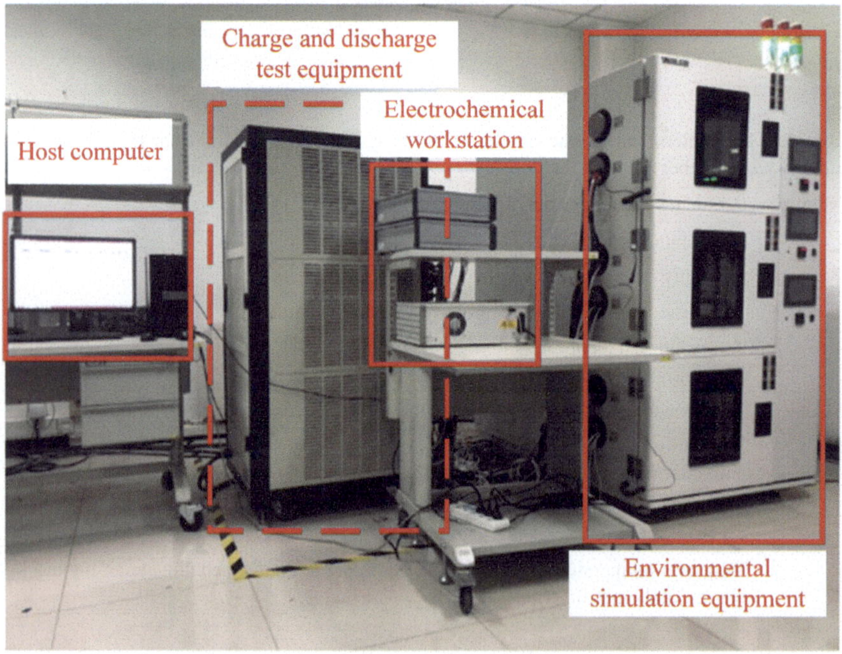

Fig. 2.8 Battery test platform

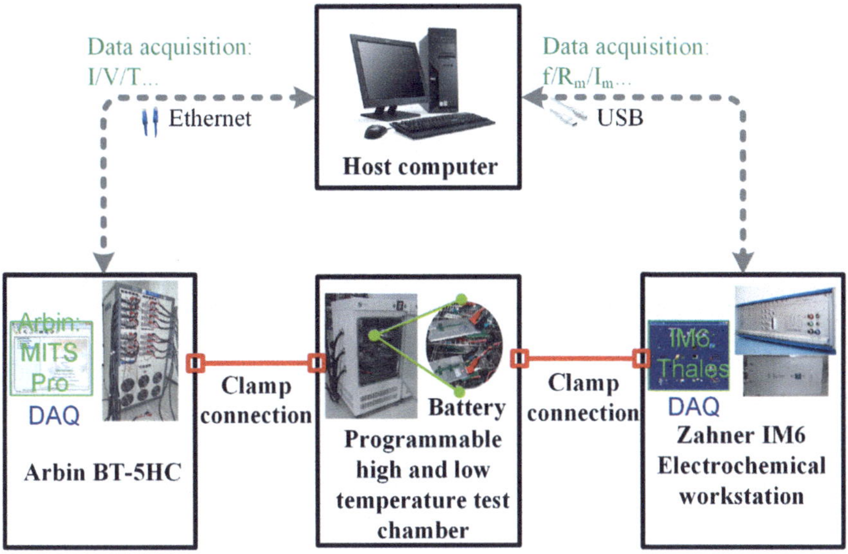

Fig. 2.9 Battery test platform structure

2.2 Battery Test Process

The Li-ion battery is a very complicated electrochemical system, and the characteristics of which are restricted by many factors such as operating conditions, ambient temperatures, humidity, aging state, vibration, and external clamping force. In order to effectively manage the battery in real time and more accurately estimate the SOC, SOH, and SOP of the battery, the dynamic relationship between the battery internal parameters and the external working environment is particularly important. Chapters 3–7 of this book introduce model-based battery state estimation methods and low-temperature heating and charging strategies. The development of these algorithms relies on a large amount of targeted battery characteristics data. Therefore, designing a set of reasonable and integrated experimental schemes for the battery becomes the primary task of developing the core algorithm of the BMS.

2.2.1 Introduction of Testing Standards

Battery testing standards include the PNGV Battery Test Manual, the USABC Electric Vehicle Battery Test Manual, Freedom CAR Battery Lifetime Test Manual released by The U.S. Department of Energy, IEC 61690 released by International Electrotechnical Commission of European Union, JIS-C-8711 released by Japan, etc. China also has specific standards for battery tests such as GB/T 18385 Electric Vehicle Power Performance Test Methods and GB/T 18386 Electric Vehicle

Energy Consumption and Driving Range Experimental Methods. The above various standards propose test methods for the battery voltage, capacity, internal resistance, charge and discharge current rate, temperature characteristics, and cycle life, and provide a lot of references in designing the comprehensive tests of the battery.

2.2.2 BMS Algorithm Development and Experimental Design

Algorithms used for SOC estimation, SOH estimation, SOP prediction, remaining useful life (RUL) prediction, low-temperature heating, and optimized charging strategy of battery are the core of BMS. In the development of an all-climate and full lifetime BMS, comprehensive testing of the battery is the best way of accurately obtaining the battery characteristics.

(1) Static capacity test is the basis for obtaining the battery SOC reference value and the SOH mapping parameter rule. The SOC is defined as the ratio of the current remaining capacity to the maximum available capacity [3]. The SOH is often defined as the ratio of the current maximum available capacity to the nominal capacity. The maximum available capacity test can get the static capacity of the battery, so it is of vital importance for the estimation of SOC and SOH. It should be pointed out that the real-time dynamic capacity is more important than the static capacity in practical applications, and the real-time dynamic capacity can better reflect the SOC and SOH levels of the battery. But accurate online estimation of dynamic capacity is a major challenge. Therefore, regular static capacity calibration becomes a viable approach;
(2) The open-circuit voltage (OCV) of the battery has a monotonic mapping relationship with SOC, and it varies with battery degradation. It is significant for improving the battery modeling and the SOC and SOH estimation accuracy by determining the OCV-SOC mapping rule [4];
(3) The internal resistance of the battery is a key parameter for calculating the heat generation rate during charging and discharging. The hybrid pulse power characteristic (HPPC) test data can be used to offline identify the parameters of the battery model. Then the parameters can be applied to the SOP estimation, low-temperature heating, and fast-charging strategies [5];
(4) Dynamic working condition test can simulate battery working condition in the actual vehicle driving process, and plays an important role in verifying the application performance of the battery key algorithms in actual working conditions;
(5) EIS can not only be used to analyze battery performance degradation and aging state characterization parameters, but also provide data support for the fractional-order modeling and the design of the low-temperature heating strategy of the battery.

In addition, due to the influence of the battery material activity, the battery characteristic data under different temperatures and aging stages are also different. In order to completely obtain the performance of the battery, each test needs to be conducted separately under different temperatures and different aging stages to achieve the efficient life cycle management of the battery in all-climate conditions.

Based on the existing battery test platform and algorithm requirements in the BMS, this book designs a comprehensive battery performance test scheme, as shown in Fig. 2.10. The test scheme mainly includes conventional electrical performance test, AC impedance test, and remaining life test [6]. The conventional electrical performance test consists of four subtests: maximum available capacity test, OCV test, HPPC test, and dynamic test. The remaining life test of the battery is composed of the charging and discharging tests at different temperatures, current rates, cutoff voltages, and SOC intervals. The battery tests involved in the different BMS algorithms are shown in Table 2.5.

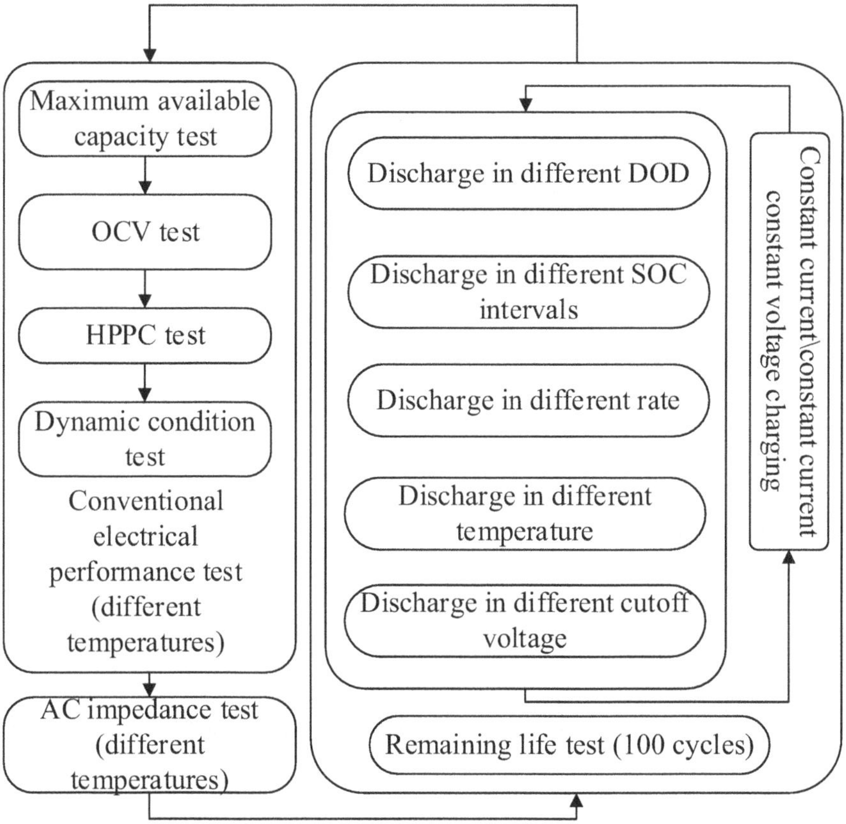

Fig. 2.10 Battery performance testing scheme

Table 2.5 Battery tests involved in different BMS algorithms

BMS algorithm	Maximum available capacity test	OCV test	HPPC test	Dynamic test	AC impedance test	Life test
SOC estimation	√	√	√	√	√	√
SOH estimation	√	√	√	√	√	√
SOP prediction	√	√	√	√		√
RUL prediction	√			√	√	√
Low-temperature heating	√		√		√	√
Optimized charging	√	√	√			√

2.2.3 *Battery Conventional Electrical Performance Test*

Maximum available capacity test

The primary purpose of this test is to calibrate the current maximum available capacity by charging and discharging the battery. As the degradation of the battery, its capacity will continue to decline. The maximum available capacity tests at different aging states and temperatures play an important supporting role in the development and evaluation of the battery SOC and SOH estimation algorithms [7].

Based on the requirements of Technical Conditions for Battery Management Systems in Electric Vehicles, it is necessary to measure the maximum discharge capacity of the battery three times in succession. The test method is to fully charge the battery at standard current with constant-current constant-voltage (CCCV). The schematic diagram of CCCV charging is shown in Fig. 2.11. After fully charging the battery, rest for a period of time and then discharge the battery with a constant current (CC) to the lower cutoff voltage. The test will repeat three times. If the deviations between the discharge capacity of three tests and the mean value of the three test results are within 2%, the maximum available capacity test result is valid. The average value of the three test results is taken as the maximum available capacity $C_{\max}$ of the battery. Otherwise, the test needs to be continued until the discharge capacity of three consecutive times meets the requirement. Generally, the charge and discharge rate of CC is 0.3C.

The current and voltage curves of a 2.4 A h ternary material battery capacity test are shown in Fig. 2.12. Unless otherwise specified, the book sets discharge current as positive and charge current as negative.

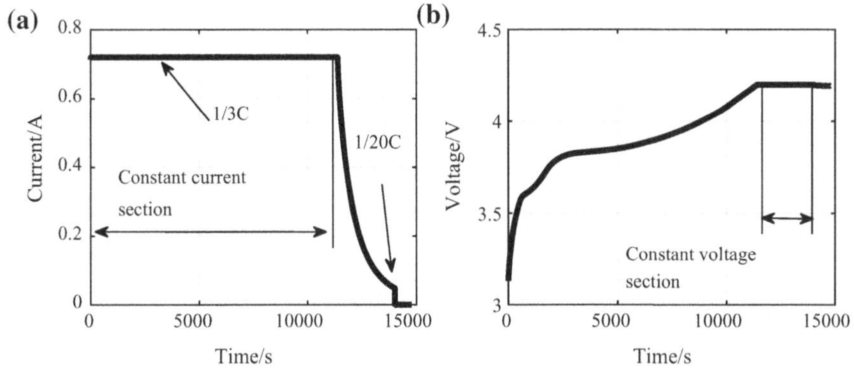

Fig. 2.11 CCCV charging diagram (charge current is positive): **a** current curve; **b** voltage curve

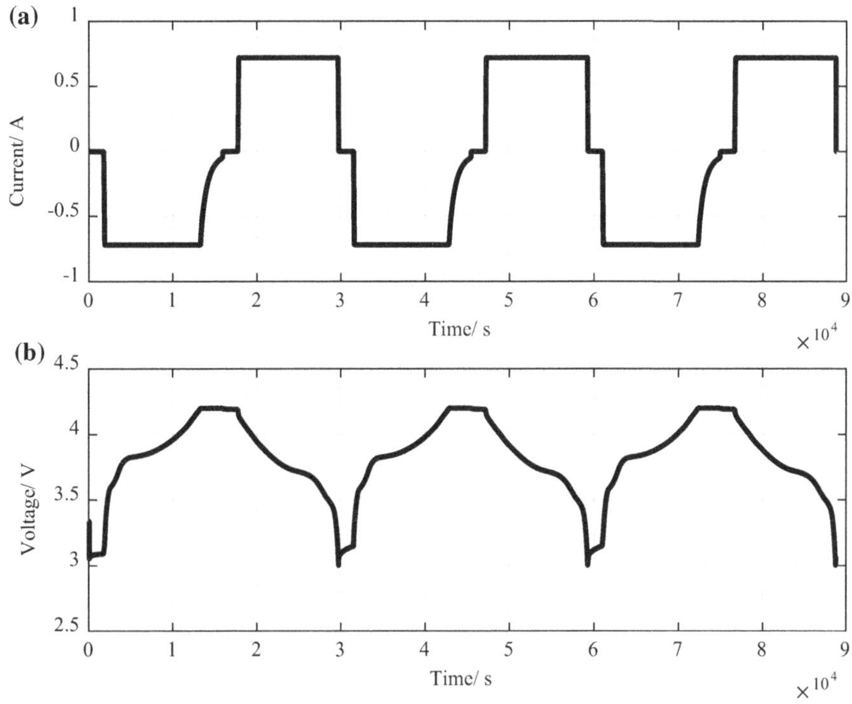

Fig. 2.12 Current and voltage curves of a 2.4 A h ternary material battery capacity test: **a** current curve; **b** voltage curve

Open-circuit voltage test

The purpose of this test is to establish the relationship between OCV, SOC, and the available capacity. Each kind of battery has its own specific OCV curve, which has a fixed relationship with the SOC at the same temperature. At the same time, the OCV is also affected by aging, which can be used for SOH estimation.

Due to the hysteresis effect, the OCV of lithium-ion batteries is different during charging and discharging. The charging process of the OCV test is as follows:

1. The battery is discharged to the cutoff voltage by a standard current. Rest for 5 h and record the terminal voltage, which is considered as the OCV when SOC = 0%;
2. The charging operation is performed on the battery by CCCV under standard current rate. The cutoff criterion is that the charging capacity is 5% of the maximum available capacity or the charging current decreases to the charging cutoff current. Then the terminal voltage is recorded after resting for 5 h;
3. Skip to Step 2 and repeat Steps 2 and 3 until the battery is fully charged.

The discharging process of the OCV test is as follows:

1. The battery is fully charged in a standard CCCV charging mode. Rest for 5 h and record the terminal voltage, which is considered as the OCV when SOC = 100%;
2. Discharge the battery at standard rate. The cutoff criterion is that the charging capacity is 5% of the maximum available capacity or the lower cutoff voltage is reached. The terminal voltage is recorded after resting for 5 h;
3. Skip to Step 2 and repeat Steps 2 and 3 until the battery reaches its discharge cutoff voltage.

OCV curves in charge and discharge process and their differences are shown in Fig. 2.13.

Hybrid pulse power characterization test

The HPPC test adopts a continuous pulse excitation sequence to charge and discharge the battery to obtain the dynamic characteristics of the battery at different SOCs. The HPPC test data is often used to identify the model parameters at different SOCs. The specific procedure are as follows:

1. The battery to be tested is fully charged in the CCCV mode with a standard current rate.
2. Rest for 5 h to make the battery terminal voltage close to equilibrium.
3. Load the hybrid pulse current excitation sequence, then perform a constant current discharge operation on the battery for a period of time. Rest for 1 h. Note that the constant current discharge process during this period is used to ensure that the SOC interval between the two pulse excitation sequence tests is 5%. So, the test data at 100, 95,..., 5% SOC can be obtained.
4. Repeat Step 3 until the battery reaches its discharge cutoff voltage.

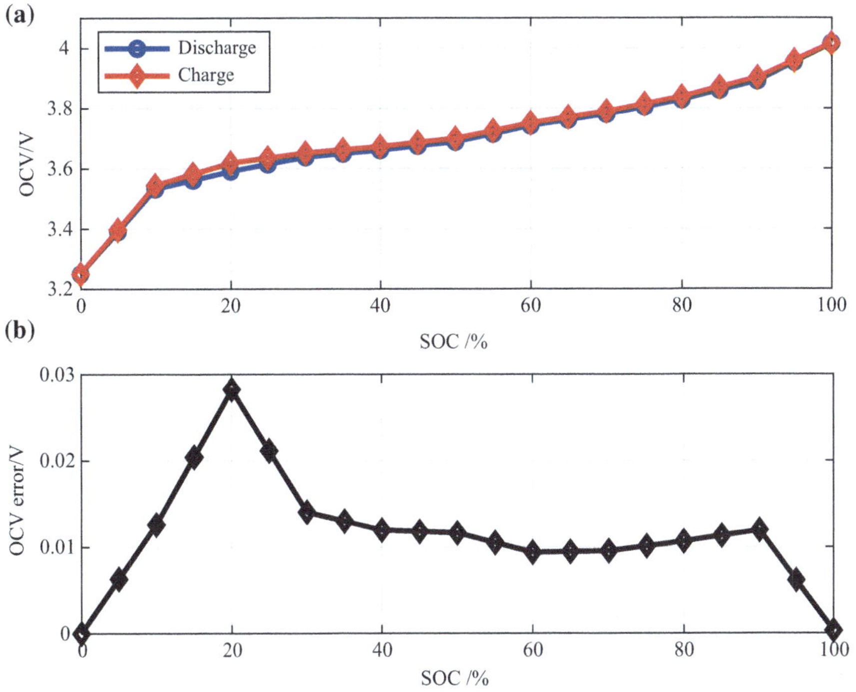

Fig. 2.13 OCV curves and their differences in the charge and discharge processes. **a** the corresponding relationship between OCV and SOC; **b** the OCV difference between the charge and discharge

In order to obtain a more comprehensive battery polarization characteristic, the experiment adopts four current rates to form a pulse excitation. In order to prevent the battery from being overcharged in the fully charged state, the pulse excitations use constant current and constant voltage. The current curve (current amplitude depends on the battery type) is shown in Fig. 2.14.

Dynamic test

Dynamic test refers to the test carried out by simulating the current excitation in the actual EV driving cycles to obtain the dynamic characteristics of the battery. Based on the dynamic test data, it is possible to simulate the applicability of the BMS core algorithms in practical applications. The commonly used dynamic tests include Dynamic Stress Test (DST), Federal Urban Driving Schedule (FUDS), Urban Dynamometer Driving Schedule (UDDS) test, New European Driving Cycle (NEDC), and China Typical City Driving Cycle (CTCDC). Figures 2.15, 2.16, 2.17, 2.18, and 2.19 show the practical current profiles of the DST, FUDS, UDDS, NEDC, and CTCDC, respectively.

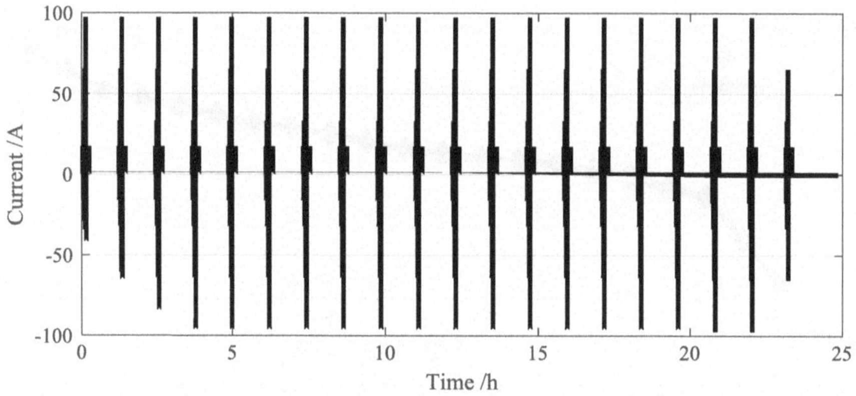

Fig. 2.14 Hybrid pulse power characteristic test current

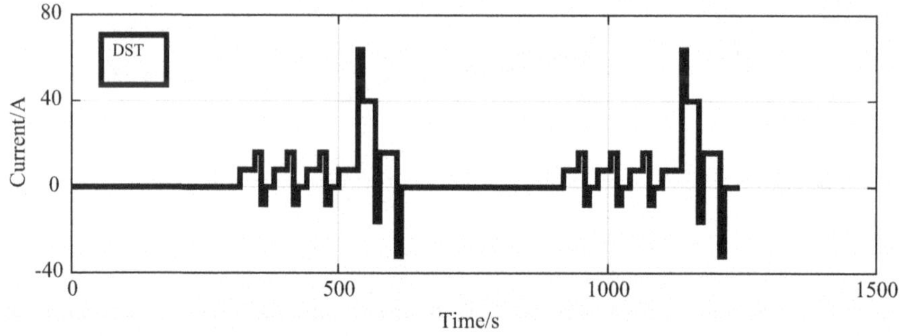

Fig. 2.15 Current profile of DST (two cycles)

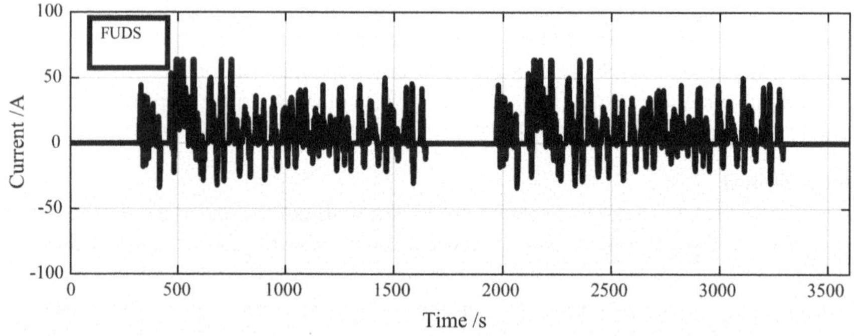

Fig. 2.16 Current profile of FUDS (two cycles)

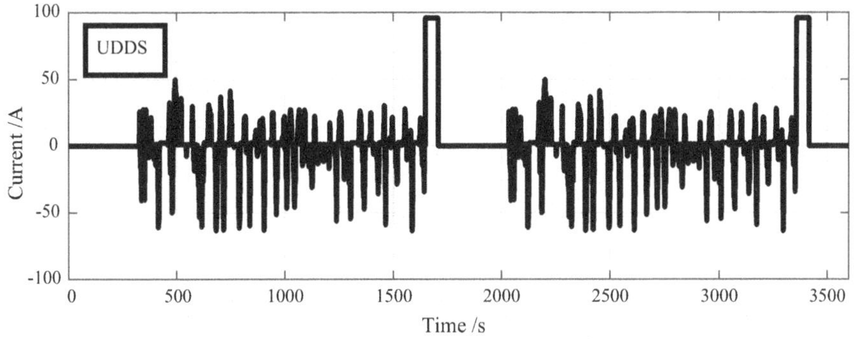

Fig. 2.17 Current profile of UDDS (two cycles)

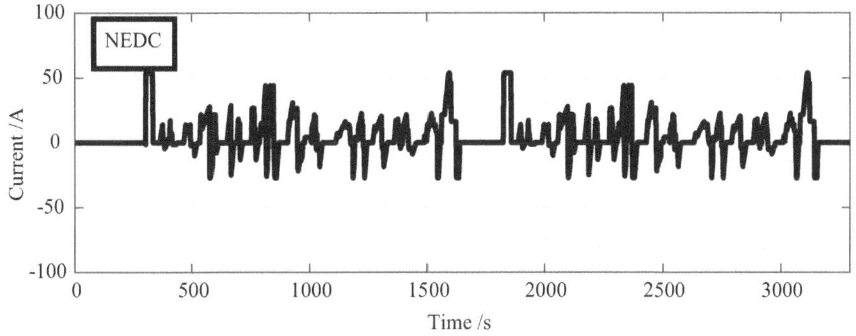

Fig. 2.18 Current profile of NEDC (two cycles)

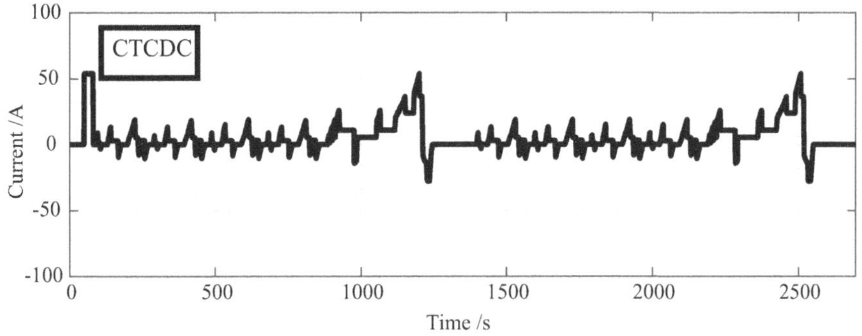

Fig. 2.19 Current profile of CTCDC (two cycles)

2.2.4 AC Impedance Test

The AC impedance test is a method of measuring the impedance spectrum of a battery system in a certain frequency range. It is achieved by using small-amplitude sinusoidal voltage signal (or current signal) as a disturbance input to generate an approximately linear relationship response of current or voltage. This "black box method" takes voltage and current as input and output, respectively, and indirectly obtains internal impedance of the battery.

In the process of studying impedance spectra, researchers found that the characteristics of the electrode interface double-layer capacitance deviate from the pure capacitance based on the electrochemical principle. Thus, the fractional-order model was established, which significantly improved the fitting accuracy of the battery characteristics in the frequency domain and time domain. In addition, there is a strong monotonic mapping relationship between the AC impedance spectrum and the aging state of the battery. After obtaining the battery AC impedance spectrum, the SOH of the battery may be calibrated by certain extracted feature parameters [8–10].

There are multiple methods to conduct EIS test for the battery by electrochemical workstations, including two-electrode system, three-electrode system, and four-electrode system, etc. [11]. Figure 2.20 shows a two-electrode system comprising a working electrode, and a counter electrode. A three-electrode system has one more reference electrode of known potential, whereby the potential of the working electrode can be obtained. Meanwhile, the current flowing through the system can be

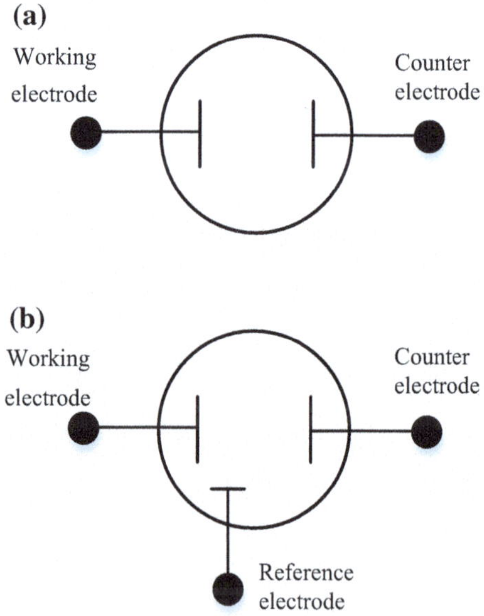

Fig. 2.20 Two-electrode system and three-electrode system for EIS testing

obtained by the working electrode and the counter electrode circuit, and the relationship between a certain electrode potential and the interface reaction of the battery can be studied. Currently, the commonly used is two-electrode system.

Figure 2.21 shows the connection mode between the battery and the electrochemical workstation in the two-electrode system and Fig. 2.22 shows the physical connection mode. The positive electrode of the battery is used as a working electrode and the negative electrode of the battery is used as a counter electrode. They are connected to the corresponding interface of the electrochemical workstation and is used

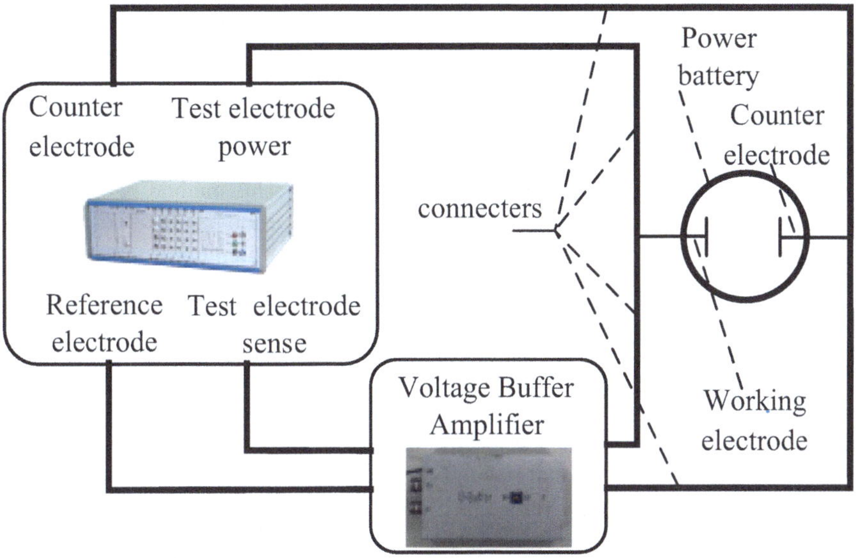

Fig. 2.21 Connection mode between the battery and the electrochemical workstation in two-electrode system

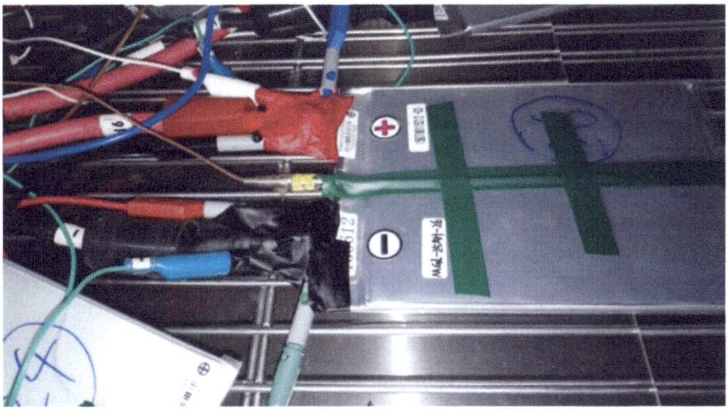

Fig. 2.22 Physical connection between the battery and the electrochemical workstation

to provide sinusoidal excitation signals. In addition, the reference electrode line is connected to the negative electrode, and the test electrode sense line is connected to the positive electrode. There are no current flows between the two to measure the voltage signal between them.

There are usually two test methods for EIS test, the galvanostatic method and the constant potential method. In theory, the measurement results of these two methods are consistent, but the application scenarios are different. The constant potential method is relatively widely used. The excitation of the battery is provided by a composite voltage signal obtained by superimposing a constant voltage and a sinusoidal signal with a definitive amplitude. At the same time, the electrochemical workstation measures the AC current response of the system and obtains impedance according to the ratio of voltage to current. The galvanostatic method is to impose the composite current signal superimposed by a direct current (can be 0) and a sinusoidal signal with a definitive amplitude on the battery. Meanwhile, the electrochemical workstation measures the voltage signal response of the battery system and calculates impedance according to the ratio of voltage to current. This method is commonly used in corrosion-related tests or fuel cell tests. For the Li-ion battery, both the constant potential in situ EIS test and the constant current in situ EIS test are used. The constant current in situ EIS test can prevent the battery from being charged or discharged during the long-term measurement process, thereby keeping the value of SOC constant.

The rest time after charging and discharging directly affects the internal stability of the Li-ion battery system, which is also an important factor affecting the reliability of the electrochemical impedance spectrum measurement results. The rest time can be determined by measuring the trend of the voltage and current during this period and comparing the measurement results of the impedance spectrum at different resting times. Under normal circumstances, the reasonable resting time is 3 h. Compared with the shorter resting time, the repeatability of the measurement results is greatly improved, and the occurrence frequency of low-frequency dead spots is significantly reduced. In addition, the EIS test is also affected by many factors, such as the connection position of the electrochemical workstation clamp, the length of the testing line on the battery tab, which must be unified in the experiment.

The EIS test results of the battery are usually presented on the negative Nyquist diagram [12]. The abscissa is the real part of the impedance, and the ordinate is the negative imaginary part of the impedance. The frequency gradually reduces from the lower left to the upper right. Figure 2.23 shows a typical impedance spectrum of a battery.

2.2.5 *Remaining Life Test*

The durability of the battery is closely related to its remaining life. It indicates the tolerance of the battery system under different working conditions, especially under extreme operational conditions. The current research on battery durability

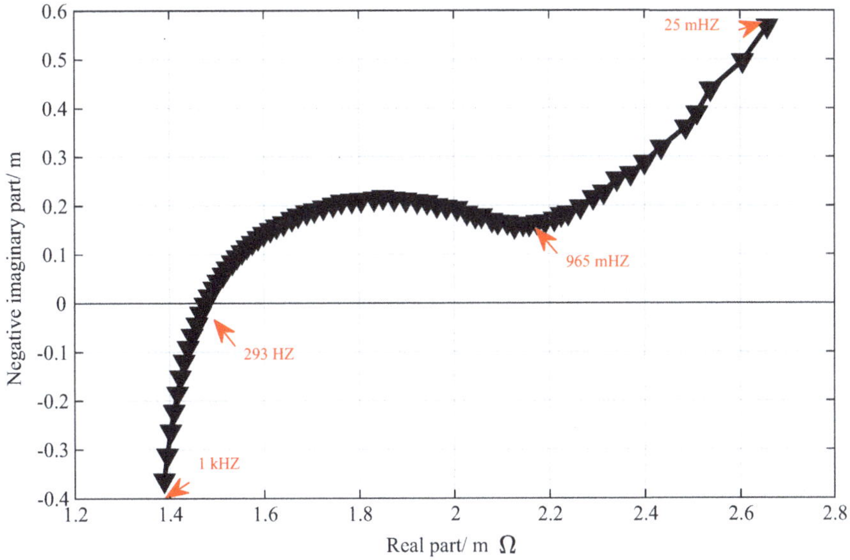

Fig. 2.23 Impedance spectrum of a battery

management focuses on battery life prediction and health status assessment under single stress or multiple stresses to alert or intervene for foreseeable battery faults and failures.

On the one hand, the battery life is usually required to last 10–15 years in the practical application of EVs, but it must meet the requirements on the minimum levels of cost and time in the relevant tests. On the other hand, considering the complex external factors in practical applications, in addition to different degradation trajectories of battery life at different stress levels, the remaining life test is required to realize decoupling of the mixed stress, of which includes the rate of charge and discharge, cutoff voltage, SOC interval, temperature, etc. Studies have shown that obtaining the degradation trajectory of the battery and the characteristics under different aging stages through the remaining life test is a feasible solution for realizing the battery cells and system RUL prediction and rapid evaluation of durability.

In summary, this book introduces a set of remaining life test plans involving different current rates, ambient temperatures, SOC intervals and discharging tests with lower cutoff voltage. Part of the test results will be used to support the research in the remaining life prediction in Chap. 6. The detailed procedures of the test are as follows:

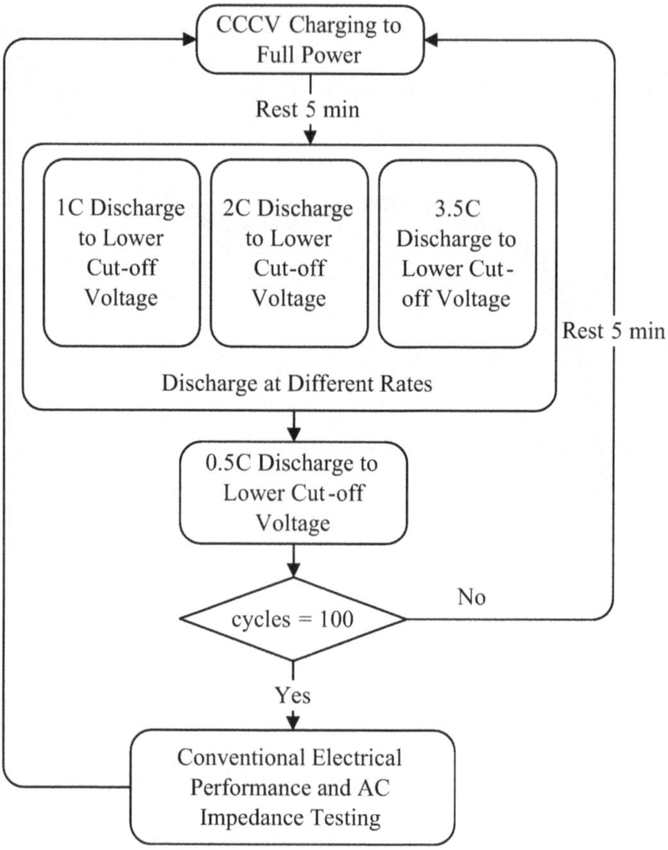

Fig. 2.24 Remaining life test with different current rates

Protocol 1: Remaining life test at different current rates (Fig. 2.24).

(1) Charge to the upper cutoff voltage with a constant current of 0.5C, and then charge to a cutoff current of 0.05C with a constant voltage.
(2) Rest for 5 min.
(3) Discharge separately at 1C, 2C, and 3.5C to the lower cutoff voltage, then discharge the battery to the lower cutoff voltage at 0.5C.
(4) Rest for 5 min and return to step (1).
(5) For every 100 cycles, a routine electrical performance test and an AC impedance test are performed once, respectively.

Protocol 2: Remaining life test at different temperatures.

Repeat the protocol 1 by setting the ambient analog device temperature at 10, 25, and 40 °C, respectively.

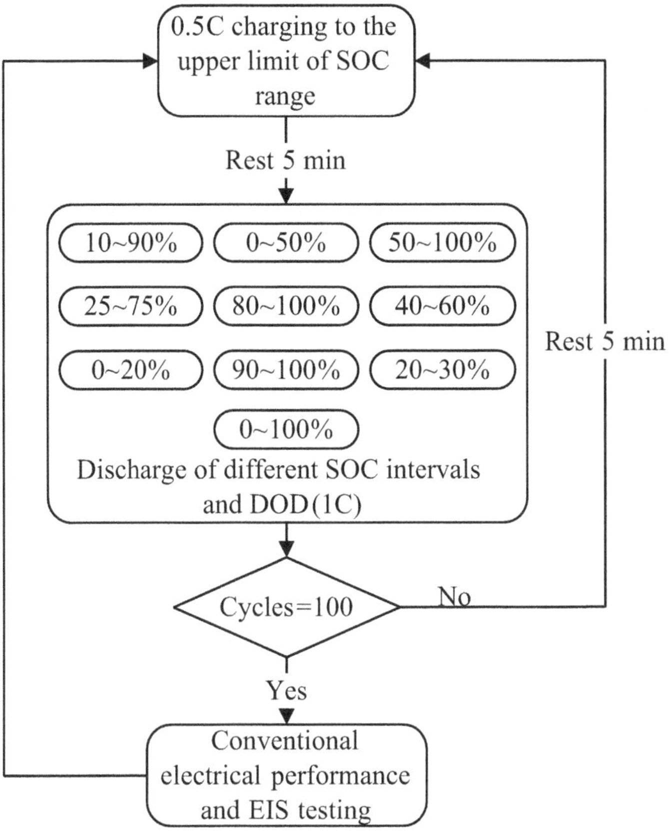

Fig. 2.25 Remaining life test with different SOC intervals

Protocol 3: Remaining battery life test for different SOC intervals (Fig. 2.25).

(1) Charge to the upper cutoff SOC with a constant current of 0.5C. If the upper cutoff SOC is 100%, it is necessary to charge the battery to the upper cutoff voltage with the CCCV mode, and then charge to the cutoff current of 0.05C in the constant voltage mode.
(2) Rest for 5 min.
(3) Discharge, respectively, at 10 SOC intervals including 0–100, 10–90, 50–100, 25–75, 0–50, 80–100, 40–60, 0–20, 90–100, 20–30%, and 5 different ΔSOC and the discharge current is 1C in all cases.
(4) Rest for 5 min and return to step (1).
(5) For every 100 cycles, a routine electrical performance test and an AC impedance test are performed once, respectively.

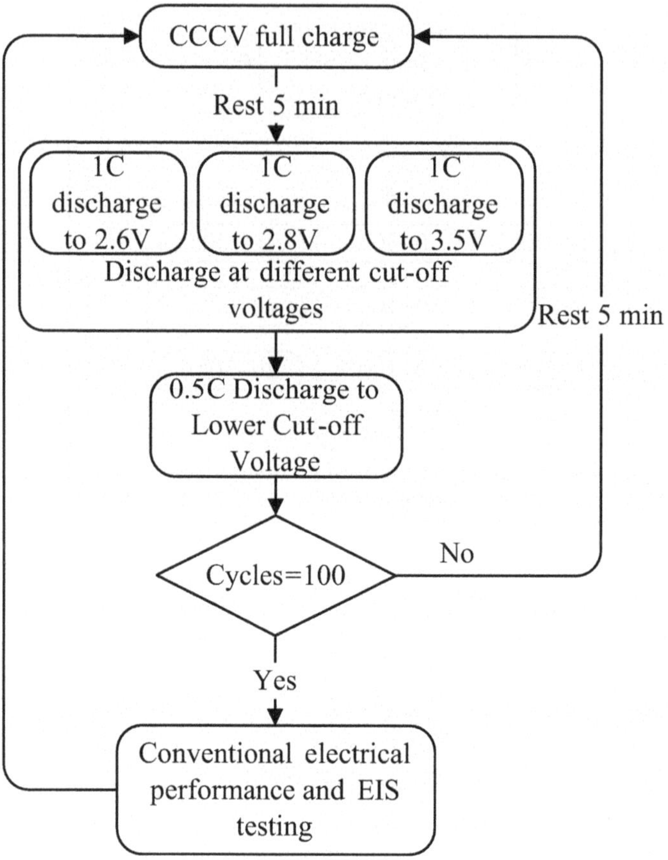

Fig. 2.26 Remaining life test with different cutoff voltages

Protocol 4: Remaining life test for different cutoff voltages (Fig. 2.26).

(1) Charge to the upper cutoff voltage with a constant current of 0.5C, then charge the battery to a cutoff current of 0.05C with a constant voltage.
(2) Rest for 5 min.
(3) Discharge at a constant current of 1C to the lower cutoff voltage of 2.6 V, 2.8 V and 3.5 V (parameters can be set based on the battery) before discharging at a constant current of 0.5C to the corresponding lower cutoff voltages of 2.6 V, 2.8 V, and 3.5 V.
(4) Rest for 5 min and return to step (1).
(5) For every 100 cycles, a routine electrical performance test and an AC impedance test are performed once, respectively.

Table 2.6 Related batteries and their basic parameters in the book

Serial number	Battery type	Nominal capacity (A h)	Upper cutoff voltage (V)	Lower cutoff voltage (V)	Shape
Battery 1	NMC	25	4.2	2.5	Square
Battery 2	NMC	32	4.05	3.0	Square
Battery 3	LMO	35	4.2	3.0	Square
Battery 4	LMO	90	4.2	3.0	Square
Battery 5	NMC	2	4.1	3.0	Cylinder
Battery 6	NCA	2.7	4.2	2.5	Cylinder
Battery 7	LFP	27	3.65	2.65	Square
Battery 8	NMC	2.5	4.2	3.0	Cylinder
Battery 9	NMC	50	4.25	2.8	Square
Battery 10	NCA	3	4.2	2.5	Cylinder

2.3 Battery Test Data

Based on the above battery testing procedures, the AESA research group have worked for many years on building up a database comprising of multiple lithium iron phosphate batteries and ternary batteries. Battery types and characteristics involved in this book are shown in Table 2.6. It should be noted that the batteries used in this book are referred to as "battery number" or "battery number-cell number". For example: "battery 1-cell 02" refers to No.2 of a ternary battery with nominal capacity of 25 A h.

The characteristics data of the battery obtained from routine electrical performance test and AC impedance test conducted at different life cycles and temperatures are also different. Understanding the working characteristics of the battery during its full lifetime cycle can ensure the stability of the BMS algorithms in the long-term use of new energy vehicles. Currently, AESA has established an all-climate, full-life battery test database, as shown in part in Table 2.7.

2.4 Characteristics Analysis of the Battery Experiments

2.4.1 Temperature Characteristics of the Battery

The effect of temperature on the capacity of batteries

The temperature has a great impact on the battery. Figure 2.27 shows the maximum available capacity of the battery 7 at different temperatures (–10, 0, 25, and 40 °C).

It can be seen that the capacity of the battery increased with temperature rising within a certain temperature range. The difference in the maximum available capacity between –10 and 40°C reaching 5 A h, which accounts for about 18.5% of the rated

Table 2.7 List of partial databases covered in this book

Battery type	Data content
$LiFePO_4$ battery	Multi-batch, multi-series power cells (a total of 120 cells)
	Aging data, characteristics, working conditions and AC impedance test data in a wide temperature range (–10 to 45 °C) and wide life span (70–100%) with the number of cycles exceeds 4000
Ternary battery (square)	Multi-batch power cells (a total of 100 cells)
	Aging data and characteristics, working condition experimental data in a wide temperature range (–10 to 45 °C), wide life span (80–100%)
Ternary battery (cylindrical)	Multi-batch and different specifications of the power cells (a total of 300 cells)
	Full-life (80–100%) capacity attenuation data and characteristic data of different magnifications, different temperatures, different cutoff voltages, and different SOC intervals
	Capacity, working conditions, AC impedance, efficiency, and other data in wide temperature (–10 to 45 °C) and under different aging time

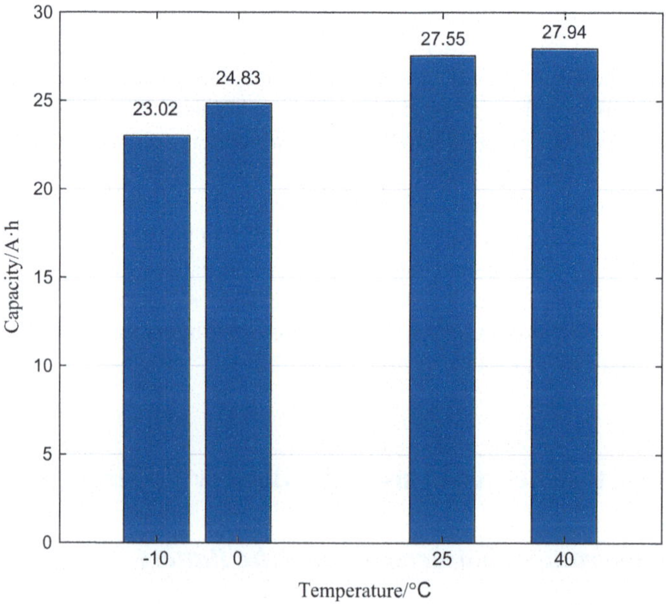

Fig. 2.27 Maximum available capacity of the battery 7 at different temperatures (–10, 0, 25 and 40 °C)

capacity. Temperature affects the battery material activity and the charge/discharge performance, which are reflected in the internal resistance and the OCV of the battery model. The terminal voltage is higher at a high temperature than at a low temperature during the discharge process. Thus, under the condition of the same discharge cutoff voltage and discharge current, the duration of battery discharge will be longer and the discharge capacity will be larger at high temperature, whereas the battery will reach its cutoff criterion earlier and the discharge capacity will be smaller at low temperature. Therefore, the operating temperature of the battery needs to be adjusted within the proper range.

The effect of temperature on open-circuit voltage

The OCV-SOC relationship of the battery is one of the most important relationships for state estimation and battery modeling. At the same temperature and testing procedures, the measured OCV-SOC curve has an excellent repeatability. However, the OCV-SOC curve will change at different temperatures. Figure 2.28 shows the OCV-SOC curves at different temperatures, the OCV-SOC curves overlap in the range of 10–100% SOC while having substantial difference in the low SOC range. The main reason is that the internal resistance of the battery increases at low temperatures, the discharge process will be faster to reach the cutoff voltage, which causes the battery not fully discharged. As a result, the feature will also affect battery modeling and state estimation. At the same time, the maximum available capacity of the battery after the temperature change will also change, which will bring uncertainties to the dynamic estimation of SOC and SOH of the battery, or even cause the non-convergence of the algorithms. Therefore, the influence of temperature must be fully considered in the development of BMS algorithms.

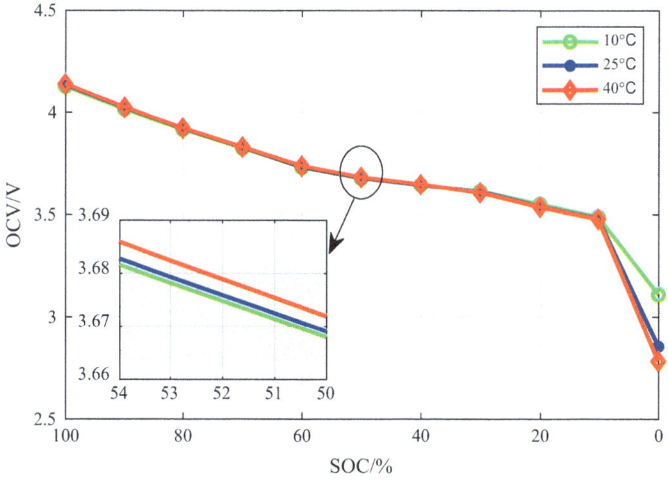

Fig. 2.28 OCV-SOC curve at different temperatures

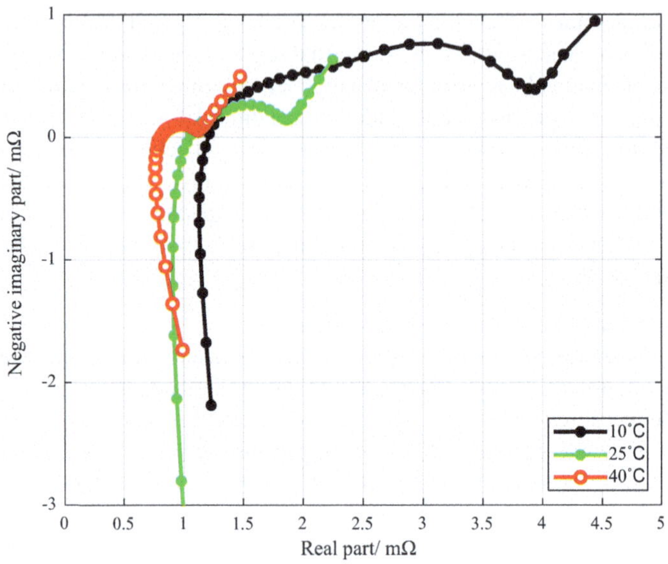

Fig. 2.29 EIS test results at different temperatures

The effect of temperature on AC impedance

Figure 2.29 indicates the EIS test results on battery 1 at different temperatures. It can be seen that the temperature has a great influence on the AC impedance of the battery. After the temperature declines, the AC impedance of the battery increases significantly. When at low temperatures, the loss of energy and capacity of the battery increases, resulting in a decrease in the maximum available capacity in the battery, which also confirms the experimental results. Therefore, it is extremely important to consider the temperature characteristics of the battery parameters to improve the prediction accuracy of the model under different application environments. The inappropriate model parameters can significantly reduce the accuracy in prediction of the battery model and the management level of the BMS. In addition, a proper and effective battery thermal management system is of great significance for the optimization and improvement of the working environment for batteries.

2.4.2 *Performance Degradation Characteristics of the Battery*

Relationship between performance degradation and the number of cycles

As the working time of the battery increases, the reduction of the active ion concentration and the loss of the electrode active material are caused, which further leads to an increase in electrolyte impurities, an increase in the charge transfer resistance, and the difficulty in Li-ion intercalation and deintercalation of the electrode. Eventually, the performance of the battery is degraded. Figure 2.30 shows the maximum available

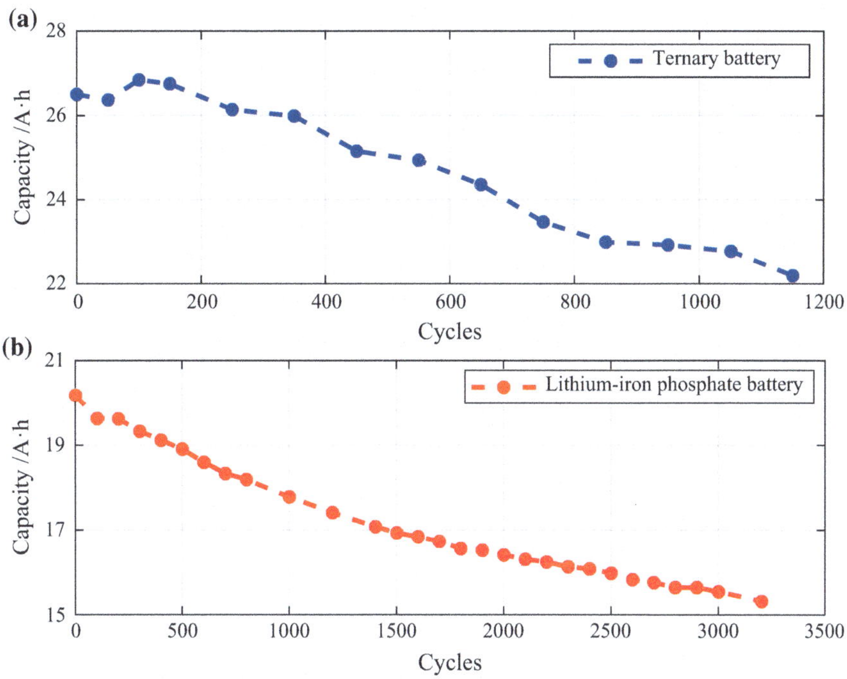

Fig. 2.30 Maximum available capacity of the battery under different aging stages

capacity of the battery at different aging stages. Consistent with the expectation, there is an approximately linear negative correlation between maximum available capacity and cycles. The cause of the specific declining trajectory will be detailed in Sect. 2.4.3. It is worth noting that, as compared to ternary batteries, $LiFePO_4$ batteries have a longer lifetime.

Effect of performance degradation on the open-circuit voltage

A decline in battery performance not only causes a reduction in the capacity, but also has an effect on the OCV-SOC curve. Figure 2.31a shows the OCV-SOC curves at different aging stages. It indicates that the OCV-SOC curve gradually changes as the number of cycles increases, which is especially noticeable in the low SOC range. Due to the different maximum available capacity of the battery at different aging stages, the electricity represented by the same SOC on the abscissa is not consistent. When the amount of discharge capacity is converted to that shown in Fig. 2.31b, the difference in the OCV of the battery will be more noticeable. In the case where the same amount of capacity is discharged, the aging of the battery is more severe, and the OCV of the battery drops more rapidly.

The effect of performance degradation on AC impedance

The crucial cause of the change in capacity and OCV under different aging stages of the battery is that the internal structure of the battery has changed. The test of the

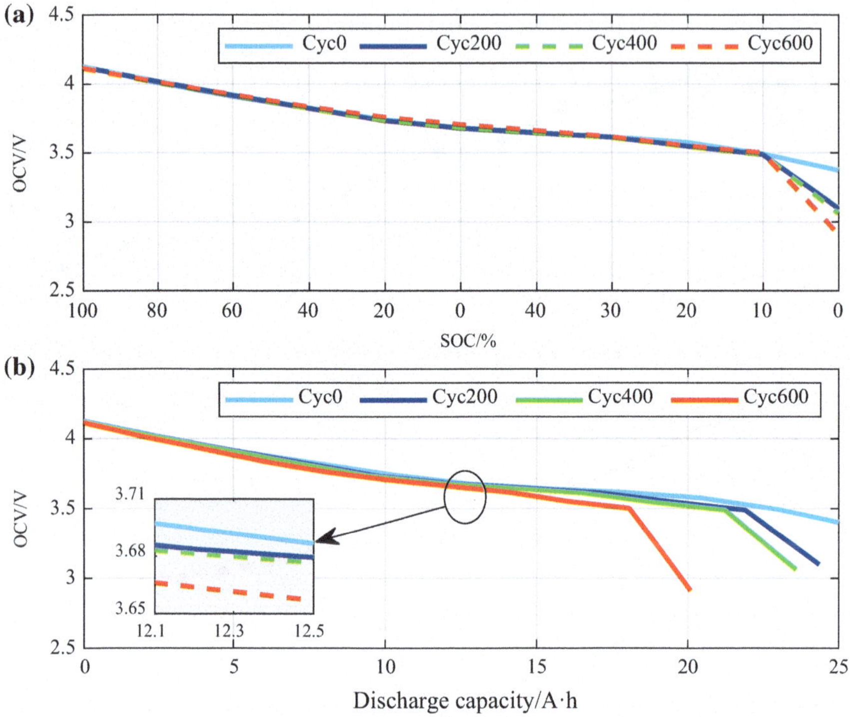

Fig. 2.31 OCV curves in different aging stages. **a** OCV-SOC curves; **b** OCV-SOC discharge capacity curves

impedance can better reflect the changes in the internal mechanism of the battery. Figure 2.32 shows that the AC impedance curves at different aging stages, which reveals that the AC impedance value is closely related to the aging stage of the battery. Overall, as the performance of the battery declines, its internal resistance gradually increases, which is also the main cause of the decline in battery lifetime. Therefore, the real-time update of the battery parameters is particularly important to improve the prediction accuracy of the model under different aging stages. Only the accurate estimation of the aging stage and internal parameters of the battery can improve the prediction performance of the battery model.

2.4.3 Cycle Characteristics of the Battery

This section provides a detailed analysis of the test data for the battery 10. It should be noted that the positive electrode material of the battery used in this section is $Li(NiCoAl)O_2$ and the negative electrode material is graphite carbon. Its specific

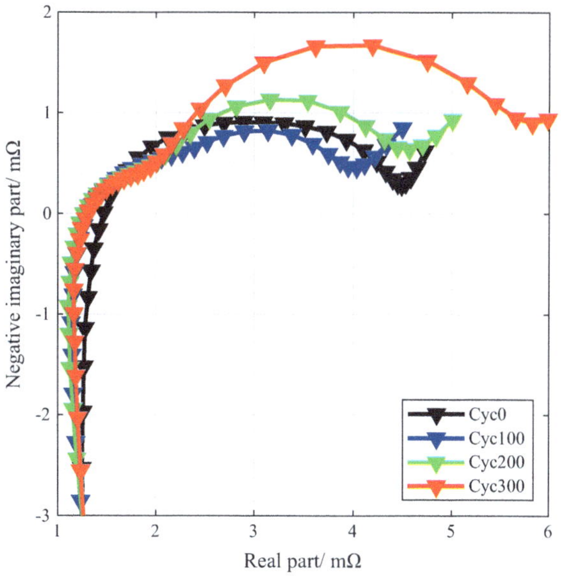

Fig. 2.32 EIS test results in different aging stages

characteristics of capacity deterioration may be different from other types of batteries, but the analysis method can still be applied for other models of batteries.

The trajectory of lifetime deterioration characteristics

Figure 2.33 shows the capacity deterioration trajectory of the battery under different discharge rates. Figure 2.33a shows the capacity decline trajectory of the battery at discharge rates of 1 and 0.5C. Figure 2.33b shows the capacity decline trajectory of the battery at discharge rates of 1.5 and 2C.

It can be seen from Fig. 2.33 that the capacity deterioration trajectory of the battery can usually be divided into two stages: stage 1 is a nonlinear capacity rapid attenuation section, while stage 2 is an approximate linear capacity deterioration section [13]. The demarcation point of these two stages is between 50 and 200 cycles. It should be noted that the exact number of segmentation cycles may not be exactly the same as shown in the figure, but has no impact on the analysis of the deterioration characteristics. The battery capacity in stage 1 experienced a relatively fast process of deterioration, the main reason of which is that the internal active material was consumed during the formation of the SEI film of the negative electrode of the battery. In stage 2, the SEI film of the battery has been completely formed, when the capacity is relatively stable and the deterioration rate is reduced, resulting in an approximately linear capacity decay trajectory [14].

Figure 2.34 describes the average capacity deterioration rate of battery at different current rates, in which the average capacity decline rate is the ratio of the percentage of its capacity deterioration to the total number of cycles. It can be seen that the average capacity deterioration rate of the battery is between 0.15% and 0.25% in

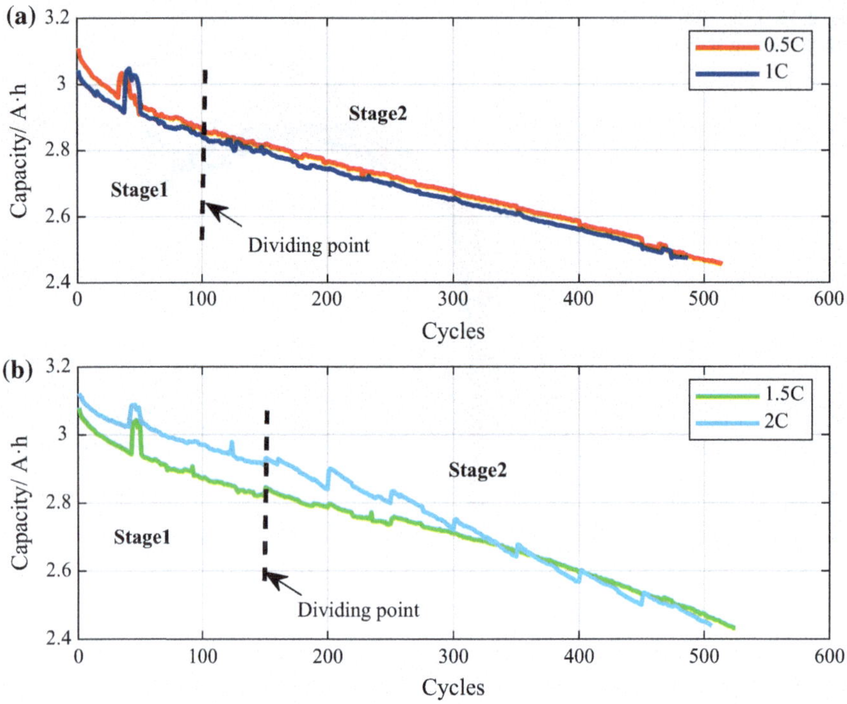

Fig. 2.33 Capacity decline characteristics at various discharge rates. **a** 0.5 and 1C; **b** 1.5 and 2C

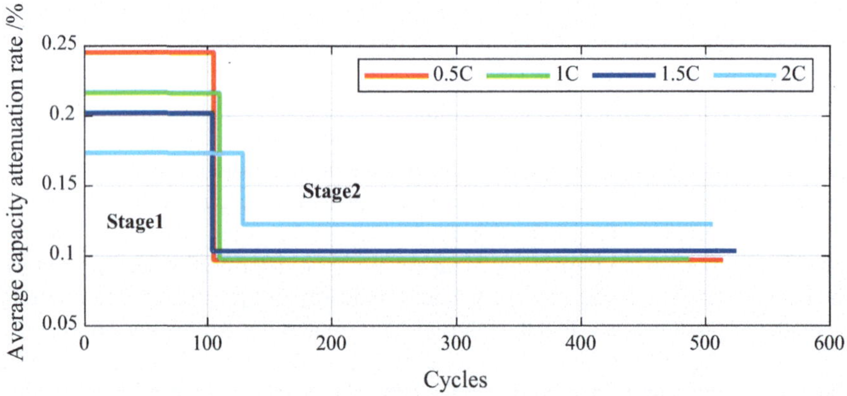

Fig. 2.34 The average capacity attenuation rate versus different current rates

stage 1, while the average rate of capacity degradation of the battery decreases to 0.1% in stage 2.

The discovery of the segmentation characteristics provides a new idea for the remaining life prediction. If the stage demarcation point can be accurately estimated, the linear model can be used to fit the capacity data of stage 2. Accordingly, the calculation process is simplified and computational burden is reduced while predicting the remaining life of the battery accurately.

Relationship between lifetime and temperature

The operating temperature has an important influence on the capacity deterioration characteristics of the battery and the lifetime. As discussed in Sect. 2.4.2, this section will specifically analyze the effect of temperature on the capacity deterioration trajectory. Figure 2.35 shows the maximum available capacity of the battery at different temperatures.

It can be seen from Fig. 2.35 that the higher the temperature is, the faster the capacity decay rate and the shorter the lifetime of the battery at discharge current rates of 1C and 2C and temperature range of 10–40 °C. Figure 2.35a also reveals the capacity declines about 0.5 A h at 10 °C after 800 cycles, and the same number of cycles results in a capacity decline 0.8 A h at 40 °C. The major cause is that the

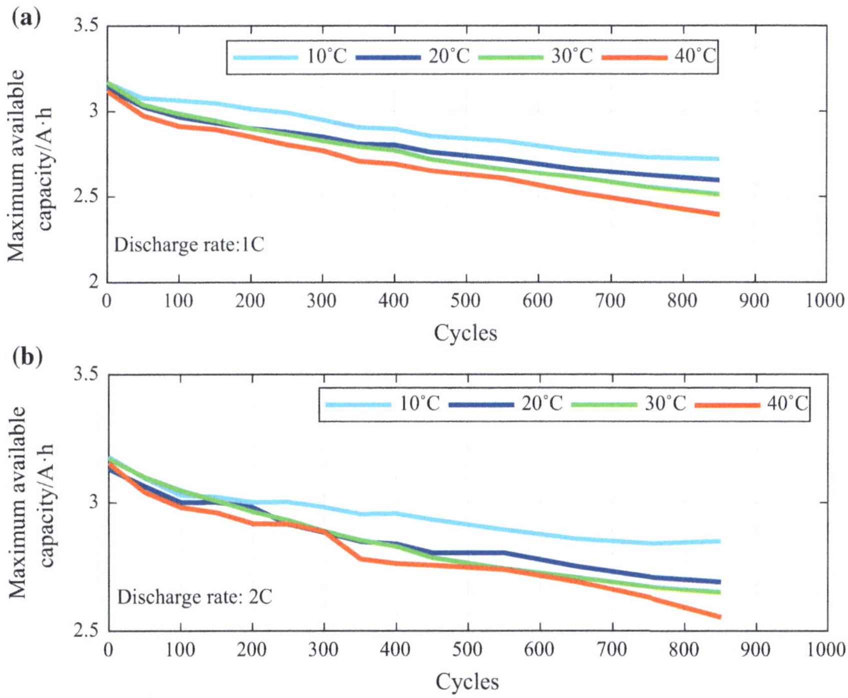

Fig. 2.35 Maximum available capacity of the battery at different temperatures: **a** discharge rate is 1C; **b** discharge rate is 2C

internal side reaction is faster at a high temperature (40 °C), consuming more active material and Li-ions, leading to a faster deterioration in capacity and shorter battery lifetime. It can be found out that the battery has a better durability at 10 °C than other temperatures tested in the experiment.

Relationship between lifetime and current rate
The battery discharge current rate exerts an important influence on its capacity deterioration characteristics and battery lifetime. Figure 2.36 shows a battery capacity deterioration trajectory at different discharge current rates (0.5, 1, 1.5, and 2C).

It is clear from Fig. 2.36 that the capacity deterioration rate of the battery declines with the increase of the current rate (within 2C) for the cylindrical ternary battery. As shown in Fig. 2.36a, the battery capacity deteriorates to 2.6 A h at 0.5C after 800 cycles, while the capacity remained above 2.8 A h at 2C. It is worth noting that under normal circumstances, the large discharge current rate will accelerate the growth of the SEI film within the battery and the rate of active material loss, resulting in an increase in the rate of the battery capacity decline. Nevertheless, the battery degradation rate is not simply positively correlated with its discharge rate. It has a significant relationship with the design of the battery materials. Therefore,

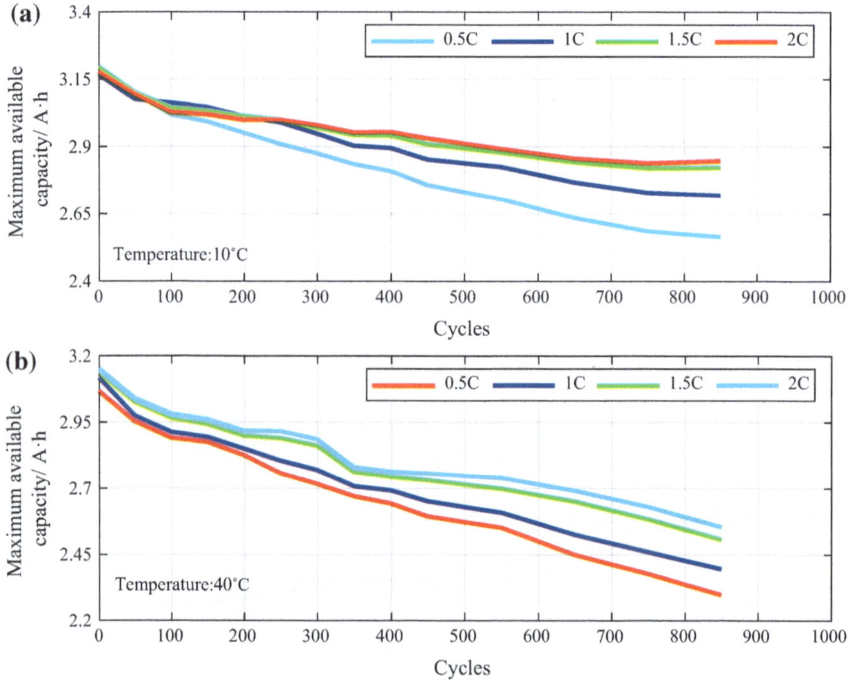

Fig. 2.36 Capacity deterioration trajectory at different discharge rates. **a** cycle temperature is 10 °C; **b** cycle temperature is 40 °C

determining the optimal working range of the battery has constructive significance for improving its durability [15, 16].

The relationship between lifetime and charge/discharge interval

The capacity loss of Li-ion batteries is mainly due to the growth of SEI film and the loss of active material inside the battery. The growth of SEI film is caused by the electrochemical side reaction inside the battery, while the loss of active material is mainly caused by substance fall off due to the change to the volume and structure of active material during the cycles [17].

SEI film growth is the major cause of capacity loss when the battery is resting and cycling over a narrow SOC range [18, 19]. Figure 2.37 shows the capacity decline trajectory of the battery in the narrower SOC interval (ΔSOC $\leq$ 20%). It can be found that the capacity loss within the narrower SOC interval is close with the loss being about 8–10% after 600 cycles.

Note: The equivalent circle here is the ratio of accumulative discharge capacity to rated capacity.

When the battery works within a wide SOC interval, Li-ions embedded in the internal active material of the battery is easy to cause damage and loss of the structure of the active material, which eventually causes the loss of Li-ions and capacity. At this time, the loss of battery capacity is the sum of the capacity loss caused by the SEI film growth along with the loss of the active material. Figure 2.38 shows the capacity loss of the battery during a wide SOC interval (ΔSOC $\geq$ 50%).

It is suggested that the capacity loss of the battery in Fig. 2.38 is greater than that in Fig. 2.37 due to the increase in active material loss. After 600 cycles, the

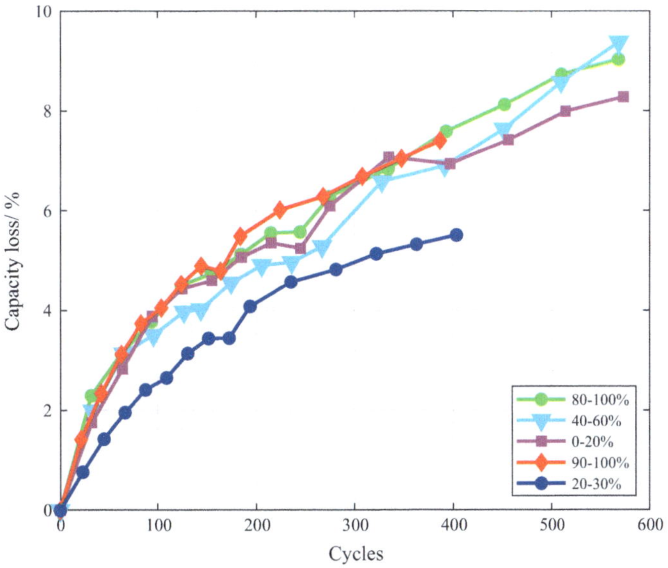

Fig. 2.37 Capacity decline trajectory with ΔSOC $\leq$ 20%

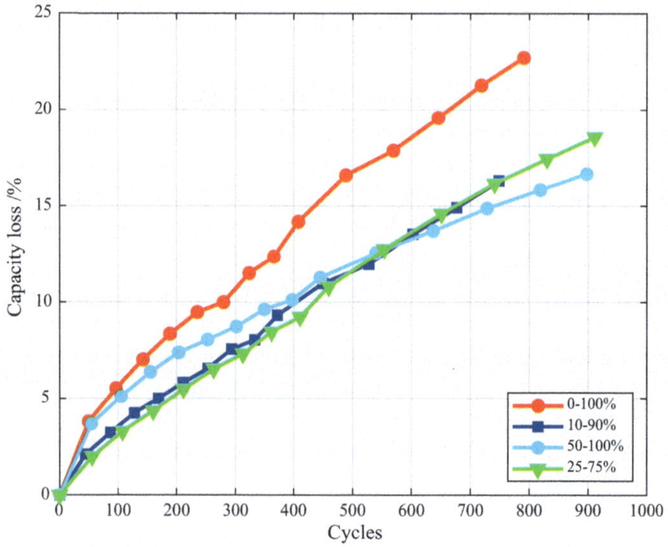

Fig. 2.38 Capacity decline trajectory with $\Delta SOC \geq 50\%$

battery capacity loss shown in Fig. 2.38 is more than 10%. In addition, the battery capacity loss under a wide SOC interval condition is quite different. For example, when $\Delta SOC = 100\%$, the battery capacity loss at 600 cycles reaches more than 20%. In contrast, when the $\Delta SOC = 80\%$ or 50%, the capacity loss remains within 20% after 900 cycles. Therefore, proper control over the charging and discharging interval of the battery can improve its durability.

2.5 Conclusion

Based on the established battery test platform consisting of battery charge/discharge equipment, frequency domain impedance characteristic test equipment, environmental simulation equipment, and connection devices, this chapter presents a systematic design of battery test plan and experimental flow, as well as establishes a database of battery characteristics including various $LiFePO_4$ batteries and ternary batteries. Meanwhile, the temperature characteristic, performance degradation and lifetime characteristics of batteries are given in detail. Experimental data shows that the maximum available capacity, the OCV and the AC impedance of the battery will be impacted by its operating temperatures and the aging stages. The deterioration trajectory of the battery lifetime depends on the SOC interval and the operation stress. Therefore, the relationships between model parameters, operating temperature, and aging stages must be considered in battery modeling. When developing the battery state estimation algorithms, consideration should also be taken to the application performance in variable stress, wide temperature range and the full lifetime. The abovementioned experimental characterization analysis presents effective data and theoretical basis for the design and verification of BMS algorithms.

References

1. Xiong R (2014) Estimation of battery pack state for electric vehicles using model-data fusion approach, Doctoral dissertation, Beijing Institute of Technology
2. Li W (2007) Arbin: an American dream for a battery business. Chinese J Power Sour 31(5):431–432
3. QC/T 897–2011 (2011). Technical conditions for battery management systems for electric vehicles. Ministry of Industry and Information Technology of the People's Republic of China
4. Xiong R, Tian J, Mu H, Wang C (2017) A systematic model-based degradation behavior recognition and health monitoring method for lithium-ion batteries. Appl Energy 207:372–383
5. Wei Z, Zhao J, Xiong R, Dong G, Pou J, Tseng KJ (2018) Online estimation of power capacity with noise effect attenuation for lithium-ion battery. IEEE Trans Ind Electron 66(7):5724–5735
6. Mu H (2016) Research on Robustness State of Charge Estimation of Lithium-ion Power Battery for Electric Vehicles, Doctoral dissertation, Beijing Institute of Technology
7. Xiong R, Cao J, Yu Q, He H, Sun F (2017) Critical review on the battery state of charge estimation methods for electric vehicles. IEEE Access 6:1832–1843
8. Tian J, Xiong R, Yu Q (2018) Fractional-order model-based incremental capacity analysis for degradation state recognition of lithium-ion batteries. IEEE Trans Ind Electron 66(2):1576–1584
9. Xi J (2012) Experimental Research on the electrochemical impedance spectroscopy of lithium iron phosphate cell. Doctoral dissertation Tsinghua University
10. Zhang Y, Wang J, He H, Peng S, Pecht M (2019) Lithium-ion battery health prognosis based on a real battery management system used in electric vehicles. IEEE Trans Veh Technol 68(5): 4110–4121
11. Orazem ME, Tribollet B (2017) Electrochemical impedance spectroscopy. Wiley, New York
12. Xiong R, Li L, Tian J (2018) Towards a smarter battery management system: A critical review on battery state of health monitoring methods. J Power Sour 405:18–29
13. Watanabe S, Kinoshita M, Nakura K (2014) Capacity fade of $LiNi_{(1-x-y)}Co_xAl_yO_2$ cathode for lithium-ion batteries during accelerated calendar and cycle life test. I. Comparison analysis between $LiNi_{(1-x-y)}Co_xAl_yO_2$ and $LiCoO_2$ cathodes in cylindrical lithium-ion cells during long term storage test. J Power Sour 247:412–422
14. Dubarry M, Truchot C, Cugnet M, Liaw BY, Gering K, Sazhin S, Michelbacher C (2011) Evaluation of commercial lithium-ion cells based on composite positive electrode for plug-in hybrid electric vehicle applications. Part I: initial characterizations. J Power Sour 196(23):10328–10335
15. Saxena S, Hendricks C, Pecht M (2016) Cycle life testing and modeling of graphite/$LiCoO_2$ cells under different state of charge ranges. J Power Sour 327:394–400
16. Ecker M, Nieto N, Käbitz S, Schmalstieg J, Blanke H, Warnecke A, Sauer DU (2014) Calendar and cycle life study of $Li(NiMnCo)O_2$-based 18650 lithium-ion batteries. J Power Sour 248:839–851
17. Zhang Y, Xiong R, He H, Pecht M (2019) Validation and verification of a hybrid method for remaining useful life prediction of lithium-ion batteries. J Clean Prod 212:240–249
18. Jin X, Vora A, Hoshing V, Saha T, Shaver G, García RE, Varigonda S (2017) Physically-based reduced-order capacity loss model for graphite anodes in Li-ion battery cells. J Power Sour 342:750–761
19. Bloom I, Cole B, Sohn J, Jones S, Polzin E, Battaglia V, Case H (2001) An accelerated calendar and cycle life study of Li-ion cells. J Power Sour 101(2):238–247

Chapter 3
Modeling Theory of Lithium-Ion Batteries

The complex electrochemical reactions inside the batteries are affected by many influencing factors and uncertainties. Establishing mathematical models of batteries is seen as a multidisciplinary problem, for which it has always been an important yet difficult problem in academia and industry. The input excitation quantity (current) and the output observation quantities (terminal voltage and temperature) constitute the finite measurable quantities in the BMS. Because of the polarizations caused by electrochemical reactions, the multi-stage characteristics of positive and negative electrode materials, etc., the terminal voltage of the battery is characterized by hysteresis, nonlinearity, and strong time variation. The battery terminal voltage is divided into dynamic and static parts. Dynamic part mainly includes the rapidly changing voltage component (ohmic polarization) and slowly changing voltage component (concentration and electrochemical polarization, etc.). Static part mainly includes open-circuit voltage (OCV) (equilibrium potential) and the hysteresis voltage (hysteresis characteristics). The hysteresis voltage component is closely associated with the current state and historical incentive trajectory of the battery.

To describe the external characteristics exhibited by batteries more accurately, reliable battery state estimation algorithms is designed and a more advanced BMS is developed for the new energy vehicles. In this process, accurate modeling is indispensable. Battery models are mainly divided into electrochemical models, equivalent circuit models (ECMs) and black box models. With the in-depth study of frequency domain characteristics of batteries, fractional-order models are established based on the ECM and AC impedance characteristics. In this chapter, the construction, parameter identification and verification of electrochemical models, ECMs and fractional-order models are introduced systematically.

R. Xiong, *Battery Management Algorithm for Electric Vehicles*,
https://doi.org/10.1007/978-981-15-0248-4_3

3.1 Electrochemical Model

3.1.1 Introduction

M. Doyle, T. F. Fuller, and J. Newman established the pseudo-two-dimensional (P2D) model based on the theory of porous electrode and concentrated solution in the middle of 1990s, which laid the foundation for the development of electrochemical models [1]. In this model, a series of partial differential equations and algebraic equations were applied to accurately describe the physical and chemical phenomena inside the battery, such as the diffusion and migration of Li-ions, the electrochemical reaction on the surface of active particles, Ohm's law, and charge conservation. So far, a majority of electrochemical models are derived and developed from this model. The electrochemical model is a first-principle model, which can not only accurately simulate the external characteristics of the battery, but also simulate the distribution and changes of the internal characteristics such as Li-ion concentration in the electrodes and electrolyte, reaction overpotential, and other physical quantities [2, 3]. Compared with other battery models, the electrochemical model has more explicit physical meanings due to its deep investigation about the reactions inside the battery.

The P2D model is universal and extensible, for which it can be applied to different types of Li-ion batteries and extended to more complex coupling models. Therefore, the P2D model plays an irreplaceable role in battery modeling [4, 5]. However, it consists of complex partial differential equations and a large number of electrochemical parameters, which bring a high computational power requirement for microcontrollers. At present, numerical methods are primarily used to solve the P2D model, such as the finite difference method, finite element method, and finite volume method.

3.1.2 Model Construction

Figure 3.1 shows a schematic diagram of P2D model for a $LiCoO_2$ battery. The internal structure of the battery is simplified into two phases (solid phase and electrolyte phase) and three regions (positive, negative, and separator). The solid phase is modeled by spherical particles, and the solid-phase diffusion process is described by the diffusion process of Li-ions along particle radius r. The electrolyte-phase diffusion process is described by the diffusion and migration of Li-ions along the thickness of the battery, namely the x-direction. In the discharge process, Li-ions diffuse from the inside of the negative active particles to the surface of the active particles. And Li-ions are released into the electrolyte through the electrochemical reactions occurring on the surface of the negative active particles. Then, Li-ions diffuse toward the positive direction, reaching the positive electrode through the separator. By the electrochemical reactions occurring on the surface of the positive active particles, Li-ions diffuse

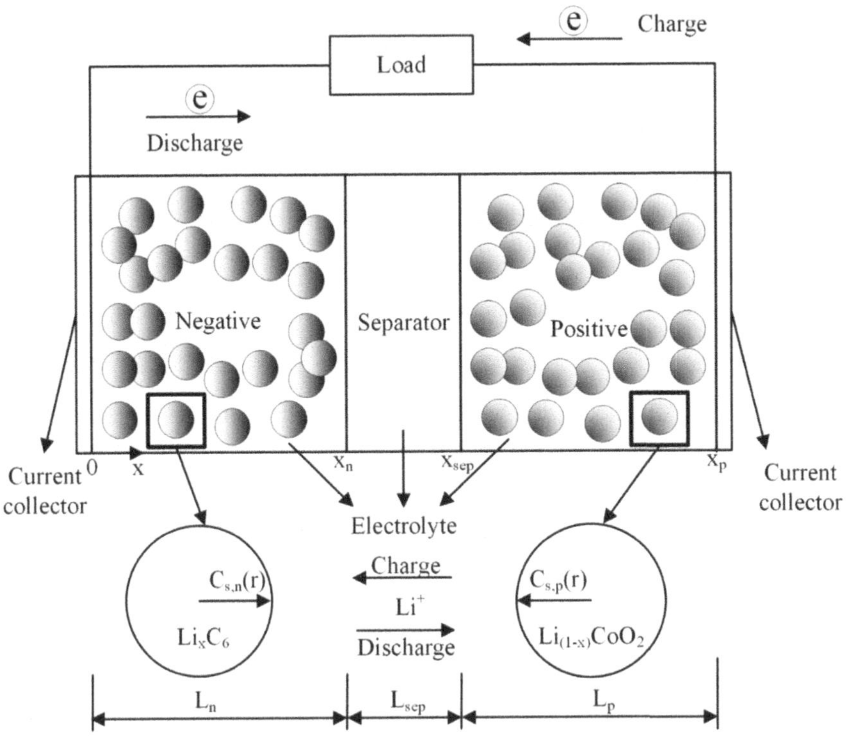

Fig. 3.1 Schematic diagram of P2D model

into the inside of the positive active particles. Meanwhile, the electrons transfer from the negative collector to the positive collector, thus generating the electric current in the external circuit.

The electrochemical reaction in the positive electrode:

$$LiCoO_2 \underset{\text{Discharge}}{\overset{\text{Charge}}{\longleftrightarrow}} Li_{(1-x)}CoO_2 + xLi^+ + xe^-$$

The electrochemical reaction in the negative electrode:

$$6C + xLi^+ + xe^- \underset{\text{Discharge}}{\overset{\text{Charge}}{\longleftrightarrow}} Li_xC_6$$

According to the working principle of the P2D model, a series of partial differential equations and algebraic equations could be established to describe the characteristics of the battery from the electrochemical perspective. Generically, the P2D model can be summarized into the following six mathematical equations [6]:

(1) Electrolyte-phase diffusion equations of Li-ions in the positive electrode, negative electrode and separator;
(2) Solid-phase diffusion equation of Li-ions in the positive electrode and negative electrode;
(3) Electrolyte-phase Ohm's law equation in the positive electrode, negative electrode and separator;
(4) Solid-phase Ohm's law equation in the positive electrode and negative electrode;
(5) Charge conservation equation in the positive electrode, negative electrode, and separator;
(6) Butler–Volmer kinetic equation on the surface of active particles in each electrode.

P2D model

Equations of the P2D model will be described in detail below. In this chapter, the parameters of the battery in different regions are differentiated by subscripts *p*, *n*, and *sep*, which represent the positive electrode, negative electrode, and separator, respectively.

(1) Li-ions diffusion in the electrolyte phase

The transfer of Li-ions in the electrolyte only includes diffusion and migration, which can be described by Fick's second law.

$$\varepsilon_{\mathrm{e}}\frac{\partial c_{\mathrm{e}}}{\partial t}=\frac{\partial}{\partial x}\left(D_{\mathrm{e}}^{\mathrm{eff}}\frac{\partial c_{\mathrm{e}}}{\partial x}\right)+a\left(1-t_{+}^{0}\right)j_{\mathrm{r}} \tag{3.1}$$

where ε_{e} represents the volume fraction of the electrolyte phase, c_{e} stands for the Li-ions concentration in the electrolyte phase, x denotes the direction along electrode thickness, $D_{\mathrm{e}}^{\mathrm{eff}}$ is the effective diffusion coefficient of Li-ions in the electrolyte phase, a refers to the specific surface area of the active particles, t_{+}^{0} is defined as the Li-ion transfer coefficient in the electrolyte phase, j_{r} stands for the molar flux at the solid–electrolyte-phase interface. On the right side of the equation, the first term describes the effect of Li-ion diffusion on the Li-ions concentration in the electrolyte phase, whereas the second term depicts the effect of Li-ions migration on the Li-ions concentration in the electrolyte phase, which is usually negligible. The boundary conditions of the diffusion equation are as follows:

$$\begin{cases}\left.\frac{\partial c_{\mathrm{e}}}{\partial x}\right|_{x=0}=\left.\frac{\partial c_{\mathrm{e}}}{\partial x}\right|_{x=x_{\mathrm{p}}}=0\\ \left.D_{\mathrm{e,sep}}^{\mathrm{eff}}\frac{\partial c_{\mathrm{e}}}{\partial x}\right|_{x=x_{\mathrm{sep}}^{-}}=\left.D_{\mathrm{e,p}}^{\mathrm{eff}}\frac{\partial c_{\mathrm{e}}}{\partial x}\right|_{x=x_{\mathrm{sep}}^{+}},\quad \left.c_{\mathrm{e}}\right|_{x=x_{\mathrm{sep}}^{-}}=\left.c_{\mathrm{e}}\right|_{x=x_{\mathrm{sep}}^{+}}\\ \left.D_{\mathrm{e,n}}^{\mathrm{eff}}\frac{\partial c_{\mathrm{e}}}{\partial x}\right|_{x=x_{\mathrm{n}}^{-}}=\left.D_{\mathrm{e,sep}}^{\mathrm{eff}}\frac{\partial c_{\mathrm{e}}}{\partial x}\right|_{x=x_{\mathrm{n}}^{+}},\quad \left.c_{\mathrm{e}}\right|_{x=x_{\mathrm{n}}^{-}}=\left.c_{\mathrm{e}}\right|_{x=x_{\mathrm{n}}^{+}}\end{cases} \tag{3.2}$$

where $x=0$ is at the junction of the negative collector and the left end of the negative electrode. $x=x_{\mathrm{n}}$ is at the junction of the separator and the right end of the negative electrode. $x=x_{\mathrm{sep}}$ is at the junction of the positive electrode and the right end of the separator. $x=x_{\mathrm{p}}$ is at the junction of the positive collector and the rightmost positive electrode.

The practical meaning of Eq. (3.2) is that at the junction of each electrode and collector, the Li-ions molar flux is 0; at the junction of each electrode and the separator, Li-ions concentration and molar flux are continuous.

(2) Li-ion diffusion in the solid phase

The Li-ion diffusion in the active materials of positive and negative electrodes is also described by Fick's second law. As the active materials in positive and negative electrodes are assumed to be spherical particles with equal radius in the P2D model, the diffusion equation of Li-ions in the solid phase is established in a spherical coordinate system. The diffusion equation is described as follows:

$$\frac{\partial c_{\mathrm{s}}}{\partial t}=\frac{1}{r^2}\frac{\partial}{\partial r}\left(D_{\mathrm{s}}r^2\frac{\partial c_{\mathrm{s}}}{\partial r}\right)=D_{\mathrm{s}}\left(\frac{2}{r}\frac{\partial c_{\mathrm{s}}}{\partial r}+\frac{\partial^2 c_{\mathrm{s}}}{\partial r^2}\right) \tag{3.3}$$

where c_{s} indicates the Li-ion concentration in the solid phase, r denotes the radius direction of solid spherical particles, and D_{s} represents the diffusion coefficient of Li-ions in the solid phase. The boundary conditions are as follows:

$$D_{\mathrm{s}}\frac{\partial c_{\mathrm{s}}}{\partial r}|_{r=0}=0,\quad D_{\mathrm{s}}\frac{\partial c_{\mathrm{s}}}{\partial r}|_{r=R_{\mathrm{s}}}=-j_{\mathrm{r}} \tag{3.4}$$

where R_{s} indicates the radius of active particles. The equation shows that the molar flux of Li-ions is 0 at the center of the active particles, and the Li-ions molar flux on the surface of the active particle is the same as the Li-ion flow participating in the electrochemical reaction.

(3) Ohm's law in the electrolyte phase

The variation rule of electrolyte-phase potential in the Li-ion battery is described by modified Ohm's law:

$$\kappa^{\mathrm{eff}}\frac{\partial \phi_{\mathrm{e}}}{\partial x}=-\frac{2RT\kappa^{\mathrm{eff}}}{F}\left(t_{+}^{0}-1\right)\frac{\partial \ln c_{\mathrm{e}}}{\partial x}-i_{\mathrm{e}} \tag{3.5}$$

where κ^{eff} represents the effective ionic conductivity of the electrolyte phase, ϕ_{e} indicates the electrolyte-phase potential, R denotes the molar gas constant, T refers to the temperature, F is defined as the Faraday constant, and i_{e} stands for the electrolyte-phase current density. On the right side of the equation, the first term describes the effect of the Li-ion concentration in the electrolyte phase on the electrolyte-phase potential and the second term describes the effect of the current in the electrolyte phase on the electrolyte-phase potential. The boundary conditions of the equation are as follows:

$$\begin{cases}\kappa^{\mathrm{eff}}\left.\frac{\partial \phi_{\mathrm{e}}}{\partial x}\right|_{x=x_{\mathrm{n}}^{-}}=\kappa^{\mathrm{eff}}\left.\frac{\partial \phi_{\mathrm{e}}}{\partial x}\right|_{x=x_{\mathrm{n}}^{+}}, & \kappa^{\mathrm{eff}}\left.\frac{\partial \phi_{\mathrm{e}}}{\partial x}\right|_{x=x_{\mathrm{sep}}^{-}}=\kappa^{\mathrm{eff}}\left.\frac{\partial \phi_{\mathrm{e}}}{\partial x}\right|_{x=x_{\mathrm{sep}}^{+}}\\ \phi_{\mathrm{e}}|_{x=x_{\mathrm{n}}^{-}}=\phi_{\mathrm{e}}|_{x=x_{\mathrm{n}}^{+}}, & \phi_{\mathrm{e}}|_{x=x_{\mathrm{sep}}^{-}}=\phi_{\mathrm{e}}|_{x=x_{\mathrm{sep}}^{+}}\end{cases} \tag{3.6}$$

(4) Ohm's law in the solid phase

The variation rule of solid-phase potential in the Li-ion battery is described by Ohm's law as

$$\sigma^{\mathrm{eff}} \frac{\partial \phi_{\mathrm{s}}}{\partial x} = -i_{\mathrm{s}} \tag{3.7}$$

where σ^{eff} indicates the effective diffusion conductivity of solid phase, ϕ_{s} denotes the solid-phase potential, and i_{s} refers to the current density of the solid phase. The boundary conditions of the equation are as follows:

$$\begin{cases} \sigma^{\mathrm{eff}} \frac{\partial \phi_s}{\partial x}\Big|_{x=0} = -i, \ \sigma^{\mathrm{eff}} \frac{\partial \phi_{\mathrm{s}}}{\partial x}\Big|_{x=x_{\mathrm{p}}} = -i \\ \sigma^{\mathrm{eff}} \frac{\partial \phi_{\mathrm{s}}}{\partial x}\Big|_{x=x_{\mathrm{n}}} = 0, \ \sigma^{\mathrm{eff}} \frac{\partial \phi_{\mathrm{s}}}{\partial x}\Big|_{x=x_{\mathrm{sep}}} = 0 \end{cases} \tag{3.8}$$

(5) Charge conservation equation

According to the law of charge conservation, the sum of electrolyte-phase current density and solid-phase current density at any position inside the battery is equal to the charge–discharge current density of the battery, which is

$$i_{\mathrm{s}} + i_{\mathrm{e}} = i \tag{3.9}$$

i can be obtained from the following equation:

$$i = \frac{i_{\mathrm{L}}}{S} \tag{3.10}$$

where i_{L} indicates the applied current and S denotes the electrode plate area.

The relationship between the Li-ion molar flux on the surface of active particles and the current density of the solid phase can be described by Faraday's law.

$$i_{\mathrm{s}} = nFj_{\mathrm{r}} \tag{3.11}$$

where n is the charge number of Li-ions, and $n = 1$.

The relationship between Li-ion molar flux and the current density of the electrolyte phase is as follows:

$$\frac{\partial i_{\mathrm{e}}}{\partial x} = ai_{\mathrm{s}} = aFj_{\mathrm{r}} \tag{3.12}$$

where a is the specific surface area of active particles, which can be obtained by

$$a = \frac{3\varepsilon_{\mathrm{s}}}{R_{\mathrm{s}}} \tag{3.13}$$

where ε_s refers to the volume fraction of the solid phase.

For the P2D model of lithium-ion batteries, the boundary conditions of Eq. (3.12) are as follows:

$$\begin{cases} i_e|_{x=0} = 0, \quad i_e|_{x=x_p} = 0 \\ \frac{\partial i_e}{\partial x}\Big|_{x=x_n} = 0, \quad \frac{\partial i_e}{\partial x}\Big|_{x=x_{sep}} = 0 \end{cases} \tag{3.14}$$

From Eqs. (3.9) and (3.12), the change rule of solid-phase current density is as follows:

$$\frac{\partial i_s}{\partial x} = -aFj_r \tag{3.15}$$

The boundary conditions of Eq. (3.15) are as follows:

$$\begin{cases} i_s|_{x=0} = i, \quad i_s|_{x=x_p} = i \\ \frac{\partial i_s}{\partial x}\Big|_{x=x_n} = 0, \quad \frac{\partial i_s}{\partial x}\Big|_{x=x_{sep}} = 0 \end{cases} \tag{3.16}$$

(6) Butler–Volmer kinetic equation

The relationship between Li-ion molar flux at the solid–electrolyte-phase interface and the surface overpotential is determined by the Butler–Volmer kinetic equation:

$$j_r = i_0\left(e^{\frac{\alpha_a F}{RT}\eta} - e^{-\frac{\alpha_c F}{RT}\eta}\right) \tag{3.17}$$

where i_0 indicates the exchange current density. α_a and α_c represent the transfer coefficients of negative and positive electrode, respectively, usually set to 0.5. η denotes the overpotential on the surface of active particles.

Exchange current density i_0 reflects the degree of difficulty in electrode reactions occurrence. Its calculation process is as follows:

$$i_0 = k_s c_e^{\alpha_a}\left(c_{s,\max} - c_{e-s}\right)^{\alpha_a} c_{e-s}^{\alpha_c} \tag{3.18}$$

where k_s indicates the electrochemical reaction constant, $c_{s,\max}$ is defined as the maximum Li-ion concentration in the active material, and c_{e-s} denotes the concentration of Li-ion at the solid–liquid phase interface.

In the positive and negative regions, the surface overpotential is related to the electrolyte-phase potential, the solid-phase potential, and the open-circuit potential [3].

$$\eta = \phi_s - \phi_e - E_{OCV} \tag{3.19}$$

where the open-circuit potential E_{OCV} of the electrode is determined by the Li-ion concentration on the surface of the active particles, which is a physical quantity only related to the material characteristics.

Although the P2D model is complex, it is still a single-input single-output system in fact. The only input of this model is the current i_L and the only output is the terminal voltage U_t. In fact, the terminal voltage of the battery is the difference of solid-phase potential between the positive and negative collectors.

$$U_t = \phi_s|_{x=x_p} - \phi_s|_{x=0} \tag{3.20}$$

By substituting Eq. (3.19) to Eq. (3.20), the expression of terminal voltage can be derived as follows:

$$U_t = \phi_e|_{x=x_p} - \phi_e|_{x=0} + \eta_p\big|_{x=x_p} - \eta_n|_{x=0} + E_{OCV,p}\big|_{x=x_p} - E_{OCV,n}\big|_{x=x_n} \tag{3.21}$$

The mathematical equations of the electrochemical model are summarized in Table 3.1.

Simplified P2D model

A total of six major equations have been introduced in detail to establish the P2D electrochemical model. Due to the complex calculation process of the P2D model, this chapter introduces a simplified P2D model to conduct the terminal voltage simulation of a battery. The flowchart of establishing the model is shown in Fig. 3.2. First, the average Li-ion molar flux along the *x*-direction is calculated. Then, the diffusion equation in the solid phase is reduced to calculate the Li-ion concentration in the solid phase inside the battery. The overpotential caused by electrochemical reactions is calculated by the Butler–Volmer kinetic equation. The SOC is calculated by the surface value of the Li-ion concentration in the solid phase, and the OCV could be derived from the relation between SOC and OCV. Finally, the simplified terminal voltage expression is obtained. The simplification steps are described in detail as follows:

(1) Solution of Li-ion molar flux on the surface of active particles

The simplified P2D model established in this chapter neglects the diffusion of solid Li-ion concentration along the electrode thickness direction. In addition, the solid active particles in each electrode are considered evenly distributed along the *x*-direction. Therefore, the electrochemical reaction intensity and the Li-ion molar flux on the surface of the active particles along the *x*-direction are consistent.

For a coordinate *x* in the negative electrode, the current density in the electrolyte phase can be written as

$$i_e(x) = i_e(0) + \int_0^x \frac{\partial i_e(x')}{\partial x'} dx' = i_e(0) + \int_0^x a_n j_{r,n}(x') F dx' \tag{3.22}$$

Combining with the boundary conditions of the phase current in the negative electrolyte, it can be derived as

Table 3.1 Mathematical expression of P2D model

	Mathematical expression	Boundary conditions
Li-ion diffusion in the electrolyte phase	$\varepsilon_{\mathrm{e}} \frac{\partial c_{\mathrm{e}}}{\partial t} = \frac{\partial}{\partial x}\left(D_{\mathrm{e}}^{\mathrm{eff}} \frac{\partial c_{\mathrm{e}}}{\partial x}\right) + a\left(1 - t_{+}^{0}\right) j_{\mathrm{r}}$	$\begin{cases} \frac{\partial c_{\mathrm{e}}}{\partial x}\Big\vert_{x=0} = \frac{\partial c_{\mathrm{e}}}{\partial x}\Big\vert_{x=x_{\mathrm{p}}} = 0 \\ D_{\mathrm{e,sep}}^{\mathrm{eff}} \frac{\partial c_{\mathrm{e}}}{\partial x}\Big\vert_{x=x_{\mathrm{sep}}^{-}} = D_{\mathrm{e,p}}^{\mathrm{eff}} \frac{\partial c_{\mathrm{e}}}{\partial x}\Big\vert_{x=x_{\mathrm{sep}}^{+}}, \quad c_{\mathrm{e}}\vert_{x=x_{\mathrm{sep}}^{-}} = c_{\mathrm{e}}\vert_{x=x_{\mathrm{sep}}^{+}} \\ D_{\mathrm{e,n}}^{\mathrm{eff}} \frac{\partial c_{\mathrm{e}}}{\partial x}\Big\vert_{x=x_{\mathrm{n}}^{-}} = D_{\mathrm{e,sep}}^{\mathrm{eff}} \frac{\partial c_{\mathrm{e}}}{\partial x}\Big\vert_{x=x_{\mathrm{n}}^{+}}, \quad c_{\mathrm{e}}\vert_{x=x_{\mathrm{n}}^{-}} = c_{\mathrm{e}}\vert_{x=x_{\mathrm{n}}^{+}} \end{cases}$
Li-ion diffusion in the solid phase	$\frac{\partial c_{\mathrm{s}}}{\partial t} = \frac{1}{r^{2}} \frac{\partial}{\partial r}\left(D_{\mathrm{s}} r^{2} \frac{\partial c_{\mathrm{s}}}{\partial r}\right) = D_{\mathrm{s}}\left(\frac{2}{r} \frac{\partial c_{\mathrm{s}}}{\partial r} + \frac{\partial^{2} c_{\mathrm{s}}}{\partial r^{2}}\right)$	$D_{\mathrm{s}} \frac{\partial c_{\mathrm{s}}}{\partial r}\vert_{r=0} = 0, \quad D_{\mathrm{s}} \frac{\partial c_{\mathrm{s}}}{\partial r}\vert_{r=R_{\mathrm{s}}} = -j_{\mathrm{r}}$
Ohm's law of the electrolyte phase	$\kappa^{\mathrm{eff}} \frac{\partial \phi_{\mathrm{e}}}{\partial x} = -\frac{2RT\kappa^{\mathrm{eff}}}{F}\left(t_{+}^{0} - 1\right) \frac{\partial \ln c_{\mathrm{e}}}{\partial x} - i_{\mathrm{e}}$	$\begin{cases} \kappa^{\mathrm{eff}} \frac{\partial \phi_{\mathrm{e}}}{\partial x}\Big\vert_{x=x_{\mathrm{n}}^{-}} = \kappa^{\mathrm{eff}} \frac{\partial \phi_{\mathrm{e}}}{\partial x}\Big\vert_{x=x_{\mathrm{n}}^{+}}, \quad \kappa^{\mathrm{eff}} \frac{\partial \phi_{\mathrm{e}}}{\partial x}\Big\vert_{x=x_{\mathrm{sep}}^{-}} = \kappa^{\mathrm{eff}} \frac{\partial \phi_{\mathrm{e}}}{\partial x}\Big\vert_{x=x_{\mathrm{sep}}^{+}} \\ \phi_{\mathrm{e}}\vert_{x=x_{\mathrm{n}}^{-}} = \phi_{\mathrm{e}}\vert_{x=x_{\mathrm{n}}^{+}}, \quad \phi_{\mathrm{e}}\vert_{x=x_{\mathrm{sep}}^{-}} = \phi_{\mathrm{e}}\vert_{x=x_{\mathrm{sep}}^{+}} \end{cases}$
Ohm's law of the solid phase	$\sigma^{\mathrm{eff}} \frac{\partial \phi_{\mathrm{s}}}{\partial x} = -i_{\mathrm{s}}$	$\begin{cases} \sigma^{\mathrm{eff}} \frac{\partial \phi_{\mathrm{s}}}{\partial x}\Big\vert_{x=0} = -i, \quad \sigma^{\mathrm{eff}} \frac{\partial \phi_{\mathrm{s}}}{\partial x}\Big\vert_{x=x_{\mathrm{p}}} = -i \\ \sigma^{\mathrm{eff}} \frac{\partial \phi_{\mathrm{s}}}{\partial x}\Big\vert_{x=x_{\mathrm{n}}} = 0, \quad \sigma^{\mathrm{eff}} \frac{\partial \phi_{\mathrm{s}}}{\partial x}\Big\vert_{x=x_{\mathrm{sep}}} = 0 \end{cases}$
Charge conservation equation	$\begin{cases} i_{\mathrm{s}} + i_{\mathrm{e}} = i \\ \frac{\partial i_{\mathrm{e}}}{\partial x} = a i_{\mathrm{s}} = aFj_{\mathrm{r}} \\ \frac{\partial i_{\mathrm{s}}}{\partial x} = -aFj_{\mathrm{r}} \end{cases}$	$\begin{cases} i_{\mathrm{e}}\vert_{x=0} = 0, \quad i_{\mathrm{e}}\vert_{x=x_{\mathrm{p}}} = 0 \\ \frac{\partial i_{\mathrm{e}}}{\partial x}\Big\vert_{x=x_{\mathrm{n}}} = 0, \quad \frac{\partial i_{\mathrm{e}}}{\partial x}\Big\vert_{x=x_{\mathrm{sep}}} = 0 \end{cases}$ $\begin{cases} i_{\mathrm{s}}\vert_{x=0} = i, \quad i_{\mathrm{s}}\vert_{x=x_{\mathrm{p}}} = i \\ \frac{\partial i_{\mathrm{s}}}{\partial x}\Big\vert_{x=x_{\mathrm{n}}} = 0, \quad \frac{\partial i_{\mathrm{s}}}{\partial x}\Big\vert_{x=x_{\mathrm{sep}}} = 0 \end{cases}$
Butler–Volmer kinetic equation	$\begin{cases} j_{\mathrm{r}} = i_{0}\left(e^{\frac{\alpha_{\mathrm{a}} F}{RT}\eta} - e^{-\frac{\alpha_{\mathrm{c}} F}{RT}\eta}\right) \\ i_{0} = k_{\mathrm{s}} c_{\mathrm{e}}^{\alpha_{\mathrm{a}}}\left(c_{\mathrm{s,max}} - c_{\mathrm{e-s}}\right)^{\alpha_{\mathrm{a}}} c_{\mathrm{e-s}}^{\alpha_{\mathrm{c}}} \end{cases}$	
Output voltage	$U_{\mathrm{t}} = \phi_{\mathrm{s}}\vert_{x=x_{\mathrm{p}}} - \phi_{\mathrm{s}}\vert_{x=0}$	

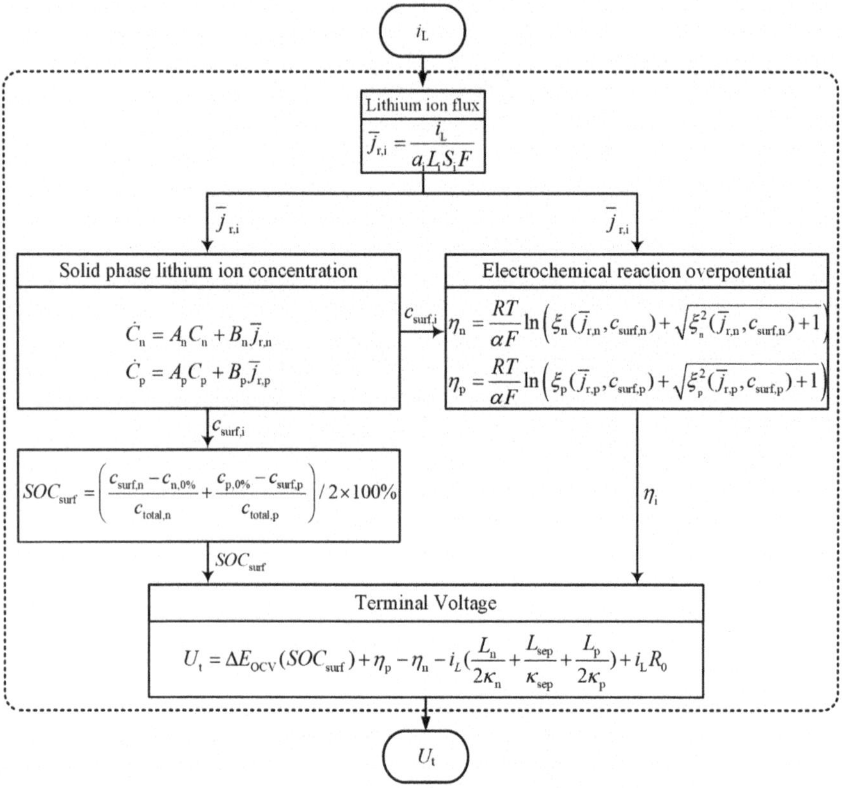

Fig. 3.2 The flowchart for establishing the simplified electrochemical model

$$\int_0^{x_n} a_n j_{r,n}(x') F dx' = \frac{i_L}{S_n} \tag{3.23}$$

The average value of Li-ion molar flux on $0 \le x \le x_n$ is used to approximate the Li-ion molar flux along the x-direction in the negative electrode.

$$\bar{j}_{r,n} = \frac{i_L}{a_n L_n S_n F} \tag{3.24}$$

Similarly, for a coordinate x in the positive electrode, the electrolyte-phase current density is calculated as

$$i_e(x) = i_e(x_{sep}) + \int_{x_{sep}}^{x} \frac{\partial i_e(x')}{\partial x'} dx' = i_e(x_{sep}) + \int_{x_{sep}}^{x} a_p j_{r,p}(x') F dx' \tag{3.25}$$

Combining the boundary conditions of positive electrolyte-phase current $i_e(x_{sep}) = \frac{i_L}{S_p}$ and $i_e(x_p) = 0$, it is derived as

$$\int_{x_{sep}}^{x_p} a_p j_{r,p}(x') F dx = -\frac{i_L}{S_p} \tag{3.26}$$

The average value of Li-ion molar flux on $x_{sep} \leq x \leq x_p$ is used to approximate the Li-ion molar flux along the x-direction in the positive electrode.

$$\bar{j}_{r,p} = -\frac{i_L}{a_p L_p S_p F} \tag{3.27}$$

Through this approximation, the relationship between Li-ion molar flux on the surface of active particles and external current can be established.

(2) Reduction and solution of solid-phase diffusion equation

The solid-phase diffusion equation is a partial differential equation. The finite difference method is adopted to mesh the active particles along the r direction. Besides, the complex partial differential equation is discretized into the ordinary differential equation to derive the state space expression of the system.

Taking the solid Li-ion concentration of negative electrode as an example, the Eq. (3.3) is discretized by the finite difference method to obtain the following equation:

$$\dot{c}_k = \frac{D_{s,n}}{(\Delta r_n)^2}\left(\frac{k+1}{k}c_{k+1} - 2c_k + \frac{k-1}{k}c_{k-1}\right),\ k = 1, 2, \ldots, m-1 \tag{3.28}$$

where $\Delta r_n = R_{s,n}/m$, and m denotes the number of discrete points.

In particular, the solid Li-ions concentration on the surface of the active particles can be obtained as

$$c_{k=m} = c_{k=m-1} - \frac{\Delta r_n}{D_{s,n}}\bar{j}_{r,n} \tag{3.29}$$

Taking $m = 10$ as an example, the state space expression of the system can be obtained as

$$\dot{c}_n = \begin{bmatrix} \dot{c}_{n,1} \\ \dot{c}_{n,2} \\ \vdots \\ \dot{c}_{n,9} \end{bmatrix} = \frac{D_{s,n}}{(\Delta r_n)^2}\begin{pmatrix} -2 & 2 & 0 & 0\,0\,0 & 0 & 0 & 0 \\ 1/2 & -2 & 3/2 & 0\,0\,0 & 0 & 0 & 0 \\ \vdots & & & & & & \vdots \\ 0 & 0 & 0 & 0\,0\,0 & 7/8 & -2 & 9/8 \\ 0 & 0 & 0 & 0\,0\,0 & 0 & 8/9 & -8/9 \end{pmatrix}\begin{bmatrix} c_{n,1} \\ c_{n,2} \\ \vdots \\ c_{n,9} \end{bmatrix} + \begin{bmatrix} 0 \\ 0 \\ \vdots \\ 0 \\ -10/9 \end{bmatrix}\frac{\bar{j}_{r,n}}{\Delta r_n} \tag{3.30}$$

Briefly written as

$$\dot{C}_{\rm n} = A_{\rm n}C_{\rm n} + B_{\rm n}\bar{j}_{\rm r,n} \tag{3.31}$$

Subsequent to the dimension reduction, the order of the state space expression stays at nine. According to the modern control theory, higher order systems are incapable of finding the exact analytic solutions, for which they need to be approximated by the numerical methods to obtain numerical solutions. Runge–Kutta method is a widely used high-precision single-step algorithm so as to solve differential equations numerically in engineering. The algorithm features high precision and can take measures to restrain the error. This chapter selects the third order Runge–Kutta method to perform an iterative calculation. The iterative process is written as

$$\begin{cases} k_1 = A_{\rm n}c_{k-1} + B_{\rm n}\bar{j}_{\rm r,n} \\ k_2 = A_{\rm n}(c_{k-1} + \frac{1}{2}k_1) + B_{\rm n}\bar{j}_{\rm r,n} \\ k_3 = A_{\rm n}(c_{k-1} - k_1 + 2k_2) + B_{\rm n}\bar{j}_{\rm r,n} \\ c_k = c_{k-1} + \frac{1}{6}(k_1 + 4k_2 + k_3) \end{cases} \tag{3.32}$$

The approach to dealing with the positive electrode Li-ion diffusion equation is the same as that for the negative electrode. In summary, the variation in Li-ion concentration for active particles of positive and negative electrodes is expressed as follows:

$$\begin{aligned} \dot{C}_{\rm n} &= A_{\rm n}C_{\rm n} + B_{\rm n}\bar{j}_{\rm r,n} \\ \dot{C}_{\rm p} &= A_{\rm p}C_{\rm p} + B_{\rm p}\bar{j}_{\rm r,p} \end{aligned} \tag{3.33}$$

The expression of Li-ion concentration on the surface of active particles is as follows:

$$\begin{aligned} c_{\rm surf,n} &= c_{\rm n,9} - \frac{\Delta r_{\rm n}}{D_{\rm s,n}}\bar{j}_{\rm r,n} \\ c_{\rm surf,p} &= c_{\rm p,9} - \frac{\Delta r_{\rm p}}{D_{\rm s,p}}\bar{j}_{\rm r,p} \end{aligned} \tag{3.34}$$

Substituting Eqs. (3.24) and (3.27) into Eq. (3.34), we obtain

$$\begin{aligned} c_{\rm surf,n} &= c_{\rm n,9} - \frac{\Delta r_{\rm n} i_{\rm L}}{a_{\rm n}L_{\rm n}S_{\rm n}D_{\rm s,n}F} \\ c_{\rm surf,p} &= c_{\rm p,9} - \frac{\Delta r_{\rm p} i_{\rm L}}{a_{\rm p}L_{\rm p}S_{\rm p}D_{\rm s,p}F} \end{aligned} \tag{3.35}$$

Therefore, the relationship between the distribution of Li-ions in the active particles and the external current has been clarified.

(3) Solution to the electrolyte-phase potential

By combining Eqs. (3.22), (3.24) and electrolyte-phase current boundary conditions, electrolyte-phase current density at any position x in the negative electrode can be obtained as

$$i_{\mathrm{e}}(x) \approx i_{\mathrm{e}}(0) + \int_{0}^{x} a_n \bar{j}_{\mathrm{r,n}} F dx' = \frac{i_{\mathrm{L}} x}{S_{\mathrm{n}} L_{\mathrm{n}}} \tag{3.36}$$

The electrolyte-phase Ohm's law shown in Eq. (3.5) can be simplified into Eq. (3.37) in the case of neglecting the influence of the Li-ion concentration distribution on the electrolyte-phase potential distribution:

$$\kappa^{\mathrm{eff}} \frac{\partial \phi_{\mathrm{e}}}{\partial x} = -i_{\mathrm{e}} \tag{3.37}$$

According to Eqs. (3.36) and (3.37), the electrolyte-phase potential at any position x in the negative electrode can be obtained as

$$\begin{aligned} \phi_{\mathrm{e}}(x) &= \phi_{\mathrm{e}}(0) + \int_{0}^{x} \frac{\partial \phi_{\mathrm{e}}(x')}{\partial x'} dx' \\ &\approx \phi_{\mathrm{e}}(0) + \int_{0}^{x} -\frac{i_{\mathrm{L}} x'}{\kappa_{\mathrm{n}}^{\mathrm{eff}} L_{\mathrm{n}} S_{\mathrm{n}}} dx' \\ &= \phi_{\mathrm{e}}(0) - \frac{x^2 i_{\mathrm{L}}}{2\kappa_{\mathrm{n}}^{\mathrm{eff}} L_{\mathrm{n}} S_{\mathrm{n}}} \end{aligned} \tag{3.38}$$

The electrolyte-phase potential difference in the anode region is as follows:

$$\phi_{\mathrm{e}}(x_{\mathrm{n}}) - \phi_{\mathrm{e}}(0) = -\frac{i_{\mathrm{L}} L_{\mathrm{n}}}{2\kappa_{\mathrm{n}}^{\mathrm{eff}} S_{\mathrm{n}}} \tag{3.39}$$

Similarly, the electrolyte-phase potential difference in the separator region is as follows:

$$\phi_{\mathrm{e}}(x_{\mathrm{sep}}) - \phi_{\mathrm{e}}(x_{\mathrm{n}}) = -\frac{i_{\mathrm{L}} L_{\mathrm{sep}}}{2\kappa_{\mathrm{sep}}^{\mathrm{eff}} S_{\mathrm{sep}}} \tag{3.40}$$

The electrolyte-phase potential difference in the cathode region is as follows:

$$\phi_{\mathrm{e}}(x_{\mathrm{p}}) - \phi_{\mathrm{e}}(x_{\mathrm{sep}}) = -\frac{i_{\mathrm{L}} L_{\mathrm{p}}}{2\kappa_{\mathrm{p}}^{\mathrm{eff}} S_{\mathrm{p}}} \tag{3.41}$$

The total electrolyte-phase potential difference in the battery is

$$\phi_e(x_p) - \phi_e(x_0) = -i_L\left(\frac{L_p}{2\kappa_p^{eff} S_p} + \frac{L_{sep}}{2\kappa_{sep}^{eff} S_{sep}} + \frac{L_n}{2\kappa_n^{eff} S_n}\right) \tag{3.42}$$

Therefore, the relationship between the electrolyte-phase potential and the input variable i_L is established.

(4) Solution of Butler–Volmer equation

From Eqs. (3.17) and (3.18), the Butler–Volmer equation of each electrode is expressed as follows:

$$\begin{aligned} j_{r,n} &= k_{s,n} c_e^{\alpha_a}\left(c_{s,max,n} - c_{e-s,n}\right)^{\alpha_a} c_{e-s,n}^{\alpha_c}\left(e^{\frac{\alpha_a F}{RT}\eta_n} - e^{-\frac{\alpha_c F}{RT}\eta_n}\right) \\ j_{r,p} &= k_{s,p} c_e^{\alpha_a}\left(c_{s,max,p} - c_{e-s,p}\right)^{\alpha_a} c_{e-s,p}^{\alpha_c}\left(e^{\frac{\alpha_a F}{RT}\eta_p} - e^{-\frac{\alpha_c F}{RT}\eta_p}\right) \end{aligned} \tag{3.43}$$

Considering that the transfer coefficients α_a and α_c are usually 0.5, they can be recorded as α. By substituting Eqs. (3.24) and (3.27), we can derive:

$$\begin{aligned} \frac{i_L}{a_n L_n S_n F} &= k_{s,n} c_e^{\alpha}\left(c_{s,max,n} - c_{e-s,n}\right)^{\alpha} c_{e-s,n}^{\alpha}\left(e^{\frac{\alpha F}{RT}\eta_n} - e^{-\frac{\alpha F}{RT}\eta_n}\right) \\ -\frac{i_L}{a_p L_p S_p F} &= k_{s,p} c_e^{\alpha}\left(c_{s,max,p} - c_{e-s,p}\right)^{\alpha} c_{e-s,p}^{\alpha}\left(e^{\frac{\alpha F}{RT}\eta_p} - e^{-\frac{\alpha F}{RT}\eta_p}\right) \end{aligned} \tag{3.44}$$

Defining the following variables:

$$\begin{aligned} \xi_n &= \frac{1}{2}\left(e^{\frac{\alpha F}{RT}\eta_n} - e^{-\frac{\alpha F}{RT}\eta_n}\right) = \sinh\left(e^{\frac{\alpha F}{RT}\eta_n}\right) = \frac{i_L}{2a_n L_n S_n F k_{s,n} c_e^{\alpha}\left(c_{s,max,n} - c_{e-s,n}\right)^{\alpha} c_{e-s,n}^{\alpha}} \\ \xi_p &= \frac{1}{2}\left(e^{\frac{\alpha F}{RT}\eta_p} - e^{-\frac{\alpha F}{RT}\eta_p}\right) = \sinh\left(e^{\frac{\alpha F}{RT}\eta_n}\right) = \frac{i_L}{2a_p L_p S_p F k_{s,p} c_e^{\alpha}\left(c_{s,max,p} - c_{e-s,p}\right)^{\alpha} c_{e-s,p}^{\alpha}} \end{aligned} \tag{3.45}$$

According to the inverse sine function, we can obtain

$$\begin{aligned} \eta_n &= \frac{RT}{\alpha F}\ln\left(\xi_n + \sqrt{\xi_n^2 + 1}\right) \\ \eta_p &= \frac{RT}{\alpha F}\ln\left(\xi_p + \sqrt{\xi_p^2 + 1}\right) \end{aligned} \tag{3.46}$$

Therefore, the relationship between the overpotential generated by the electrochemical reaction and the input variable i_L has been established.

(5) Solution to open-circuit potential

The open-circuit potential of an electrode is determined by the Li-ion concentration on the surface of the active particle, a physical quantity solely related to the characteristics of the electrode material. At present, the most commonly adopted method to obtain the open-circuit potential of each electrode is to charge and discharge the half-battery at a very small current rate. However, this method requires relevant experimental equipment to make positive and negative half-batteries and takes a long time. Therefore, in this chapter, the relationship between the open-circuit potential and the Li-ion concentration on the surface of active particles is substituted by the relationship between the OCV and SOC.

The available Li-ion concentration refers to the absolute value of Li-ion concentration variation in the active particles during the entire discharge process of the battery from 100% SOC to 0% SOC. The available Li-ion capacity is proportional to the maximum available capacity of the battery, and its expression is as follows:

$$\begin{aligned} c_{\text{total,n}} &= c_{\text{n,100\%}} - c_{\text{n,0\%}} = \frac{C_{\max}}{S_{\text{n}} L_{\text{n}} \varepsilon_n F} \\ c_{\text{total,p}} &= c_{\text{p,0\%}} - c_{\text{p,100\%}} = \frac{C_{\max}}{S_{\text{p}} L_{\text{p}} \varepsilon_p F} \end{aligned} \tag{3.47}$$

where $c_{\text{total,n}}$ is the available Li-ion concentration in the negative electrode, and $c_{\text{total,p}}$ is the available Li-ion concentration in the positive electrode.

In the electrochemical model, SOC of the battery is defined as the ratio of the remaining available Li-ion concentration to the maximum available Li-ion concentration.

$$SOC = \frac{\bar{c}_{\text{n}} - c_{\text{n,0\%}}}{c_{\text{total,n}}} \times 100\% = \frac{c_{\text{p,0\%}} - \bar{c}_{\text{p}}}{c_{\text{total,p}}} \times 100\% \tag{3.48}$$

where $\bar{c}_{\text{p}}$ and $\bar{c}_{\text{n}}$ are the average solid-phase Li-ion concentration in the positive and negative electrodes. However, the open-circuit potential of the electrode is not related to the average Li-ion concentration in the active particles, but only to the Li-ion concentration on the surface of the active particles. Therefore, the concept of surface SOC is introduced as

$$SOC_{\text{surf}} = \frac{c_{\text{surf,n}} - c_{\text{n,0\%}}}{c_{\text{total,n}}} \times 100\% = \frac{c_{\text{p,0\%}} - c_{\text{surf,p}}}{c_{\text{total,p}}} \times 100\% \tag{3.49}$$

where SOC_{surf} is the battery surface SOC. The surface SOC calculated from positive and negative electrodes should be consistent theoretically. Considering the calculation error, the surface SOC can be obtained as follows:

$$SOC_{\text{surf}} = \frac{\left(\frac{c_{\text{surf,n}} - c_{\text{n,0\%}}}{c_{\text{total,n}}} + \frac{c_{\text{p,0\%}} - c_{\text{surf,p}}}{c_{\text{total,p}}}\right)}{2} \times 100\% \tag{3.50}$$

The OCV of the battery is a function of SOC and available capacity, which is usually fitted by the following polynomial function:

$$OCV(SOC, C_{\max}) = \alpha_0 + \alpha_1 SOC + \alpha_2 SOC^2 + \alpha_3 SOC^3 + \alpha_4 SOC^4 + \alpha_5 SOC^5 + +\alpha_6 SOC^6 \tag{3.51}$$

where α_0–α_6 represent the coefficients of fitting function, whose value is related to the battery available capacity.

In the OCV test, the terminal voltage is recorded after the battery rests for a long time when the Li-ion concentration in the active particles tends to be uniform along with the continuous Li-ions diffusion. At this time, the current stays at 0. The electrode is at an equilibrium potential and the overpotential is 0. Li-ions concentration on the surface of the active particles is nearly equal to the average Li-ions concentration. We can derive the relations as

$$\begin{aligned} c_{\text{surf,n}} &= \bar{c}_{\text{n}} \\ c_{\text{surf,p}} &= \bar{c}_{\text{p}} \\ OCV &= E_{\text{OCV,p}} - E_{\text{OCV,n}} = \Delta E_{\text{OCV}} \end{aligned} \tag{3.52}$$

Combining Eqs. (3.48), (3.49), and (3.52), we obtain

$$SOC = SOC_{\text{surf}} \tag{3.53}$$

Combining Eqs. (3.51) and (3.53), we obtain

$$\Delta E_{OCV}(SOC_{\text{surf}}) = \alpha_0 + \alpha_1 SOC_{\text{surf}} + \alpha_2 SOC_{\text{surf}}^2 + \alpha_3 SOC_{\text{surf}}^3 + \alpha_4 SOC_{\text{surf}}^4 + \alpha_5 SOC_{\text{surf}}^5 + +\alpha_6 SOC_{\text{surf}}^6 \tag{3.54}$$

Therefore, the relationship between the -circuit potential of the electrode and the Li-ions concentration on the surface of the active particles can be established.

Substituting the results of Steps (1)–(5) into Eq. (3.21) ,we obtain the simplified terminal voltage expression.

$$U_{\text{t}} = \Delta E_{\text{OCV}}(SOC_{\text{surf}}) + \eta_{\text{p}} - \eta_{\text{n}} - i_L\left(\frac{L_{\text{n}}}{2\kappa_{\text{n}} S_{\text{n}}} + \frac{L_{\text{sep}}}{\kappa_{\text{sep}} S_{\text{sep}}} + \frac{L_{\text{p}}}{2\kappa_{\text{p}} S_{\text{p}}}\right) + i_{\text{L}} R_0 \tag{3.55}$$

where $i_{\text{L}} R_0$ is used to compensate for the voltage drop. As this simplified electrochemical model ignores the distribution of solid–electrolyte-phase potential and Li-ions molar flux along the x-direction, the voltage drop caused by SEI film generated during battery use is not discussed. Accordingly, $i_{\text{L}} R_0$ is used to approximate the neglected voltage drop due to the simplification of the model.

This section introduces a series of mathematical equations of the simplified P2D model, and the method of solving partial differential equation via the finite difference method. So far, each part of the voltage drop in Eq. (3.55) has been calculated. The simulation of the terminal voltage can be easily achieved.

3.1.3 Parameter Identification of the Model

There are up to 23 internal parameters in the simplified P2D model. Some parameters are difficult or even impossible to obtain through measurement and some parameters change along with battery aging. Accordingly, identifying the model parameters is an inevitable work. However, this model contains several partial differential equations and has strong nonlinearity. It is difficult to identify the parameters by traditional methods such as the least square (LS) method and Kalman filter (KF). Genetic algorithm (GA) is a method of searching the optimal solution by simulating the natural evolutionary process. It is capable of solving the parameter identification problem of strongly nonlinear systems and is quite suitable for large-scale optimization of complex models [7].

GA starts with a population of solution sets where the optimal solution to the problem is possible to exist. After the initial population generation, according to the law of survival of the fittest, there's an increasingly superior approximate solution to evolve gradually. In each generation, individuals are selected according to their fitness in the problem domain prior to being crossed and mutated using genetic operators to generate a population of offspring that represents a new set of solutions. In this process, the progeny population will adapt better to the environment than their parents like natural evolution. The optimization objective of the GA applied in this chapter is the square sum of the terminal voltage error. The flowchart is shown in Fig. 3.3.

The specific steps of GA are as follows:

(1) Determine the boundary conditions of the parameters of battery model and set the running parameters of GA. Set the numbers of individuals and maximum generation to 80 and 200, respectively. The crossover rate and the mutation rate are set to 0.5 and 0.01, respectively.
(2) Generate the initial population.
(3) Calculate the individual fitness of the population and determine whether it meets the requirements. If the requirements are satisfied, the identification ends. Otherwise, the next step is taken.
(4) Inheritance, crossover, and mutation of the population are done to obtain offspring.
(5) Return to Step (3).

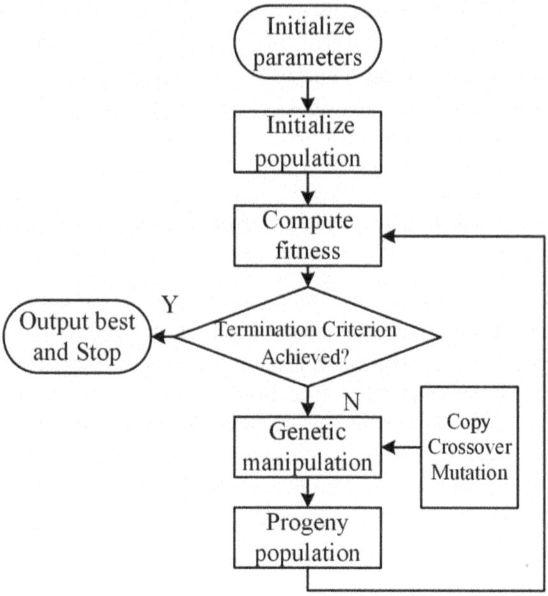

Fig. 3.3 Flowchart of genetic algorithm

3.1.4 Case Study

Some parameters used in the electrochemical model have been determined in advance to the battery design and will not change obviously during the whole life cycle. Some parameters are not sensitive to the aging state of the battery or have no obvious correlation with the aging state. During the use of the battery, complex and irreversible side reactions occur inside the battery. These side reactions can consume Li-ions and generate byproducts depositing on the surface of the active material to form SEI films. With the aging of the battery, the thickness of the SEI film will increase, leading to an increase in the internal resistance. In the meantime, it will be more difficult for the Li-ions to intercalate and de-intercalate from the active particles, which is manifested in the decrease in the Li-ions solid-phase diffusion coefficients. The irreversible side reactions also permanently consume Li-ions, resulting in a decrease in the initial Li-ions concentration of the electrodes. In addition, with the aging of the battery, the electrochemical reactions will be more difficult to occur, resulting in a decrease in the electrochemical reaction rate constant. Therefore, the parameters, which are closely related to the battery aging process might exhibit a certain law include: maximum Li-ions concentration in active particles of each electrode $c_{s,max,p}$ and $c_{s,max,n}$, positive and negative solid-phase diffusion coefficient $D_{s,p}$ and $D_{s,n}$, positive and negative electrochemical reaction rate constant $k_{s,p}$ and $k_{s,n}$, cell resistance R_0.

Taking the above seven variables as parameters to be identified, GA is used to identify the parameters. The number of the battery is 1, the identification is carried out using the DST profile at 25 °C. Positive and negative solid-phase diffusion coefficient $D_{s,p}$ and $D_{s,n}$, positive and negative electrochemical reaction rate constant $k_{s,p}$ and

$k_{s,n}$, and cell resistance R_0 are not only related to the aging state, but also to SOC. Therefore, in order to get better model accuracy, these five parameters are identified in each SOC interval, respectively. To balance the parameter identification process complexity and the model accuracy, the cell discharge interval is equally divided into 10% intervals from 100 to 0%. The parameter identification results are shown in Table 3.2.

Figure 3.4a presents a comparison between the simulation results and the experimental results under DST test. Figure 3.4b is the terminal voltage error in the SOC range of 100–20%.

From Fig. 3.4, it can be seen that the electrochemical model has an excellent simulation accuracy in the middle and high SOC region. The simulation error is usually restricted to within 20 mV. Although the accuracy decreases at a large rate current, it is still maintained within 50 mV. The above results show that GA is effective in identifying the parameters used in the model, and the simplified P2D model can accurately simulate the characteristics of Li-ion batteries.

3.2 The Equivalent Circuit Models

3.2.1 Model Introduction

Based on the external characteristics of the battery, the ECM was developed by using resistors, capacitors, and voltage sources to form a circuit network. The ECM uses a voltage source to represent the thermodynamic equilibrium potential and RC network to describe the dynamic characteristics of battery [8]. It has ideal applicability to a variety of working conditions for the battery, and the state equation of the model can be derived for analysis and application. Therefore, it has been extensively used in modeling and simulation of new energy vehicles and the model-based BMS. Figure 3.5 shows a typical ECM consisting of n RC networks, known as the n-RC model for short. The model is composed of three parts [9]:

(1) Voltage source: U_{OC} indicates the OCV of the battery;
(2) Ohmic internal resistance: R_i represents the resistance of electrode material, electrolyte, and diaphragm as well as the contact resistance of each part of the battery;
(3) RC network: The dynamic characteristics of the battery, including polarization characteristics and diffusion effects, are described by the polarization resistance R_{Di} and the polarization capacitance C_{Di}, where $i = 0, \ldots, n$.

In Fig. 3.5, U_{Di} represents the polarization voltage of the battery. According to the Kirchhoff's voltage law and the Kirchhoff's current law, as well as the relationship between the change of the capacitor voltage and its current, the state equation of the n-RC model can be expressed as

Table 3.2 Identification results of battery 1

Parameters	100–90%	90–80%	80–70%	70–60%	60–50%	50–40%	40–30%	30–20%	20–10%	10–0%
$c_{s,max,p}$ (mol/m^3)	31423									
$c_{s,max,p}$ (mol/m^3)	22754									
$D_{s,p}$ (10^{-15}m^2/s)	20.35	21.01	34.06	41.46	43.40	58.10	54.17	34.68	25.87	19.91
$D_{s,n}$ (10^{-15}m^2/s)	12.76	13.03	33.37	30.72	49.82	24.65	20.79	22.80	20.58	9.90
$k_{s,p}$ (10^{-10}m$^{2.5}$/mol$^{0.5}$s)	5.735	0.061	0.058	1.356	0.291	1.561	5.194	5.307	0.139	2.443
$k_{s,p}$ (10^{-10}m$^{2.5}$/mol$^{0.5}$s)	5.091	5.162	5.603	5.591	5.635	5.613	4.721	5.363	5.507	6.344
R_0 (mΩ)	3.308	3.313	3.636	3.314	4.153	4.349	4.173	5.215	6.964	10.103

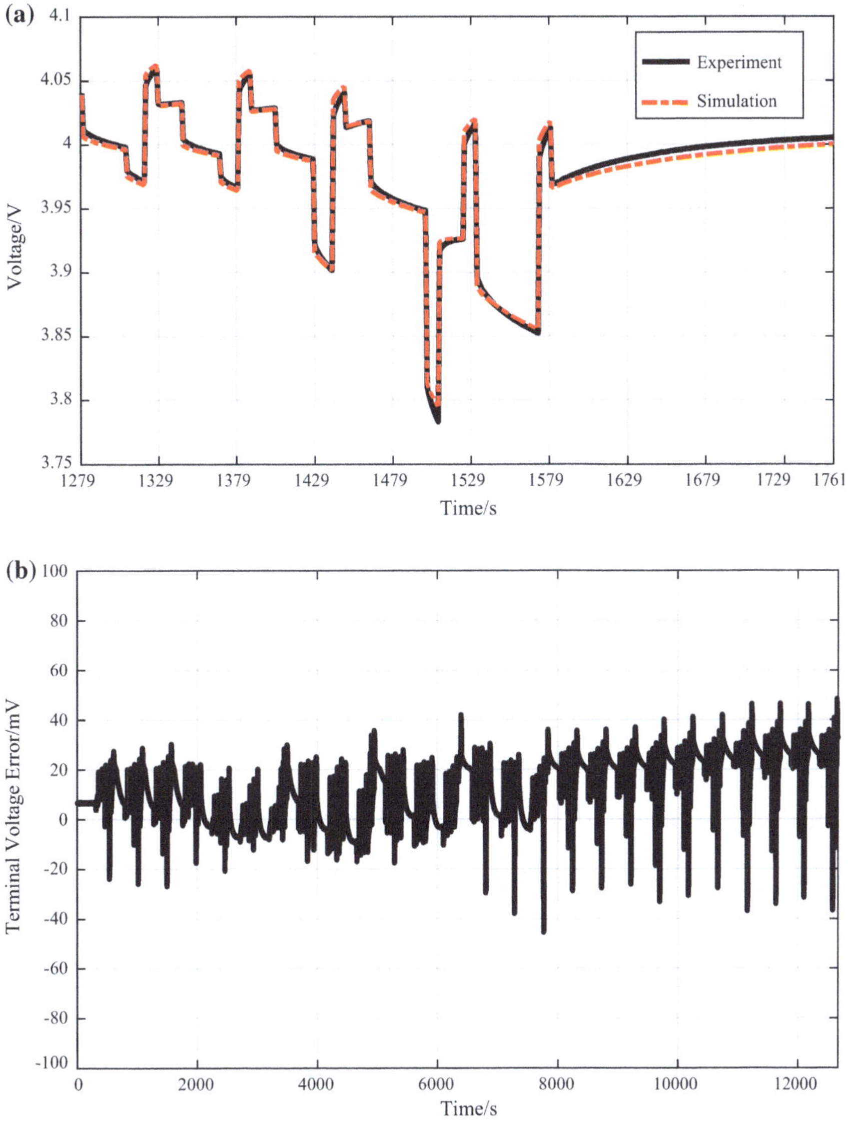

Fig. 3.4 **a** Modeling results at 25 °C **b** simulation error

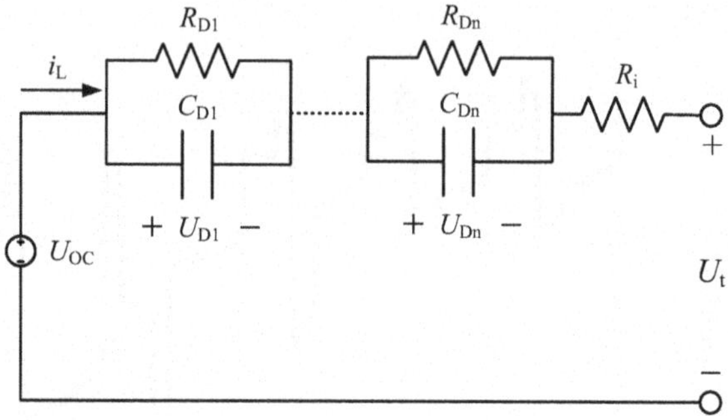

Fig. 3.5 Schematic diagram of n-RC model

$$\begin{cases} \dot{U}_{D1} = -\frac{U_{D1}}{R_{D1}C_{D1}} + \frac{iL}{C_{D1}} \\ \dot{U}_{D2} = -\frac{U_{D2}}{R_{D2}C_{D2}} + \frac{iL}{C_{D2}} \\ \cdots \\ U_t = U_{OC} - U_{D1} - U_{D2} - \cdots - iLR_i \end{cases} \tag{3.56}$$

The Rint model, the Thevenin model, and the Dual Polarization (DP) model are special cases for the n-RC equivalent circuit model when $n = 0$, $n = 1$, and $n = 2$, respectively. They have already been extensively used for battery state estimation and management algorithms. Then, three commonly used equivalent circuit models will be introduced in detail.

Rint model ($n = 0$)

The Rint model is shown in Fig. 3.6. The OCV is described by an ideal voltage source U_{OC}, and the ideal resistance R_i describes the ohmic internal resistance of the battery. R_i and U_{OC} are functions of SOC and temperature [10]. The relationship between output voltage and input current of Rint model is as follows:

$$U_t = U_{OC} - i_L R_i \tag{3.57}$$

The Thevenin model ($n = 1$)

Figure 3.7 shows the circuit structure of the Thevenin model [11–13]. R_D and C_D represent polarization internal resistance and polarization capacitance, respectively, and U_D denotes the voltage drop of the parallel RC branch, which is used to simulate

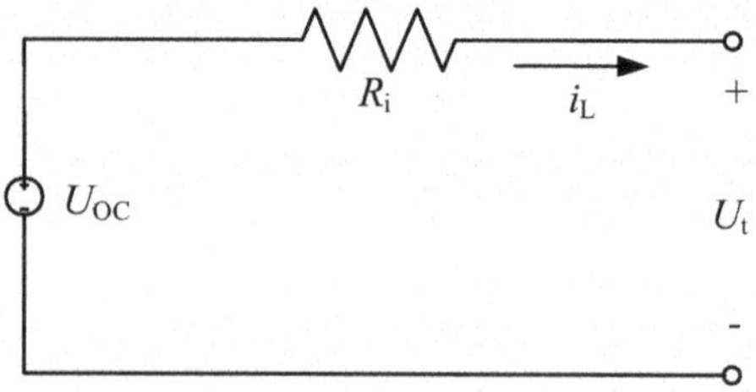

Fig. 3.6 Schematic diagram of the Rint model

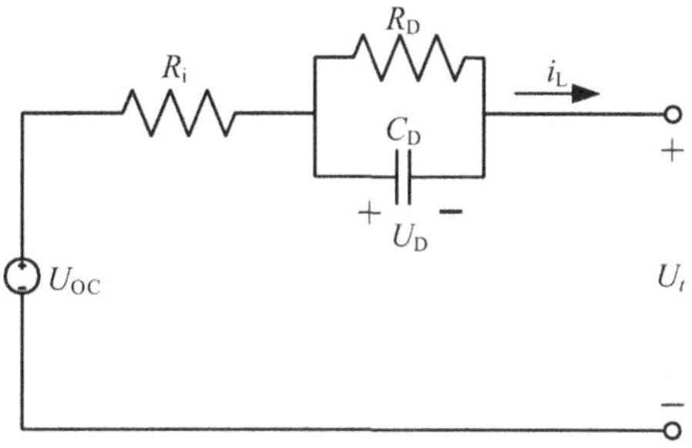

Fig. 3.7 Schematic diagram of the Thevenin model

polarization voltage. Compared with the Rint model, this model adds an RC branch to describe thee polarization characteristics during charing and discharging. The state space equation of the circuit model is as follows:

$$\begin{cases} \dot{U}_{\mathrm{D}} = \frac{i_L}{C_{\mathrm{D}}} - \frac{U_{\mathrm{D}}}{R_{\mathrm{D}} C_{\mathrm{D}}} \\ U_{\mathrm{t}} = U_{\mathrm{OC}} - U_{\mathrm{D}} - i_{\mathrm{L}} R_{\mathrm{i}} \end{cases} \tag{3.58}$$

Dual polarization model ($n = 2$)

Figure 3.8 shows the circuit structure of the Dual Polarization model.

The model uses two parallel RC branches to describe the polarization characteristics shown by the battery [14]. The equation of the this model is

$$\begin{cases} \dot{U}_{\mathrm{D1}} = \frac{i_L}{C_{\mathrm{D1}}} - \frac{U_{\mathrm{D1}}}{R_{\mathrm{D1}} C_{\mathrm{D1}}} \\ \dot{U}_{\mathrm{D2}} = \frac{i_L}{C_{\mathrm{D2}}} - \frac{U_{\mathrm{D2}}}{R_{\mathrm{D2}} C_{\mathrm{D2}}} \\ U_{\mathrm{t}} = U_{\mathrm{OC}} - U_{\mathrm{D1}} - U_{\mathrm{D2}} - i_{\mathrm{L}} R_{\mathrm{i}} \end{cases} \tag{3.59}$$

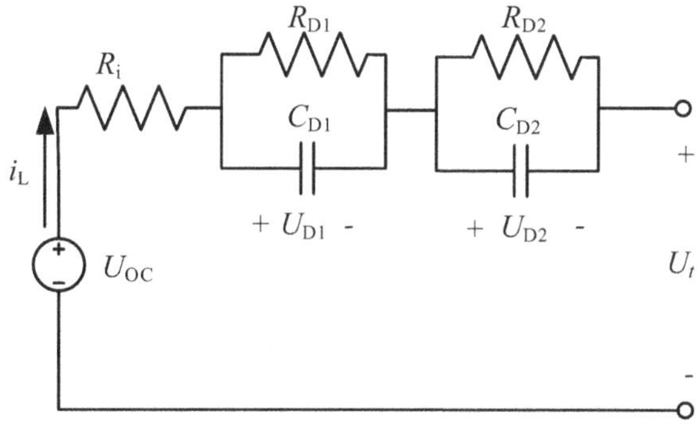

Fig. 3.8 Schematic diagram of the DP model

3.2.2 Model Constructing

In order to facilitate the application of BMS, the model needs to be discretized, and the n-RC model is taken as an example to deduce the discrete form of the model. The transfer function of the n-RC model can be written as

$$G(s)=\frac{U_{\mathrm{t}}(s)-U_{\mathrm{oc}}(s)}{i_{\mathrm{L}}(s)}=-\left(R_{\mathrm{i}}+\frac{R_{\mathrm{D1}}}{1+R_{\mathrm{D1}}C_{\mathrm{D1}}s}+\frac{R_{\mathrm{D2}}}{1+R_{\mathrm{D2}}C_{\mathrm{D2}}s}+\cdots+\frac{R_{\mathrm{D}n}}{1+R_{\mathrm{D}n}C_{\mathrm{D}n}s}\right) \tag{3.60}$$

Defining $E_{\mathrm{L}}(s)=U_{\mathrm{t}}(s)-U_{\mathrm{oc}}(s)$, then,

$$E_{\mathrm{L}}(s)=-i_{\mathrm{L}}(s)\left(R_{\mathrm{i}}+\frac{R_{\mathrm{D1}}}{1+R_{\mathrm{D1}}C_{\mathrm{D1}}s}+\frac{R_{\mathrm{D2}}}{1+R_{\mathrm{D2}}C_{\mathrm{D2}}s}+\cdots+\frac{R_{\mathrm{D}n}}{1+R_{\mathrm{D}n}C_{\mathrm{D}n}s}\right) \tag{3.61}$$

The bilinear transformation method is frequently adopted to map the system from the s-plane to the z-plane. A bilinear transformation method shown in Eq. (3.62) is employed for the discretization calculation [15, 16].

$$s=\frac{2}{\Delta t}\frac{1-z^{-1}}{1+z^{-1}} \tag{3.62}$$

where Δt represents sampling interval time of the system, and the equation based on the z-plane is

$$G(z^{-1})=\frac{c_{n+1}+\cdots+c_{2n+1}z^{-2n+1}}{1-c_1z^{-1}-\cdots-c_nz^{-n}} \tag{3.63}$$

where $c_i (i=1,2,\ldots 2n+1)$ is the coefficient related to the parameters of the model. It can be converted to the discrete time domain, and the result is shown in Eq. (3.64):

$$E_{\mathrm{L},k}=c_1E_{\mathrm{L},k-1}+\cdots+c_nE_{\mathrm{L},k-n}+c_{n+1}i_{\mathrm{L},k}+c_{2n+1}i_{\mathrm{L},k-n} \tag{3.64}$$

Since the OCV is coupled with the SOC, the operating temperature T and the aging state A_{ge}, the $U_{\mathrm{OC},k}$ is defined as the function of the z_k, the temperature T_k and the aging state $A_{\mathrm{ge},k}$ at time k, which is

$$U_{\mathrm{OC},k}=f(z_k,T_k,A_{\mathrm{ge},k}) \tag{3.65}$$

Differentiate $U_{\mathrm{OC},k}$ in Eq. (3.65) with respect to time k.

$$\frac{\mathrm{d}U_{\mathrm{OC}}}{\mathrm{d}t}=\frac{\partial U_{\mathrm{OC}}}{\partial z}\frac{\mathrm{d}z}{\mathrm{d}t}+\frac{\partial U_{\mathrm{OC}}}{\partial T}\frac{\mathrm{d}T}{\mathrm{d}t}+\frac{\partial U_{\mathrm{OC}}}{\partial A_{ge}}\frac{\mathrm{d}A_{ge}}{\mathrm{d}t} \tag{3.66}$$

For the sake of simplifying the calculation, the following assumptions are made.

Hypothesis 1 The influence of the amount of electricity consumed or regained at each sampling interval on its SOC is approximate to zero, namely, $\mathrm{d}z/\mathrm{d}t \approx 0$.

The influence of the variety of electricity at every sampling interval on its SOC can be expressed as follows:

$$\frac{\mathrm{d}z}{\mathrm{d}t} = \frac{\eta_{\mathrm{i}} i_{\mathrm{L}} \times \Delta t}{C_{\max} \times 3600} \tag{3.67}$$

It is assumed that the average operating current rate of the battery is 1C. The maximum available capacity is $C_{\max}$, the sampling interval Δt is 1 s, and the efficiency of charge–discharge rate η_{i} of the battery is 1. Then, the above equation can be calculated as

$$\frac{\mathrm{d}z}{\mathrm{d}t} = \frac{1 \times C_{\max} \times 1}{C_{\max} \times 3600} = \frac{1}{3600} < 0.03\% \tag{3.68}$$

Generally speaking, the actual sampling time is less than 1 s, and the value will be smaller, for which the hypothesis is valid.

Hypothesis 2 The temperature of the battery remains unchanged within the sampling interval, that is $\mathrm{d}T/\mathrm{d}t \approx 0$.

Relying on the temperature management of BMSs, the temperature rise/decrease of batteries ought to be slow, and then $\mathrm{d}T/\mathrm{d}t \approx 0$ holds for the normal operating conditions.

Hypothesis 3 The aging state of the battery remains unchanged within the sampling interval, that is $\mathrm{d}A_{\mathrm{ge}}/\mathrm{d}t \approx 0$.

Battery aging is a long-term and slow process, for which the aging state can be considered as a constant value at each time of sampling.

Based on the above assumptions, Eq. (3.66) can be simplified into:

$$\frac{\mathrm{d}U_{\mathrm{OC}}}{\mathrm{d}t} = \frac{\partial U_{\mathrm{OC}}}{\partial z}\frac{\mathrm{d}z}{\mathrm{d}t} + \frac{\partial U_{\mathrm{OC}}}{\partial T}\frac{\mathrm{d}T}{\mathrm{d}t} + \frac{\partial U_{\mathrm{OC}}}{\partial A_{ge}}\frac{\mathrm{d}A_{ge}}{\mathrm{d}t} = 0 \tag{3.69}$$

In the discrete time domain, that is

$$\Delta U_{\mathrm{OC},k} = U_{\mathrm{OC},k} - U_{\mathrm{OC},k-1} \approx 0 \tag{3.70}$$

Then Eq. (3.64) can be simplified into:

$$\begin{aligned} U_{\mathrm{t},k} = {} & \left(1 - \sum_{i=1}^{n} c_i\right) U_{\mathrm{OC},k} + c_1 U_{\mathrm{t},k-1} + c_2 U_{\mathrm{t},k-2} + \ldots + c_{\mathrm{n}} U_{\mathrm{t},k-n} \\ & + c_{\mathrm{n}+1} i_{\mathrm{L},k} + c_{\mathrm{n}+2} i_{\mathrm{L},k-1} + \cdots + c_{2\mathrm{n}+1} i_{\mathrm{L},k-n} \end{aligned} \tag{3.71}$$

Defining the data matrix and parameter matrix of the system as

$$\begin{cases} \mathbf{\Phi}_{\mathrm{n},k} = \begin{bmatrix} 1 & U_{\mathrm{t},k-1} & U_{\mathrm{t},k-2} \ldots U_{\mathrm{t},k-\mathrm{n}} & i_{\mathrm{L},k} & i_{\mathrm{L},k-1} & i_{\mathrm{L},k-2} \ldots i_{\mathrm{L},k-\mathrm{n}} \end{bmatrix} \\ \mathbf{\theta}_{\mathrm{n},k} = \begin{bmatrix} \left(1 - \sum\limits_{i=1}^{n} c_i\right) U_{\mathrm{OC},k} & c_1 & c_2 & c_3 \ldots c_{2\mathrm{n}+1} \end{bmatrix}^{\mathrm{T}} \end{cases} \tag{3.72}$$

Then Eq. (3.60) can be simplified to:

$$y_k = \mathbf{\Phi}_{\mathrm{n},k} \mathbf{\theta}_{\mathrm{n},k} \tag{3.73}$$

Based on the identification results of the parameter matrix, the model parameters could be identified. For example, the OCV can be interpreted as

$$U_{\mathrm{oc},k} = \frac{\mathbf{\theta}_{\mathrm{n},k}(1)}{1 - \sum_{i=1}^{n} c_i} \tag{3.74}$$

3.2.3 Parameters Identification

The ECM use online or offline data to identify battery parameters. The online parameter identification method utilizes the data measured in real time such as current, voltage, and temperature to identify parameters, thereby realizing online updating of model parameters. The offline parameter identification method uses the data stored by the BMS to update and calibrate parameters, which can be automatically updated periodically according to the designer's intention and the needs of BMS. This book will introduce two typical parameter identification methods.

Online parameter identification

Recursive least squares (RLS) algorithm based on adaptive filtering is a common method of model parameter identification and data mining [17, 18]. For the case that the model and parameters are easy to be affected by the uncertain application environment, the RLS can overcome the uncertainty of model parameters through periodic correction and update, so as to accurately capture the real-time characteristics of the system. However, it is difficult for the conventional RLS method to obtain stable and reliable estimation results for the heuristic system with continuous and slow changes. To address this problem, the RLS method with a forgetting factor can be used to achieve a reliable identification of the parameters. An optimal forgetting factor is employed to assign less weight to old data and more weight to recent data. The system identification is realized as follows.

Consider the following system:

$$y_k = \mathbf{\Phi}_k \mathbf{\theta}_k + e_{\mathrm{Ls},k} \tag{3.75}$$

where y_k denotes the vector of output values, $e_{\mathrm{Ls},k}$ indicates the white noise information mainly from the measurement, $\boldsymbol{\theta}_k$ and $\boldsymbol{\Phi}_k$ represent the parameter vector and the data matrix, respectively, and subscript k indicates that the data value is the kth sampling instant. The detailed computation processes of the RLS algorithm are presented in Eq. (3.76).

$$\begin{cases} \mathbf{K}_{\mathrm{Ls},k} = \mathbf{P}_{\mathrm{Ls},k-1}\boldsymbol{\Phi}_k^{\mathrm{T}}\left[\boldsymbol{\Phi}_k\mathbf{P}_{\mathrm{Ls},k-1}\boldsymbol{\Phi}_k^{\mathrm{T}} + \mu\right]^{-1} \\ \hat{\theta}_k = \hat{\theta}_{k-1} + \mathbf{K}_{\mathrm{Ls},k}[y_k - \boldsymbol{\Phi}_k\hat{\theta}_{k-1}] \\ \mathbf{P}_{\mathrm{Ls},k} = \frac{1}{\mu}[\mathbf{I} - \mathbf{K}_{\mathrm{Ls},k}\boldsymbol{\Phi}_k]\mathbf{P}_{\mathrm{Ls},k-1} \end{cases} \tag{3.76}$$

where $\mathbf{K}_{\mathrm{Ls},k}$ and $\mathbf{P}_{\mathrm{Ls},k}$ denote the gain and error covariance matrixes of the RLS algorithm at the kth sampling interval, and μ indicates the forgetting factor. When $\mu = 1$, the above formula degenerates into the traditional RLS method. Based on the Eqs. (3.72), (3.75), and (3.76), the calculation flow of online parameter identification method using the RLS is shown in Fig. 3.9.

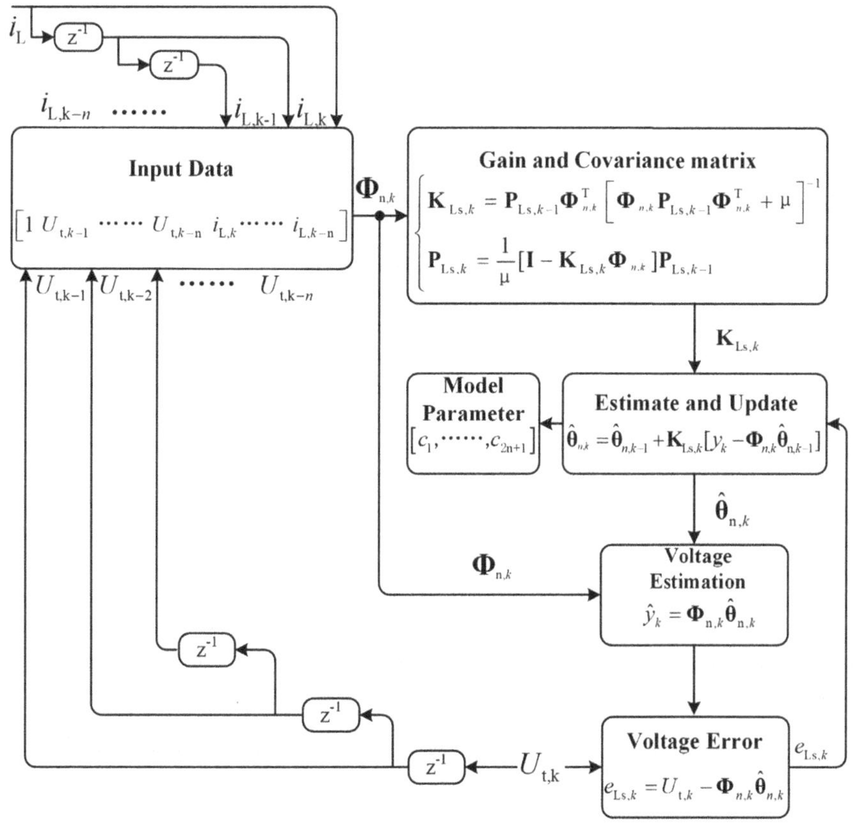

Fig. 3.9 Flowchart of model parameter identification based on RLS algorithm

The input data variables are constructed based on the real-time measured data such as current, voltage and temperature. Then, the corresponding least squares gain and error covariance are calculated to realize the online identification and update.

The MATLAB code for online parameter identification of the Thevenin model is as follows.

```
clear;
load DSTT40.mat;              %Import data
Ut=DSTT40(:,2);               %Terminal voltage
I=DSTT40(:,1);                %The discharge current is positive
T=length(Ut);                 %Length of data
Uoc=zeros(T,1);               %Store OCV for different models
Rs1=zeros(T,1);               %Store ohmic resistance for the Thevenin model
tau1=zeros(T,1);              %Store time constant for Thevenin model
Rt1=zeros(T,1);               %Store polarization resistance for the Thevenin model
e=zeros(T,1);                 %Store the error of the terminal voltage
%Initial RLS parameters
u=0.97;                       %Forgetting factor
Phi=zeros(1,4);               %Data vector
theta=zeros(1,4);             %Parameter vector
P=10^6*eye(4);                %Covariance matrix
K=zeros(4,1);                 %Gain
for t=1:T
if t<1
        %RLS recursive process
        Phi=[1,Ut(t-1),I(t),I(t-1)];
        K=P*Phi'/(Phi*P*Phi'+u);
        theta=theta+K*(Ut(t)-Phi*theta);
        P=(eye(4)-K*Phi)*P/u;
        %Parameter   analysis
        Uoc(t,2)=theta(1)/(1-theta(2));
        e(t)=(Ut(t)-Phi*theta(3))/(1+theta(2));
        Rs1(t)=(theta(4)-theta(3))/(1+theta(2));
        tau1(t)=(-1-theta(2))/(2*(theta(2)-1));
        Rt1(t)=-0.5*(theta(4)+theta(3))*(1+2*tau1(t))-Rs1(t);
    end
end
```

Offline parameter identification

In order to realize the automatic updating and optimization of the parameters of the battery models while avoiding the complicated and cumbersome calibration, a five-step offline parameter identification method shown in Fig. 3.10 is introduced in this section, which is easily applied to parameterize and evaluate ECMs of different orders in the BMS.

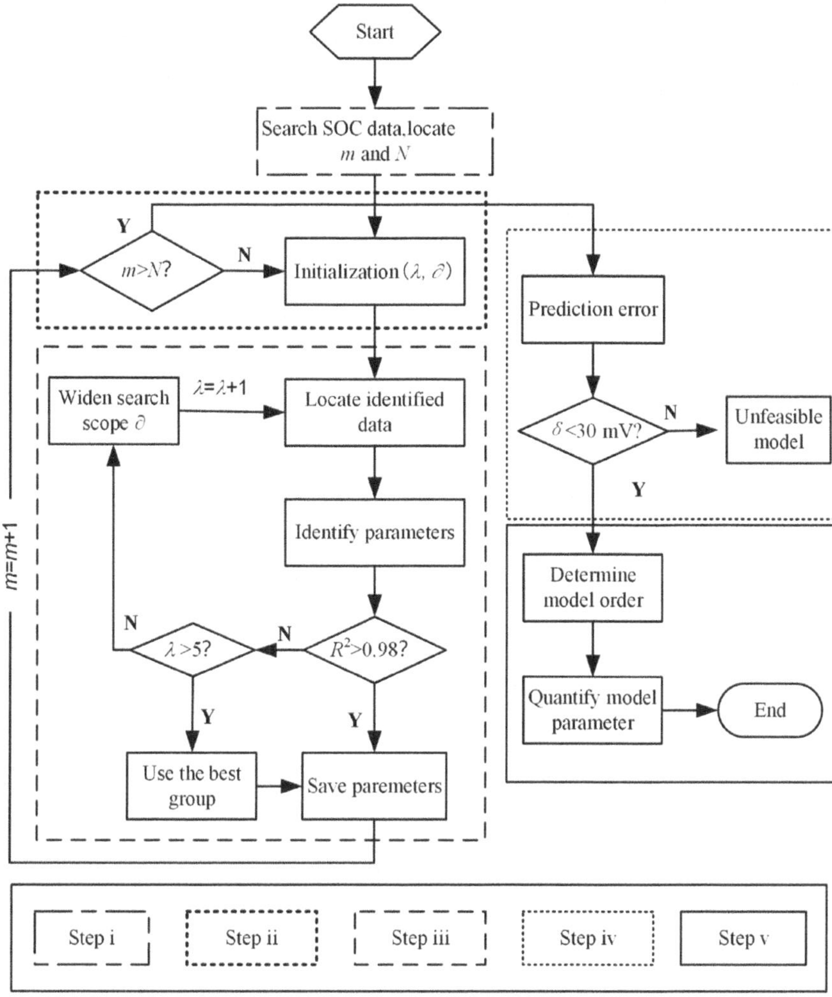

Fig. 3.10 Five-step parameter identification approach for battery

The specific implementation steps of this method are as follows [19]:

(1) *Region determination*. Complete the clearing of registers and variables, and extract the measured signals of the battery from the model calibration data stored in the BMS. Then, search the range of SOC (z_s, z_e), determine the parameter updating area and calculate the array number N of the parameter table. For example, with the 5% of the SOC interval of adjacent parameter group (the whole SOC range 10–100%), we can calculate that the number N is 19, the initial SOC z_s is 10%, and the end SOC z_e is 100%. At the beginning of the program, set the current order of the data group $m = 1$.

(2) *Initialization*. Make judgment whether the parameter identification is complete or not, and start the model evaluation if it has been completed. Otherwise, initialize the system initial value. ∂ denotes the SOC range of each group and λ represents the allowable times to expand the data range during parameter identification.
(3) *Parameter identification*. First, a calculation is performed in the SOC interval for the parameter identification. For example, based on the SOC interval ($z_s + (m-1) \times 5\%$ $z_s + (m-1) \times 5\% + \partial$), the current and voltage used for system identification are determined. Second, the multiple linear regression method is applied to identifying parameters of ECMs with different RC branch numbers on the basis of mathematical equations. Then, based on the R^2 to judge the reliability of the obtained parameters, with the minimal value of R^2 being set to 0.98. If $R^2 > 0.98$, it is considered that the parameters are reliable and the parameter identification is completed. If $R^2 < 0.98$, the value of ∂ will be enlarged to obtain more data for further identification. However, the maximum allowable times k is 5. If all the R^2 in the 5 times is less than 0.98, the parameter group with the biggest R^2 will be selected as model parameters. Finally, save the parameters and prepare for parameter identification or model evaluation at the next moment.
(4) *Model parameter evaluation and optimization*. The accuracy of the model is determined with a threshold of 30 mV. If the error δ is smaller than this value, the model is considered to be reasonable. If the error δ is greater than 30 mV, the model is considered to be unusable. The remaining battery models will be prepared for order evaluation.
(5) *Model order determination and system parameter updating*. Based on the Akaike information criterion (AIC), the accuracy and complexity of the model in different orders (except for the invalid order model) are evaluated. Since the model order n is much smaller than the data length L_D and the average of the residual squares $\hat{s}_k^2$ under the optimal parameters is less than 1 V, the AIC criterion can be simplified as follows:

$$\mathrm{AIC} = 2\log \hat{s}_k^2 + 2n \tag{3.77}$$

where the calculation method of $\hat{s}_k^2$ is as follows:

$$\hat{s}_k^2 = \frac{1}{L_D}\sum_{k=1}^{L_D}\{y_k - \hat{y}_k\}^2 \tag{3.78}$$

where $\hat{y}_k$ represents the estimated terminal voltage of the model.

Different from the traditional modeling methods, the above methods can be used not only to identify model parameters, but also to balance of model structure and complexity, and optimize the imprecise model to improve the accuracy of the model.

3.2.4 Case Study

Online parameter identification results

Based on the mathematical equations of the Thevenin model and the RLS algorithm, the parameter identification and model verification of the Thevenin model are realized by the hybrid pulse power characterisation (HPPC) test. The experimental object is the battery 2-cell 1. Figure 3.11 presents the OCV curve identified by the online data-driven method and Fig. 3.12 shows the comparison of voltage errors of different ECMs.

Figure 3.11a shows the variation in OCV with time as obtained by online identification of ECM with different orders under the HPPC test. When SOC is higher, the identification results of each model are similar. With the decrease in SOC, the difference of identification results between Rint model and other models increases gradually. Figure 3.11b shows that the OCV obtained by online estimation has a monotonic relationship with SOC, which gradually increases with the rise in SOC, and this characteristic can be applied to SOC estimation. The change step of SOC in the HPPC test is 5%, and the OCV remains constant in the same SOC region. It is revealed that the identification result of OCV based on online data-driven method is stable and reliable.

Figure 3.12 depicts the voltage error with time. In addition to the Rint model, other ECMs are demonstrated that can be reliably identified using the online data-driven parameter identification method. The maximum error is within 30 mV.

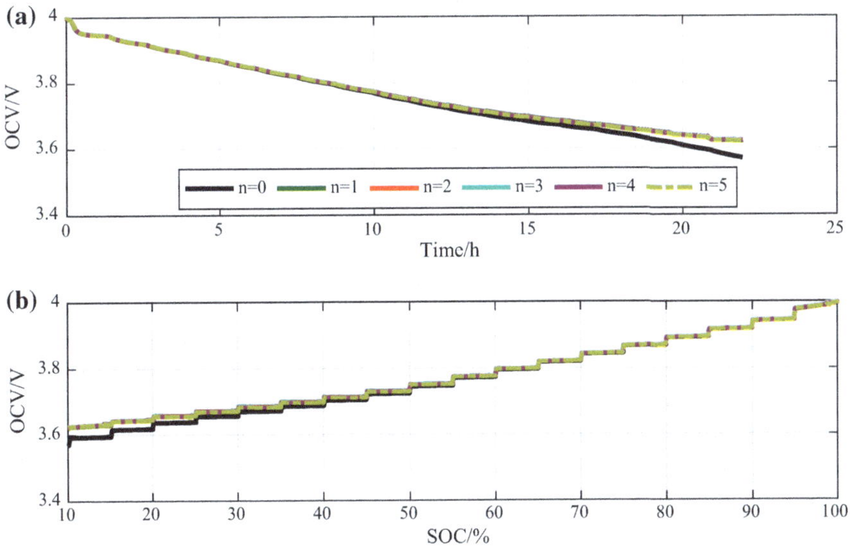

Fig. 3.11 The OCV results identified by HPPC test

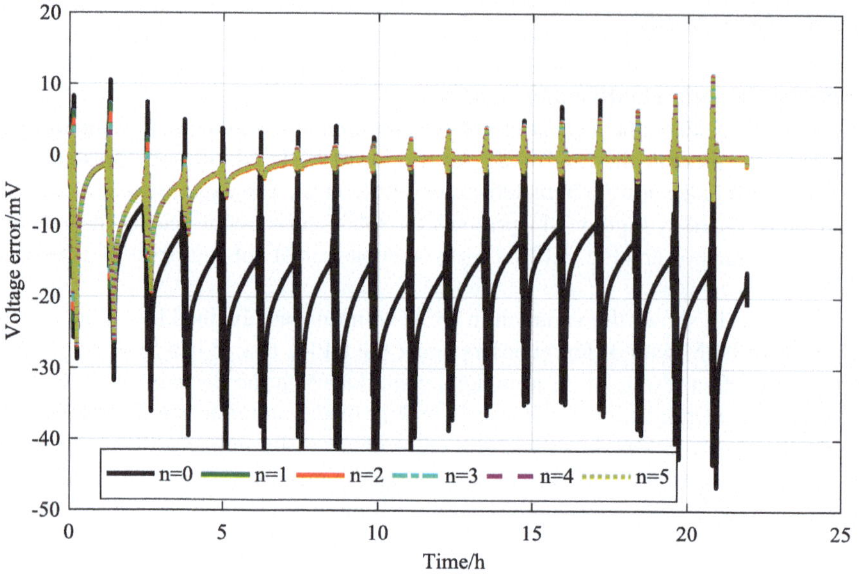

Fig. 3.12 The comparison curve of the Thevenin model voltage error

Offline parameter identification results

Figures 3.13 and 3.14 show the simulation results of battery 2-cell 1 based on the five-step parameter identification approach. Figure 3.13 shows the offline identification results of the OCV and the partially enlarged figure and Fig. 3.14 shows the comparative analytical results of the battery voltage prediction error.

OCV is one of the most important parameters of battery models. Figure 3.13 shows that the OCV estimation results of ECMs with different number of RC networks are relatively close, especially when considering the RC network, the deviation of the OCV estimates of the model is within 0.5 mV. Figure 3.14 indicates that the prediction accuracy of the Rint model does not meet the design requirements, and the maximum prediction error is 63 mV, which exceeds the set threshold. Therefore, the Rint model will be regarded as an invalid model. In contrast, the prediction errors of other models meet the requirements. For convenience in comparing the accuracy and calculation cost of each order model, in this section, the accuracy of ECMs with different number of RC networks will be evaluated from five aspects: correlation degree coefficient R^2, the maximum and average value of absolute error, calculation time using look-up table method or fitting method, as shown in Table 3.3. It is worth noting that the R^2 selects the minimum set of values in the parameter identification process of each model. In order to ensure objectivity in evaluating the error value, all the error statistics in this section use the absolute value of the prediction error. For example, the voltage error in Table 3.3 is the absolute value, and then the corresponding maximum, average and standard errors are calculated.

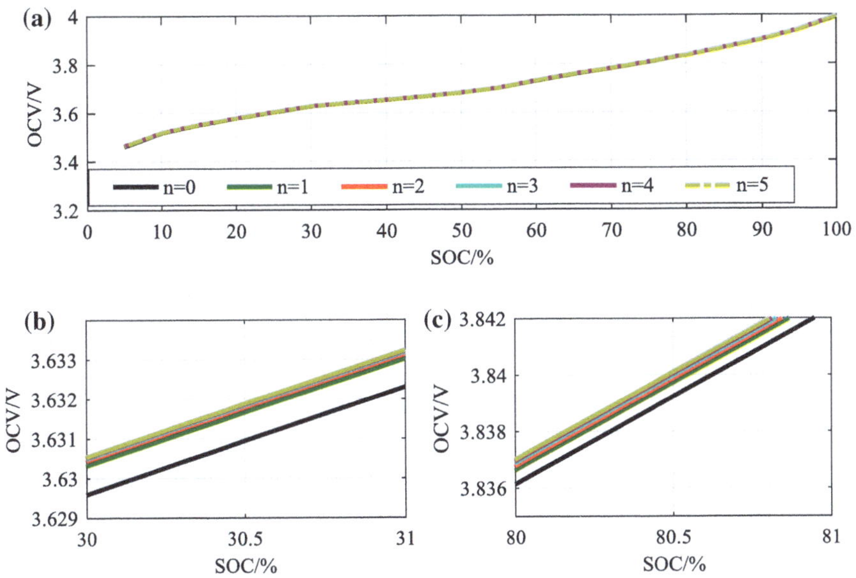

Fig. 3.13 The offline identification results of the OCV and the partial enlarged view

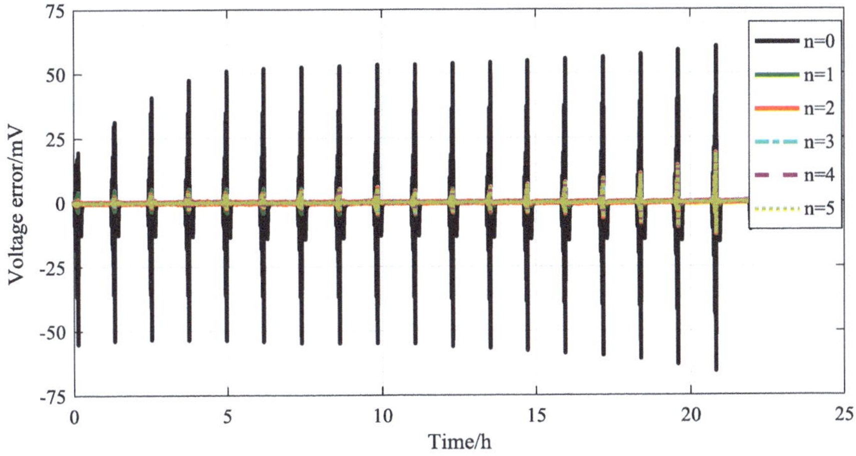

Fig. 3.14 The comparative analysis result of the battery voltage prediction error

Table 3.3 The analysis results of the absolute value of voltage prediction error based on the HPPC test of battery 2-cell 1

RC network	R^2	Maximum error (mV)	Mean error (mV)	Computation time look-up table (s)	Computation time-fitting method (s)
n = 0	0.926	63.789	1.701	19.61	0.61
n = 1	0.991	19.002	0.116	36.28	1.12
n = 2	0.992	18.508	0.114	71.08	3.56
n = 3	0.992	18.131	0.112	94.32	7.13
n = 4	0.993	19.115	0.112	119.53	15.32
n = 5	0.991	19.079	0.111	147.63	27.21

Table 3.3 shows that R^2 is less than the threshold and model prediction error is greater than the threshold only when $n = 0$. Therefore, it is difficult to accurately describe the dynamic characteristics of batteries without an RC branch. When $n = 3$, the prediction accuracy of the model is observed to be the best. When $n = 5$, the prediction accuracy of the model decreases, and the maximum prediction error is larger than the model with $n = 3$. It can be seen that the number of RC networks is not so large as possible.

In this case, both look-up table and fitting methods are compared. The look-up table method is to establish the data table of SOC against the parameters based on the 19 sets of model parameters obtained through identification. Then, the subsequent model parameters are obtained through interpolation. The fitting method is that the parameters are fitted by a sixth-order polynomial with SOC as the variable. It is revealed by the calculation time that the method of parameter fitting is basically consistent with the look-up table method in terms of accuracy. However, it can reduce the calculation cost.

AIC is an indicator of model performance. If the minimum AIC value of the model is regarded as the best model, which is the model with the best balance between accuracy and structural complexity, then the model with $n = 1$ is the most ideal. In this case, the Thevenin model is more suitable for NMC batteries (Table 3.4). It has been indicated by the AIC calculation results that, when the accuracy of the battery model varies little, the model complexity is a major influencing factor in the AIC value, and that, when the model accuracy varies significantly, the model accuracy is a major indicator influencing the AIC value. To validate the adaptability of the model, the analytical results based on the UDDS dynamic working condition are listed in Table 3.5.

Table 3.4 The AIC evaluation results based on HPPC test of battery 2-cell 1

Number of RC branches	$n = 1$	$n = 2$	$n = 3$	$n = 4$	$n = 5$
AIC	2.73	10.46	14.68	17.79	20.53

Table 3.5 The statistical results of the predication voltage error under the UDDS condition

Number of RC branches	Maximum error (mV)	Mean error (mV)	AIC
n = 1	27.966	0.828	6.31
n = 2	27.694	0.796	8.21
n = 3	27.791	0.775	10.16
n = 4	27.913	0.760	12.12
n = 5	27.762	0.748	14.08

It is clear that the ECMs with different RC networks based on the HPPC test exhibit satisfactory prediction accuracy in the UDDS profile, and the AIC results also prove that the Thevenin model is ideal.

3.3 The Fractional-Order Model

3.3.1 Model Introduction

Figure 3.15 demonstrates the electrochemical impedance spectroscopy (EIS) test results for the battery 1-cell 2. In the middle-frequency range, there is a semicircle whose center is below the horizontal axis. This phenomenon is generally considered to be related to the double layer at the interface between the electrode and

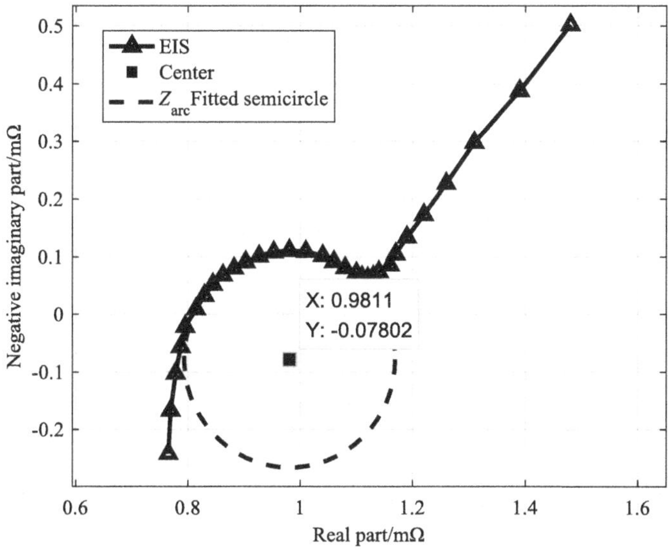

Fig. 3.15 The dispersion effect in electrochemical impedance spectroscopy

the electrolyte. When fitting impedance spectra, the researchers discovered that the ECM using ideal capacitances often failed to achieve the desired fitting accuracy. The phenomenon that the measured frequency response characteristics of the double layer are inconsistent with the pure capacitance is called the dispersion effect. This deviation from the pure capacitance can be fitted by a constant phase element (CPE). Its impedance expression is as follows [20]:

$$Z'_{\mathrm{CPE}}(\omega) = \frac{\omega^{-\alpha}}{Y}\cos\left(\frac{\alpha\pi}{2}\right),\ Z''_{\mathrm{CPE}}(\omega) = \frac{\omega^{-\alpha}}{Y}\sin\left(\frac{\alpha\pi}{2}\right),\quad 0 < \alpha < 1 \tag{3.79}$$

where ω indicates the angular frequency. The CPE has two parameters. The first parameter is Y, with its unit being $\mathrm{s}^n\Omega^{-1}$. The second parameter, α, is a dimensionless exponent which is used to measure how far the CPE deviates from the pure capacitor element. If $\alpha = 0$, the CPE is a pure resistance element. If $\alpha = 1$, it converts into a pure capacitance element.

It can be seen from Eq. (3.79) that the phase angle of the CPE satisfies

$$\tan\phi = \tan\left(\frac{\alpha\pi}{2}\right),\ \phi = \frac{\alpha\pi}{2} \tag{3.80}$$

Therefore, the phase angle of the element is the independent of frequency, for which it is known as the constant phase angle element.

In the impedance spectra fitting, the CPE is often used in parallel with a pure resistance, and its impedance is recorded as Z_{arc}. It is expressed as

$$Z_{\mathrm{arc}}(\omega) = \frac{\frac{1}{R} + Y\omega^{\alpha}\cos\left(\frac{\alpha\pi}{2}\right) - jY\omega^{\alpha}\sin\left(\frac{\alpha\pi}{2}\right)}{\left(\frac{1}{R}\right)^2 + \left(\frac{2}{R}\right)Y\omega^{\alpha}\cos\left(\frac{\alpha\pi}{2}\right) + (Y\omega^{\alpha})^2} \tag{3.81}$$

$$Z'_{\mathrm{arc}}(\omega) = \frac{\frac{1}{R} + Y\omega^{\alpha}\cos\left(\frac{\alpha\pi}{2}\right)}{\left(\frac{1}{R}\right)^2 + \left(\frac{2}{R}\right)Y\omega^{\alpha}\cos\left(\frac{\alpha\pi}{2}\right) + (Y\omega^{\alpha})^2} \tag{3.82}$$

$$Z''_{\mathrm{arc}}(\omega) = \frac{-Y\omega^{\alpha}\sin\left(\frac{\alpha\pi}{2}\right)}{\left(\frac{1}{R}\right)^2 + \left(\frac{2}{R}\right)Y\omega^{\alpha}\cos\left(\frac{\alpha\pi}{2}\right) + (Y\omega^{\alpha})^2} \tag{3.83}$$

By eliminating Y, we get:

$$\left(Z'_{\mathrm{arc}}(\omega) - \frac{R}{2}\right)^2 + \left[Z''_{\mathrm{arc}}(\omega) - \frac{R\cot\left(\frac{\alpha\pi}{2}\right)}{2}\right]^2 = \left[\frac{R}{2\sin\left(\frac{\alpha\pi}{2}\right)}\right]^2 \tag{3.84}$$

It is indicated that the arc of the intermediate frequency in Fig. 3.15 can be fitted by the Eq. (3.84). The larger the parameter α is, the greater distance from the center of the arc to the real axis, which also proves that the pure capacitor ($\alpha = 1$) is incapable to fit the dispersion effect [21, 22].

However, the CPE is difficult to handle in the time domain and needs to be processed by fractional calculus theory. There are the following three commonly used definitions of fractional calculus [23]:

(1) The definition of Grünwald–Letnikov (G-L)
The G-L definition is derived from the traditional integer calculus and is defined as

$$ {}_a\mathrm{D}_t^{\alpha} f(t) = \lim_{h \to 0} h^{-\alpha} \sum_{j=0}^{[(t-a)/h]} (-1)^j \binom{\alpha}{j} f(t - jh) \tag{3.85} $$

The above formula is a unified expression of fractional differential and integral, where ${}_a\mathrm{D}_t^{\alpha}$ indicates a fractional calculus operator. The positive and negative of α represent the fractional differential and integral, respectively. a and t denote the lower and upper limits of calculus. h is defined as the step size. $[(t-a)/h]$ presents the rounding up of $(t-a)/h$. $\binom{\alpha}{j}$ indicates the binomial coefficient:

$$ \binom{\alpha}{j} = \begin{cases} 1 & j = 0 \\ \frac{\alpha(\alpha-1)\cdots(\alpha-(j-1))}{j!} & j > 0 \end{cases} \tag{3.86} $$

(2) The definition of Riemann–Liouville (R-L)
R-L integral definition:

$$ {}_a\mathrm{D}_t^{-\alpha} f(t) = \frac{1}{\Gamma(-\alpha)} \int_a^t (t - \tau)^{-\alpha-1} f(\tau) d\tau \tag{3.87} $$

R-L differential definition:

$$ {}_a\mathrm{D}_t^{\beta} f(t) = \frac{1}{\Gamma(n-\beta)} \frac{d^n}{dt^n} \left[\int_a^t (t - \tau)^{n-\beta-1} f(\tau) d\tau \right] \tag{3.88} $$

where $0 < \alpha \le 1, n - 1 < \beta \le n, n \in \mathrm{N}$.

(3) The definition of Caputo
Caputo integral definition:

$$ {}_a\mathrm{D}_t^{-\alpha} f(t) = \frac{1}{\Gamma(-\alpha)} \int_a^t (t - \tau)^{-\alpha-1} f(\tau) d\tau \ (\alpha > 0) \tag{3.89} $$

Caputo differential definition:

$$
{}_{a}\mathrm{D}_{t}^{\beta} f(t) = \frac{1}{\Gamma(n-\beta)} \left[\int_{a}^{t} (t-\tau)^{n-\beta-1} f^{(n)}(\tau) d\tau \right] \tag{3.90}
$$

where $n - 1 < \beta < n, n \in \mathrm{N}$.

It can be confirmed that the definitions of G-L and R-L are completely equivalent to most functions. R-L definitions are most commonly used in the theoretical analysis. Caputo definition is more suitable for describing and discussing initial value problems of fractional differential equations, while G-L definition provides the most direct form and method for discretization approximation.

3.3.2 Model Constructing

By replacing the pure capacitive element in the Thevenin model with the CPE, the fractional-order model can be established as shown in Fig. 3.16.

Similar to the ECM, its transfer function is expressed as [24, 25]

$$
G(s) = \frac{U_t(s) - U_{oc}(s)}{i_{\mathrm{L}}(s)} = -\left(R_i + \frac{R_{ct}}{1 + R_{ct} Y s^{\alpha}} \right) = -\frac{(R_{ct} + R_i) + R_{ct} R_i Y s^{\alpha}}{1 + R_{ct} Y s^{\alpha}} \tag{3.91}
$$

$$
U_t(s) - U_{oc}(s) + R_{ct} Y s^{\alpha} (U_t(s) - U_{oc}(s)) = -(R_{ct} + R_i) i_{\mathrm{L}}(s) - R_{ct} R_i Y s^{\alpha} i_{\mathrm{L}}(s) \tag{3.92}
$$

The inverse Laplace transform of the upper formula can be derived at the kth sampling time.

$$
U_t(k) = -(R_{ct} + R_i) i_{\mathrm{L}}(k) - R_{ct} R_i Y \mathrm{D}^{\alpha} i_{\mathrm{L}}(k) + U_{oc}(k) + R_{ct} Y \mathrm{D}^{\alpha} (U_{oc}(k) - U_t(k)) \tag{3.93}
$$

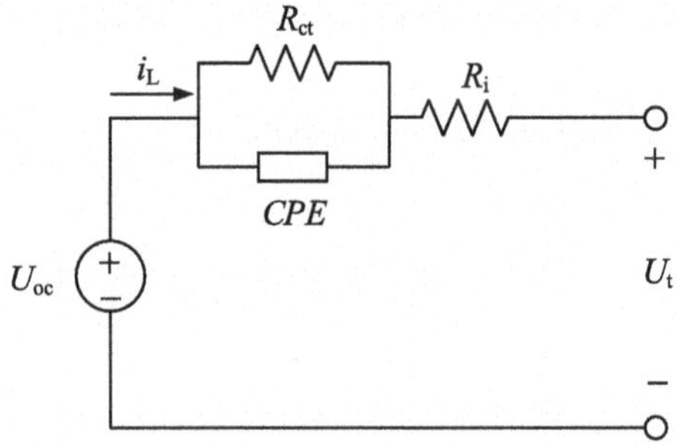

Fig. 3.16 A commonly used fractional model

3.3.3 Parameter Identification

Equation (3.93) is rewritten as

$$U_t(k) = \mathbf{\Phi}_k^T \boldsymbol{\theta}_k = \left[-i_L(k) \ -\mathrm{D}^\alpha i_L(k) \ 1 \ -\mathrm{D}^\alpha U_t(k) \right] \\ \left[R_i + R_{ct} \ R_{ct} R_i Y \ (1 + R_{ct} Y \mathrm{D}^\alpha) U_{oc}(k) \ R_{ct} Y \right]^\mathrm{T} \tag{3.94}$$

where D^α is the fractional-order calculus, which can be calculated using the G-L definition. To reduce the amount of calculation, the G-L definition can be approximated as

$${}_a\mathrm{D}_t^\alpha f(t) \approx \lim_{h \to 0} \frac{1}{h^\alpha} \sum_{j=0}^{L} (-1)^j \binom{\alpha}{j} f(t - jh) \tag{3.95}$$

where L is the memory length selected according to the short memory criterion.

Under the condition of fixed α, the parameter identification using Eq. (3.94) is a linear problem, which can be realized by the RLS method. Each parameter can be resolved by the value of parameter vector $\boldsymbol{\theta}$:

$$\begin{cases} U_{oc}(k) = \dfrac{\hat{\theta}_k(3) - \hat{\theta}_k(4)\left(\frac{1}{h^\alpha} \sum\limits_{j=1}^{L} (-1)^j \binom{\alpha}{j} U_{oc}(k-j) \right)}{\left(1 + \frac{1}{h^\alpha} \hat{\theta}_k(4)\right)} \\ R_i(k) = \dfrac{\hat{\theta}_k(2)}{\hat{\theta}_k(4)} \\ R_{ct}(k) = \hat{\theta}_k(1) - R_i(k) \\ Y(k) = \dfrac{\hat{\theta}_k(4)}{R_{ct}(k)} \end{cases} \tag{3.96}$$

3.3.4 Case Study

It is assumed that $\alpha = 0.8$. The model parameters of the battery 1-cell 2 are identified under the DST condition at 40 °C. The terminal voltage error is shown in Fig. 3.17.

The maximum terminal voltage error is found to be within 8 mV. It is revealed that the fractional-order model is also capable of accurately estimating the terminal voltage.

As a significant parameter in the model, OCV can be used to reflect the aging state of the battery and calculate the SOC. Figure 3.18 shows the comparison between the OCV obtained by online identification of the fractional-order model and experimental values after fitting with SOC polynomial. When SOC ranges from 10 to 100%, the identified OCV can be well matched with the experimental value, with the maximum error of 15 mV.

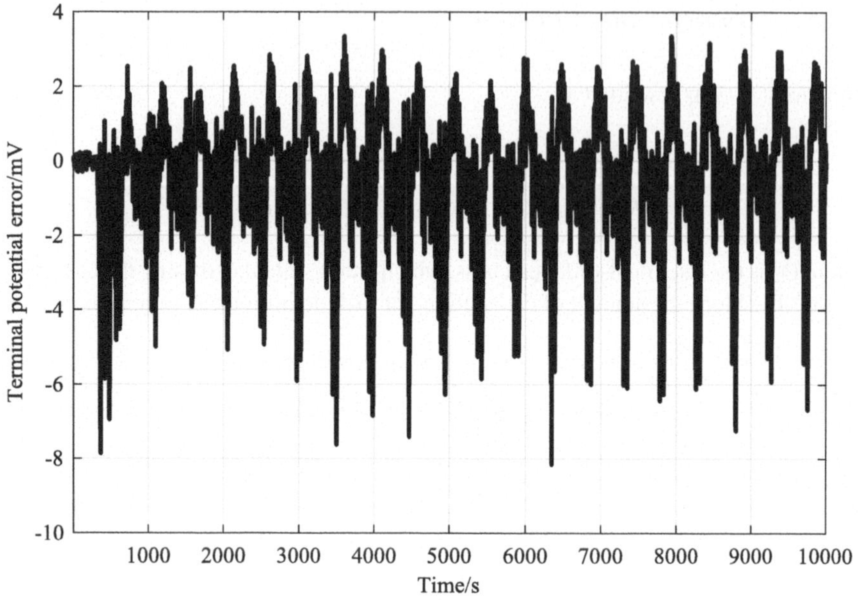

Fig. 3.17 The terminal voltage error under the DST condition

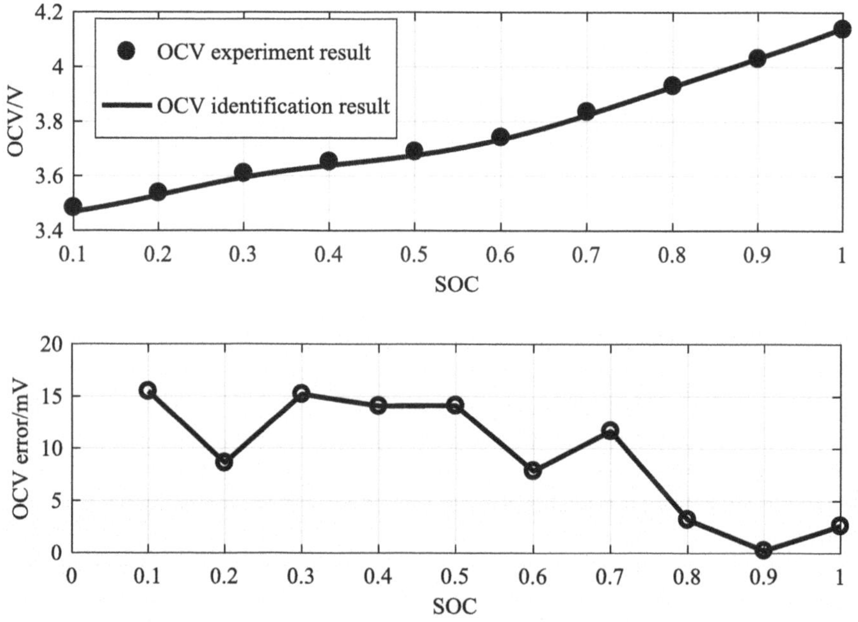

Fig. 3.18 Comparison of OCV identification value and experimental value

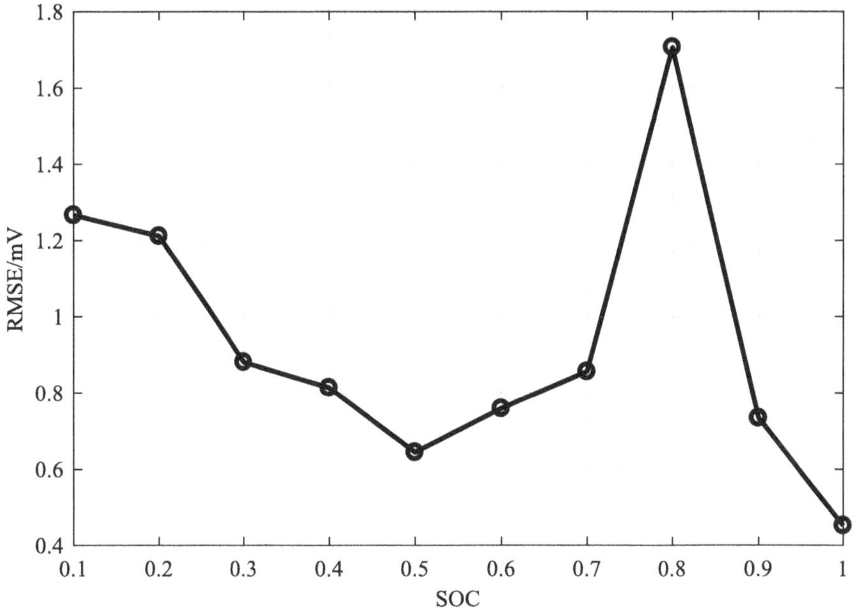

Fig. 3.19 The root mean square error of the battery under offline identification

It is noteworthy that the differential order is fixed in the above studies, and the global offline identification of differential order and other parameters can be achieved by GA. Taking the HPPC data on the same aging state and the same temperature of the battery as an example, all parameters are optimized by GA, and the root mean square error (RMSE) is shown in Fig. 3.19. The RMSE of the fractional-order model is less than 1.7 mV, suggesting that the fractional-order model can also accurately predict the battery terminal voltage and be used for offline identification.

Figure 3.20 indicates the relationship between the differential order α of CPE and the SOC in the fractional-order model. It can be seen that the optimization result of α fluctuates within the range of 0.61–0.93, and that in most cases, it is below 0.9, which further confirms that the double layers of the battery deviate from the pure capacitance.

3.4 Conclusion

In this chapter, three types of battery models are introduced, namely the electrochemical model, equivalent circuit model, and fractional-order model. The modeling approaches and parameter identification methods differ significantly. Among them, the electrochemical model derives from the internal reaction principle of the battery, which has high accuracy. However, the computation cost is significant, and

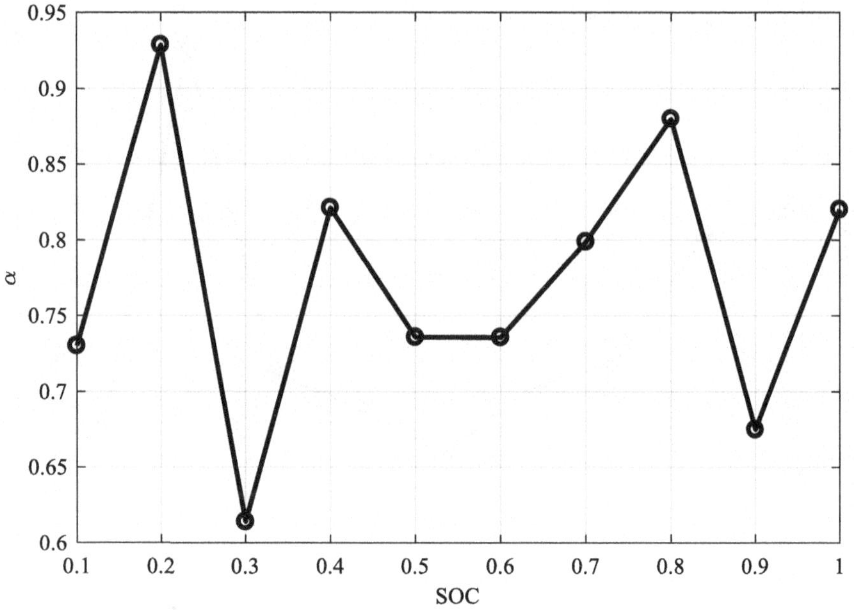

Fig. 3.20 Differential order identification of the CPE

the dimensionality reduction processing is required in the calculation process. The advantage shown by the electrochemical model lies in its clear physical meaning, which can reflect the mapping relationship between the electrochemical parameters of batteries and the terminal voltage. Besides, it has some obvious advantages in terms of the battery performance degradation mechanism analysis, aging modeling, SOH estimation, and fault diagnosis.

The equivalent circuit model describes the voltage response through ideal electrical components, the static voltage characteristics displayed by a constant voltage source, and the dynamic voltage characteristics such as polarization and hysteresis described by parallel RC branches. It is widely applied to SOC, SOH estimation, and SOP prediction of various BMS, as well as energy management algorithm for the advantages of small calculation and excellent real-time performance. However, the identified parameters deviate significantly from reality, so that the model lacks real physical significance due to a lack of the actual constraints on various parameters. As a result, this model is difficult to reflect the internal characteristics of the battery. Besides, it is difficult to be used for the thermal–electrical coupling modeling and mechanism analysis of the battery.

As compared to the equivalent circuit model, the fractional-order model is based on the measurement results of the electrochemical impedance spectroscopy, and is combined with the fractional-order theory in the time domain. The accuracy of voltage simulation and SOC estimation is higher than that of the equivalent circuit model, but the computation is often larger. This model can be regarded as the generalization of the traditional equivalent circuit model, and its application in SOH estimation has great research value as the fractional-order model in the frequency domain is often used to fit EIS and obtain plenty of information on aging mechanism of batteries.

References

1. Fuller TF, Doyle M, Newman J (1994) Simulation and optimization of the dual Lithium ion insertion cell. J Electrochem Soc 141:1–10
2. Lin C, Tang A (2016) Simplification and efficient simulation of electrochemical model for Li-ion battery in EVs. Energy Procedia 104:68–73
3. Lin C, Tang A, Xing J (2017) Evaluation of electrochemical models based battery state-of-charge estimation approaches for electric vehicles. Appl Energy 207:394–404
4. Cai L, White RE (2009) Reduction of model order based on proper orthogonal decomposition for Lithium-Ion battery simulations. J Electrochem Soc 156(3):154–161
5. Xiong R, Li L, Tian J (2018) Towards a smarter battery management system: a critical review on battery state of health monitoring methods. J Power Sources 405:18–29
6. Xiong R, Li L, Li Z, Yu Q, Mu H (2018) An electrochemical model based degradation state identification method of lithium-ion battery for all-climate electric vehicles application. Appl Energy 219:264–275
7. Ye M, Gong H, Xiong R, Mu H (2018) Research on the battery charging strategy with charging and temperature rising control awareness. IEEE Access 6:64193–64201
8. Nejad S, Gladwin DT, Stone DA (2016) A systematic review of lumped-parameter equivalent circuit models for real-time estimation of lithium-ion battery states. J Power Sources 316:183–196
9. Lin C, Mu H, Xiong R, Cao J (2017) Multi-model probabilities based state fusion estimation method of lithium-ion battery for electric vehicles: state-of-energy. Appl Energy 194:560–568
10. Zhang Y, Xiong R, He H, Shen W (2017) Lithium-Ion battery pack state of charge and state of energy estimation algorithms using a hardware-in-the-loop validation. IEEE Trans Power Electron 32:4421–4431
11. Yu Q, Xiong R, Lin C, Shen W, Deng J (2017) Lithium-ion battery parameters and state-of-charge joint estimation based on H-infinity and unscented Kalman filters. IEEE Trans Veh Technol 66:8693–8701
12. Xiong R, Yu Q, Shen W, Lin C, Sun F (2019) A sensor fault diagnosis method for a lithium-ion battery pack in electric vehicles. IEEE Trans Power Electron 34(10):9709–9718
13. Xiong R, Yu Q, Wang LY, Lin C (2017) A novel method to obtain the open circuit voltage for the state of charge of lithium ion batteries in electric vehicles by using H-infinity filter. Appl Energy 207:346–353
14. Zhao X, Cai Y, Yang L, Deng Z, Qiang J (2017) State of charge estimation based on a new dual-polarization-resistance model for electric vehicles. Energy 135:40–52
15. Zhang DH, Zhu GR, Bao J, Ma Y, He SJ, Qiu S, Chen W (2015) Research on parameter identification of battery model based on adaptive particle swarm optimization algorithm. J Comput Theor Nanosci 12:1362–1367
16. Remmlinger J, Buchholz M, Soczka-Guth T, Dietmayer K (2013) On-board state-of-health monitoring of lithium-ion batteries using linear parameter-varying models. J Power Sources 239:689–695
17. Deng Z, Yang L, Cai Y, Deng H, Sun L (2016) Online available capacity prediction and state of charge estimation based on advanced data-driven algorithms for lithium iron phosphate battery. Energy 112:469–480
18. Wei Z, Zou C, Leng F, Soong BH, Tseng K-J (2018) Online model identification and state-of-charge estimate for Lithium-ion battery with a recursive total least squares-based observer. IEEE Trans Ind Electron 65:1336–1346

19. Sun F, Xiong R, He H (2016) A systematic state-of-charge estimation framework for multi-cell battery pack in electric vehicles using bias correction technique. Appl Energy 162:1399–1409
20. Huang J, Li Z, Yann B, Zhang J (2016) Graphical analysis of electrochemical impedance spectroscopy data in Bode and Nyquist representations. J Power Sources 309:82–98
21. Wang B, Liu Z, Li SE, Moura SJ, Peng H (2017) State-of-charge estimation for Lithium-ion batteries based on a nonlinear fractional model. IEEE Trans Control Syst Technol 25:3–11
22. Yang R, Xiong R, He H, Chen Z (2018) A fractional-order model-based battery external short circuit fault diagnosis approach for all-climate electric vehicles application. J Clean Prod 187:950–959
23. Ivo P (2010) Fractional-order nonlinear systems (modeling, analysis and simulation), 2nd ed. Springer
24. Xiong R, Tian J, Shen W, Sun F (2018) A novel fractional order model for state of charge estimation in lithium ion batteries. IEEE Trans Veh Technol 68(5):4130–4139
25. Tian J, Xiong R, Yu Q (2019) Fractional-order model-based incremental capacity analysis for degradation state recognition of Lithium-ion batteries. IEEE Trans Ind Electron 66:1576–1584

Chapter 4
Battery SOC and SOH Estimation

Battery SOC and SOH estimation are core functions performed by the BMS. Accurate SOC and SOH estimation can ensure the safe and reliable operation of the battery system, and provide the basis for energy management and safety management of EVs. However, batteries exhibit the characteristics of limited measurable parameters, coupling feature, degradation with time, strong time-varying, and nonlinearity. The vehicle applications are also encountering the requirements of series-parallel group of inconsistent complex system, various operation conditions (wide rate charge and discharge), and all-climate (–30 to 55 °C temperature range). Battery SOC and SOH estimation with high precision and strong robustness are extremely challenging, and they have been the industry's technical difficulties and hotspots in the international academic research. This chapter will systematically describe the basic theory and application of battery SOC and SOH estimation, discuss the performance of online SOC estimation with the known static capacity and dynamic capacity as well as the necessity of SOH and SOC collaborative estimation. A detailed algorithm flow for the practical application of BMS will also be provided.

4.1 SOC Estimation

The battery SOC of new energy vehicle is equivalent to the oil meter of traditional fuel vehicle. As one of the significant factors in energy management, SOC plays a crucial role in optimizing vehicle energy management, improving battery capacity and energy utilization, preventing batteries from overcharging and overdischarging, as well as ensuring the safety and long service lifetime of batteries. This section will elaborate on the SOC estimation method under the condition that the static capacity of the battery is known.

R. Xiong, *Battery Management Algorithm for Electric Vehicles*,
https://doi.org/10.1007/978-981-15-0248-4_4

4.1.1 Classification

The electrochemical reaction process and stage of the battery are complex and difficult to determine, and the operating conditions are harsh and variable for vehicles. Therefore, it is quite difficult to obtain accurate SOC as it is a hidden state. The battery SOC estimation methods could be divided into four categories [1]: the estimation method based on characteristic parameters, the ampere-hour integral estimation method, the model-based estimation method, and the data-driven estimation method, as shown in Fig. 4.1.

The estimation method based on characteristic parameters
The method is mainly divided into two steps:

- Establishing an offline relationship between the characteristic parameters and SOC.
- Calculating the parameter values in real time, and calibrating the SOC of battery.

The application of this method needs to meet two premises: the offline relationship between the established characteristic parameters and the SOC ought to be relatively stable, and the selected characteristic parameters should be easily available. Optional characteristic parameters include the remaining capacity, impedance spectrum, OCV, etc.

The available capacity can be obtained using the discharge experiment method, which is considered to be the most direct method for determining the battery SOC. However, it is difficult to determine the available capacity by constant current discharge for a long time during actual driving, which limits this method only applicable

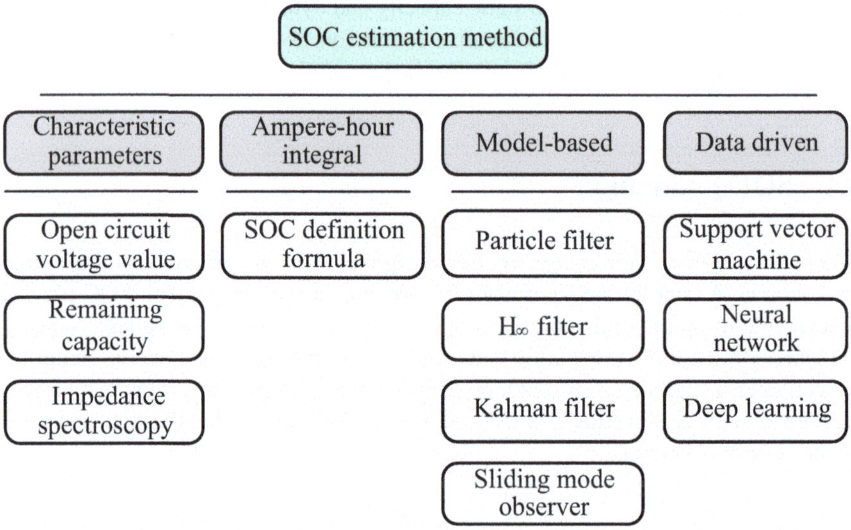

Fig. 4.1 Classification of SOC estimation methods

to some specific environments such as laboratories. The method based on electrochemical impedance spectroscopy [2] needs the electrochemical workstation to attain the impedance under different SOC and establish the mapping relationship between SOC and parameters, then calibrate SOC by looking up tables. The stable OCV-SOC relationship [3] is often used in the industry to calibrate battery SOC, and a large number of BMS products also rely on this relationship to calibrate the initial battery SOC. However, the battery is required to rest for a long time so that the accurate OCV can be acquired. As a result, it often needs to be combined with online OCV identification methods in practical applications.

Ampere-hour integral estimation method

This method is also known as the coulomb counting method [4], which is based on the definition of SOC, as shown in Eq. (4.1).

$$z(t) = z(t_0) - \frac{\int_{t_0}^{t} \eta_i i_L(\tau) d\tau}{C_{\max}} \tag{4.1}$$

where $z(t)$ represents the estimated value of the battery SOC at time t, $z(t_0)$ indicates the initial value of the battery SOC. η_i denotes the coulomb efficiency of the battery, which is determined by experiments. For lithium-ion batteries, the discharge efficiency is generally regarded as 1, the charging efficiency ranges between 0.98 and 1 (within 3C charging current). $i_L(\tau)$ stands for the charging and discharging current at time τ and $C_{\max}$ refers to the maximum available capacity of the battery under current conditions.

Despite the ampere-hour integral method is classical and widely used for SOC estimation, it has the following three major defects:

(1) The accurate value of the initial SOC is difficult to obtain.
(2) The method requires accurate current sensors. However, the accuracy of current sensors is usually affected by noise, temperature drift, and other unknown random disturbances in practical applications. In integral calculation, these random variables are easy to cause accumulated errors, and the rounding calculation carried out by the controller also has a certain impact.
(3) The deterioration of the battery static capacity affects the SOC estimation accuracy.

The three factors mentioned above mutually affect each other, which further reduce the reliability of this method. In order to eliminate the restriction imposed by the above factors and improve the estimation accuracy, complex and tedious periodic calibration is required. For this reason, this method is often combined with other methods to develop a fusion method. For example, OCV is used to determine the initial SOC of the battery, and the SOC trajectory is calculated using the ampere-hour integral method.

Model-based estimation method

The method uses the model and state estimation algorithm to complete the SOC estimation of the battery. Therefore, the method first needs to establish a reliable model. This section mainly introduces and illustrates the model-based SOC estimation method with the ECM. Based on the established ECM and its state-space equation, the filter and the observer algorithms are applied to build a model-based SOC estimation framework. The specific implementation processes (Fig. 4.2) are summarized as following:

(1) Based on the SOC of previous moment or the initial SOC and the measured current, the SOC estimation at the current time is calculated by using the ampere-hour integral method;

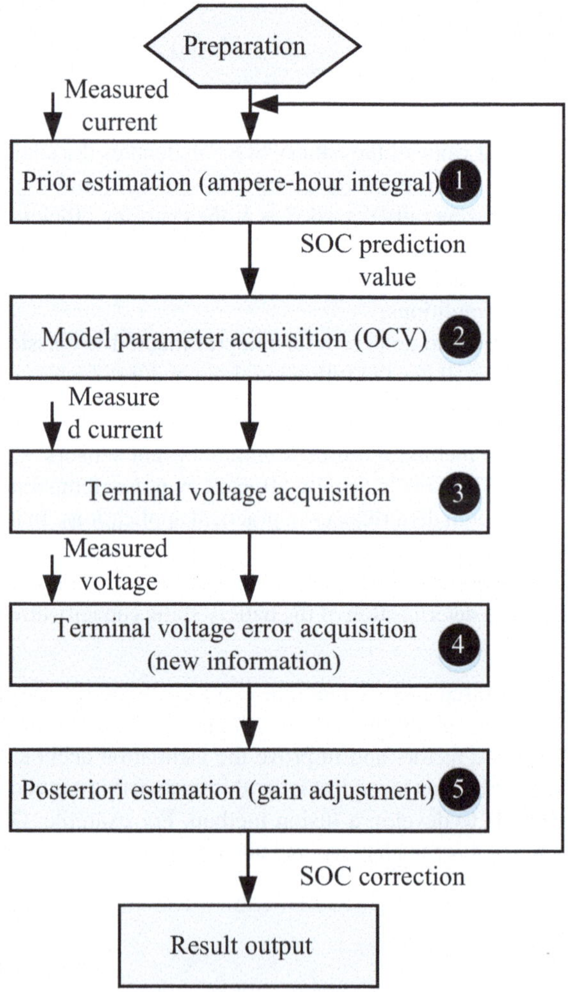

Fig. 4.2 Flowchart of the model-based estimation method

(2) Calculate the model parameters based on the relation between parameter and SOC, such as the OCV-SOC relationship;
(3) Calculate the terminal voltage based on the state-space equation;
(4) Calculate the voltage error, that is, the innovation with the measured voltage;
(5) Modify the estimated SOC with a certain gain of the innovation, so as to obtain the final SOC correction value and use it as the input of the next moment.

The form of the gain in Step (5) depends on the state estimation algorithm. Obviously, the model-based method is a closed-loop method, which makes the algorithm have certain robustness by continuously modifying the SOC results. In general, the estimation accuracy of the model-based method is determined by the prior estimation process and posteriori process. In case that the estimation result of the ampere-hour integral is reliable (the prior SOC estimation value is accurate), the gain can be appropriately reduced; otherwise, it will be increased. However, the excessive correction will make the SOC value vary significantly, the gain should be adjusted according to the actual situation.

The performance of the model-based estimation method relies on the model and estimation algorithm. Kalman filter (KF) algorithm [5] is the most widely used algorithm for battery SOC estimation. KF is an optimal estimation method with minimum variance proposed by American scholar Kalman in the early 1960s, which is convenient for real-time processing. It provides a solution to deal directly with random noise. The KF treats the parameter error as noise and treats the parameters to be estimated as space state variables. The system noise and random noise are filtered out by recursive method to obtain the accurate state. However, the original KF is only applicable to linear systems, and the proposed extended Kalman filter (EKF) [5] extends it to the nonlinear systems. EKF linearizes the battery model with Taylor expansion. However, in the process of linearization, truncation error will arise, which will increase SOC estimation error, and even cause divergence under improper initial settings. Therefore, it is necessary to improve and optimize the battery model, or use the improved KF algorithm to guarantee the accuracy and robustness of the system. Although EKF considers the noise in the actual process as much as possible, it still encounters two problems.

(1) It assumes that the noise remains unchanged, which obviously does not match with the actual. The noise covariance algorithm is proposed to address this problem. It updates adaptively the statistical characteristics of noise in the filtering algorithm as the change of estimated results, such as the adaptive extended Kalman filter (AEKF) [6].
(2) It assumes the noise as white noise. When this assumption fails to meet, the H_∞ filter (HIF) [7] based on the principle of minimizing the maximum estimation error can be used to complete the SOC estimation of battery. It acknowledges that the statistical characteristics of noise are unknown in the actual process. Based on the principle of HIF, the optimal solution to the state estimation in the worst case is acquired.

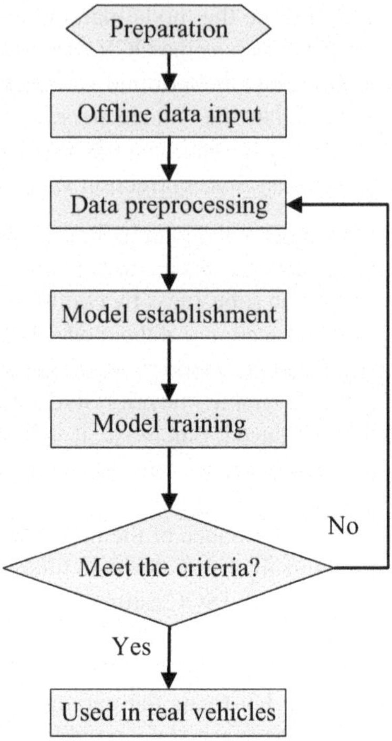

Fig. 4.3 Flowchart of the data-driven estimation method

Data-driven estimation method

Based on a large amount of offline data, this method establishes and trains a mapping relationship model between the battery current, voltage, temperature and SOC [8]. The flowchart is shown in Fig. 4.3, which is divided into three steps:

(1) Preprocessing of the offline data. To obtain the format that meets the input and output requirements of the model, offline data needs to be preprocessed such as data cleaning, normalization, data segmentation. Data segmentation refers to dividing the normalized data into the training set, verification set, and test set according to a certain proportion.
(2) Modeling and training. According to the size of the data, the structure of the model is preliminarily determined. Afterward, the training set is adopted to train the model, and the verification result of the verification set is used as the training cutoff criteria.
(3) Model test. The test set is used to determine whether or not the accuracy meets the requirements. If so, the training is completed. Otherwise, it will return to Step (1) to design and plan again.

The data-driven method has special advantages in solving strongly nonlinear problems and has high estimation accuracy, but it often requires a large amount of experimental data as prior knowledge. In addition, the experimental data should fully reflect the characteristics of the battery; otherwise, it will easily lead to over-fitting of the model. Meanwhile, the complexity of the model, the selected training function, and the training cutoff criteria also have a direct impact on the estimation accuracy and generalization ability of the model.

The typical representative of this method is the neural network model [9], which requires little consideration given to the internal chemical reaction details of the battery. Meanwhile, it displays a strong fitting ability, which makes it suitable for SOC estimation of any kind of batteries in theory. However, in recent years, it has been discovered that increasing the number of neurons in the neural network often contributes to a rapid increase in the parameters of the model, which further leads to the over-fitting of the model. Therefore, the study of neural networks has gradually shifted to the deep learning network [10] with stronger generalization ability. Meanwhile, the structure of the trained neural network model is more complex and the calculation cost is substantial, thus high-performance chips are often needed in vehicle applications. Therefore, a large number of chips used for neural network/deep learning are gradually launched in the market.

Table 4.1 summarizes the advantages and disadvantages of four SOC estimation methods as well as their performances in accuracy and robustness [11].

The model-based estimation method has great application prospects in battery management systems for EV. In this chapter, the model-based estimation method will be discussed in detail.

4.1.2 Model-Based SOC Estimation Method

This method integrates the ampere-hour integral method and the characterization parameter mapping method. The relationship between SOC and OCV is the most commonly used characterization parameter mapping. The state estimation algorithm is used to achieve the optimal fusion of the two methods. The inaccurate initial SOC will directly lead to an incorrect OCV, thus increasing the voltage deviation of the model. Only by constantly adjusting the estimated SOC, could the OCV at the next moment be an accurate value, which can minimize the voltage error and realize the closed-loop correction of the SOC. The specific implementation details are as follows:

(1) Establishment and discretization of battery model

According to the Thevenin model established in Chap. 3, the corresponding differential equation is shown in Eq. (4.2):

$$\begin{cases} \dot{U}_{\mathrm{D}} = -\frac{1}{C_{\mathrm{D}}R_{\mathrm{D}}}U_{\mathrm{D}} + \frac{1}{C_{\mathrm{D}}}i_{\mathrm{L}} \\ U_{\mathrm{t}} = U_{\mathrm{oc}} - U_{\mathrm{D}} - i_{\mathrm{L}}R_{\mathrm{i}} \end{cases} \tag{4.2}$$

It is assumed that the parameters of the battery model can be regarded as fixed values within a sampling interval, and that the battery model can be linearized and simplified

Table 4.1 Advantages and disadvantages of four estimation methods and the evaluation of estimation accuracy and robustness

Methods	Advantages	Disadvantages	Accuracy	Robustness
Estimation method based on characteristic parameters	• Easy to implement • Low computing cost • Good real-time performance	• Easily affected by the uncertainty factors, such as temperature, working condition, and degree of aging • Regular calibration of OCV or EIS information is required • Precision measuring instruments are required	Poor	Good
Ampere-hour integral estimation method	• Easy to implement • Low computing cost • Excellent real-time performance	• Dependence on the accurate SOC initial value • The open-loop calculation method needs periodic revision • It is susceptible to current drift, noise, and aging	Ordinary	Poor
Model-based estimation method	• High estimation accuracy • Closed-loop feedback control • Good real-time performance • Strong adaptability	• It depends strongly on the accuracy of the model • Computing costs are relatively high • Improper initial values diverge the estimated results	Excellent	Excellent
Data-driven estimation method	• High estimation accuracy • Good at dealing with nonlinear problems	• High algorithm complexity • High dependence on the training data	Excellent	Poor

as a time-invariant system in a sampling interval. The basic solution of Eq. (4.2) is as follows:

$$U_{\mathrm{D}}(t) = \mathrm{e}^{-\frac{1}{C_{\mathrm{D}}R_{\mathrm{D}}}(t-t_0)}U_{\mathrm{D}}(t_0) + \int\limits_{t_0}^{t} \mathrm{e}^{-\frac{1}{C_{\mathrm{D}}R_{\mathrm{D}}}(t-t_0)}\frac{1}{C_{\mathrm{D}}}i_{\mathrm{L}}(\tau)\,d\tau \tag{4.3}$$

where t indicates the current time and t_0 represents the initial time. Define $t_0 = k\Delta t$, $t = (k+1)\Delta t$, $k = 0, 1, 2, \ldots$, Δt denotes the current collection interval, then:

$$\begin{aligned} U_{\mathrm{D}}[(k+1)\Delta t] &= \mathrm{e}^{-\frac{1}{C_{\mathrm{D}}R_{\mathrm{D}}}(t-t_0)}U_{\mathrm{D}}(k\Delta t) \\ &+ \int\limits_{k\Delta t}^{(k+1)\Delta t} \mathrm{e}^{-\frac{1}{C_{\mathrm{D}}R_{\mathrm{D}}}[(k+1)\Delta t-\tau]}\frac{1}{C_{\mathrm{D}}}i_{\mathrm{L}}(\tau)\,d\tau \end{aligned} \tag{4.4}$$

where τ is between $k\Delta t$ and $(k+1)\Delta t$. The current is considered as a constant value in a sampling interval, namely $i_{\mathrm{L}}(\tau) = i_{\mathrm{L}}[(k+1)\Delta t]$. In this case, the input can be placed outside the integral symbols, and Eq. (4.4) can be formulated as

$$\begin{aligned} U_{\mathrm{D}}[(k+1)\Delta t] &= \mathrm{e}^{-\frac{1}{C_{\mathrm{D}}R_{\mathrm{D}}}(t-t_0)}U_{\mathrm{D}}(k\Delta t) \\ &+ \int\limits_{k\Delta t}^{(k+1)\Delta t} \mathrm{e}^{-\frac{1}{C_{\mathrm{D}}R_{\mathrm{D}}}[(k+1)\Delta t-\tau]}\,d\tau\frac{1}{C_{\mathrm{D}}}i_{\mathrm{L}}[(k+1)\Delta t] \end{aligned} \tag{4.5}$$

where $t = (k+1)\Delta t - \tau$, and $dt = -d\tau$. The lower limit of the integral is $\tau = k\Delta t$, and $t = (k+1)\Delta t - k\Delta t = \Delta t$. When the upper limit of the integration is $\tau = (k+1)\Delta t$, the $t = (k+1)\Delta t - \tau = 0$. Therefore, Eq. (4.5) can be simplified as follows:

$$U_{\mathrm{D}}[(k+1)\Delta t] = \mathrm{e}^{-\frac{1}{C_{\mathrm{D}}R_{\mathrm{D}}}(t-t_0)}U_{\mathrm{D}}(k\Delta t) + \int\limits_{0}^{\Delta t} \mathrm{e}^{-\frac{1}{C_{\mathrm{D}}R_{\mathrm{D}}}t}\,dt\frac{1}{C_{\mathrm{D}}}i_{\mathrm{L}}[(k+1)\Delta t] \tag{4.6}$$

Then, the polarization voltage in the battery model can be calculated as

$$U_{\mathrm{D}}[(k+1)\Delta t] = \mathrm{e}^{-\frac{\Delta t}{\tau}}U_{\mathrm{D}}(k\Delta t) + R_{\mathrm{D}}i_{\mathrm{L}}[(k+1)\Delta t]\left[1 - \mathrm{e}^{-\frac{\Delta t}{\tau}}\right] \tag{4.7}$$

where $\tau = R_{\mathrm{D}} \times C_{\mathrm{D}}$. For the sake of simplicity, $U_{\mathrm{D}}[(k+1)\Delta t]$ and $i_{\mathrm{L}}[(k+1)\Delta t]$ are denoted by $U_{\mathrm{D},k+1}$ and $i_{\mathrm{L},k+1}$ respectively. Then, we have

$$U_{\mathrm{D},k} = \mathrm{e}^{-\frac{\Delta t}{\tau}} \times U_{\mathrm{D},k-1} + \left[1 - \mathrm{e}^{-\frac{\Delta t}{\tau}}\right] \times i_{\mathrm{L},k}R_{\mathrm{D}} \tag{4.8}$$

where the subscript k represents the time t_k.

Meanwhile, the discrete calculation equation of battery SOC can be derived using the ampere-hour integral method:

$$z_k = z_{k-1} - \eta_i i_{L,k} \Delta t / C_{max} \tag{4.9}$$

(2) The determination of gain feedback

The model-based state estimation algorithm modifies the state or parameter primarily by establishing the mapping relationship between the state and the measured signals of the system and implementing the feedback correction of the observation error. The inaccurate state observations will cause significant observation errors, which can be used to adjust the gain of the filter to achieve accurate estimation of state or parameters. Figure 4.4 shows the main parameters of cell 1. OCV is the experimental value, and other parameters are the identification results.

It can be seen that only the OCV is monotonically increasing along with SOC in the above four parameters. Therefore, the mapping relationship between OCV and SOC can be used to correct the SOC estimation error. Through the monotonically increasing function of OCV and SOC, the monotonic relationship between the battery terminal voltage and the SOC is obtained, thereby improving the optimization efficiency and convergence speed in the state feedback of the filter. To accurately characterize the relationship between the above four parameters and SOC, Fig. 4.5 illustrates the relationship between the change of parameters and SOC.

In order to accurately characterize the corresponding relationship between OCV and SOC, a simplified electrochemical model [12] is used to analyze the battery

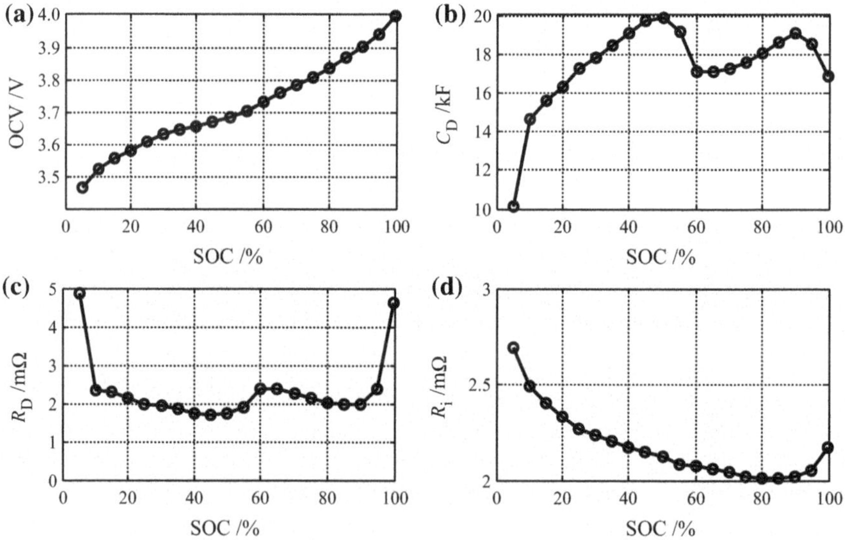

Fig. 4.4 Model parameters: **a** OCV; **b** polarization capacitance; **c** polarization internal resistance; **d** ohm internal resistance

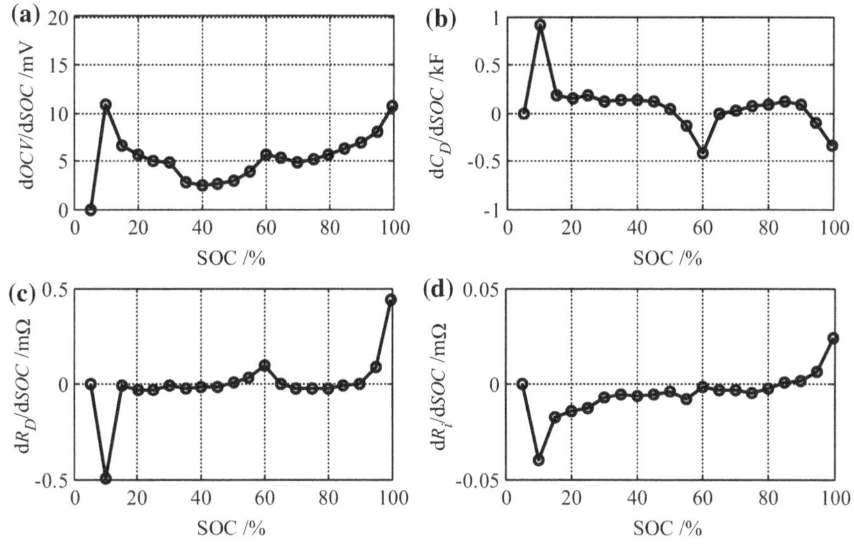

Fig. 4.5 The variation of model parameters with respect to SOC: **a** d*OCV*/d*SOC*; **b** dC_{D}/d*SOC*; **c** dR_{D}/d*SOC*; **d** dR_{i}/d*SOC*

OCV, as shown in Eq. (4.10). The model can enhance the correlation between SOC and OCV of battery in addition to improve the convergence speed of filter algorithm.

$$U_{\mathrm{oc}}(z) = \alpha_0 + \alpha_1 z + \alpha_2 z^2 + \alpha_3 z^3 + \alpha_4 / z + \alpha_5 \ln z + \alpha_6 \ln(1 - z) \quad (4.10)$$

where α_i ($i = 0, 1, \ldots, 6$) refers to the fitting coefficient, which is used to fit the mapping relationship between the OCV and SOC of the battery.

The above is the linearization equation of the Thevenin model, which can be combined with a filtering algorithm to realize the real-time estimation of SOC. Its general operation flow is shown in Fig. 4.6. This chapter takes AEKF and HIF as examples to illustrate the implementation and the estimation effect of the model-based SOC estimation algorithm.

4.1.3 Battery SOC Estimation Based on AEKF Algorithm

(1) Adaptive extended Kalman filtering algorithm

For any nonlinear discrete system, take $f(\boldsymbol{x}_{k-1}, \boldsymbol{u}_{k-1})$ as the system state equation. $h(\boldsymbol{x}_k, \boldsymbol{u}_k)$ is the system observation equation, and the general form of the state equation and observation equation is

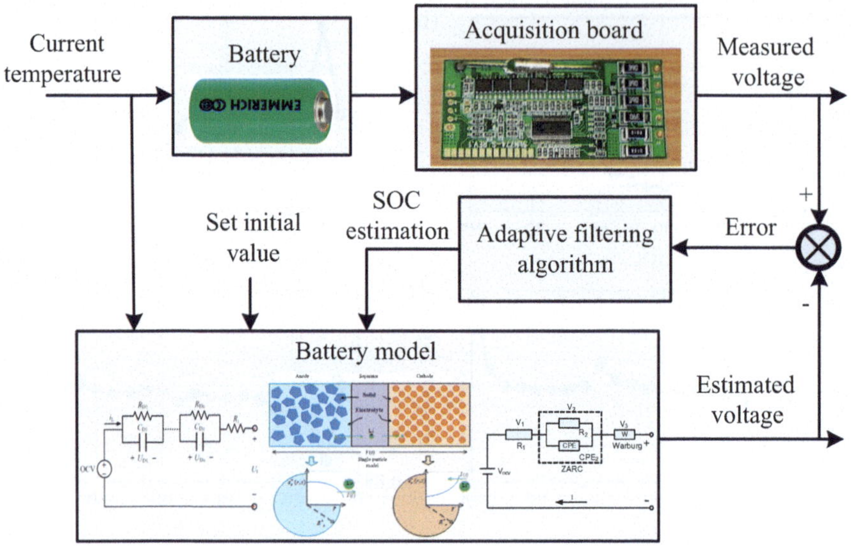

Fig. 4.6 Flowchart of model-based SOC estimation method

$$\begin{cases} \text{state equation:} & \boldsymbol{x}_k = f(\boldsymbol{x}_{k-1}, \boldsymbol{u}_{k-1}) + \boldsymbol{\omega}_{k-1} \\ \text{observation equation:} & \boldsymbol{y}_k = h(\boldsymbol{x}_k, \boldsymbol{u}_k) + \boldsymbol{v}_k \end{cases} \tag{4.11}$$

where $\boldsymbol{x}$ represents the state vector of the n-dimensional system, $\boldsymbol{u}$ indicates the input vector, and $\boldsymbol{y}$ denotes the output vector (or observation value). $\boldsymbol{\omega}_{k-1}$ is defined as the white noise of the system with zero mean and the covariance is $\boldsymbol{Q}_k$. $\boldsymbol{v}_{\mathrm{k}}$ indicates the white noise sequence of the measurement with zero mean and the covariance is $\mathbf{R}_k$. $\boldsymbol{\omega}$ and v are independent of each other.

At each moment, the $f(\boldsymbol{x}_k, \boldsymbol{u}_k)$ and $h(\boldsymbol{x}_k, \boldsymbol{u}_k)$ are linearized with the first-order Taylor expansion. That is

$$\begin{cases} f(\boldsymbol{x}_k, \boldsymbol{u}_k) \approx f(\hat{\boldsymbol{x}}_k, \boldsymbol{u}_k) + \frac{\partial f(\boldsymbol{x}_k, \boldsymbol{u}_k)}{\partial \boldsymbol{x}_k}\Big|_{\boldsymbol{x}_k = \hat{\boldsymbol{x}}_k} (\boldsymbol{x}_k - \hat{\boldsymbol{x}}_k) \\ h(\boldsymbol{x}_k, \boldsymbol{u}_k) \approx h(\hat{\boldsymbol{x}}_k, \boldsymbol{u}_k) + \frac{\partial h(\boldsymbol{x}_k, \boldsymbol{u}_k)}{\partial \boldsymbol{x}_k}\Big|_{\boldsymbol{x}_k = \hat{\boldsymbol{x}}_k} (\boldsymbol{x}_k - \hat{\boldsymbol{x}}_k) \end{cases} \tag{4.12}$$

where $\hat{\boldsymbol{x}}_k$ is the estimated value of $\boldsymbol{x}_k$.

Define $\boldsymbol{A}_k \triangleq \frac{\partial f(\boldsymbol{x}_k, \boldsymbol{u}_k)}{\partial \boldsymbol{x}_k}\Big|_{\boldsymbol{x}_k = \hat{\boldsymbol{x}}_k}$, $\boldsymbol{C}_k \triangleq \frac{\partial h(\boldsymbol{x}_k, \boldsymbol{u}_k)}{\partial \boldsymbol{x}_k}\Big|_{\boldsymbol{x}_k = \hat{\boldsymbol{x}}_k}$. Substituting Eq. (4.12) into Eq. (4.11) to obtain the linearized state equation, and the observation equation can be transformed into:

$$\begin{cases} \boldsymbol{x}_k \approx \boldsymbol{A}_{k-1}\boldsymbol{x}_{k-1} + \left[f(\hat{\boldsymbol{x}}_{k-1}, \boldsymbol{u}_{k-1}) - \boldsymbol{A}_{k-1}\hat{\boldsymbol{x}}_{k-1}\right] + \boldsymbol{\omega}_{k-1} \\ \boldsymbol{y}_k \approx \boldsymbol{C}_k\boldsymbol{x}_k + \left[h(\hat{\boldsymbol{x}}_k, \boldsymbol{u}_k) - \boldsymbol{C}_k\hat{\boldsymbol{x}}_k\right] + \boldsymbol{v}_k \end{cases} \tag{4.13}$$

Based on the above system equations, the optimal state estimation process of the nonlinear discrete system using AEKF algorithm is established, as shown in Fig. 4.7.

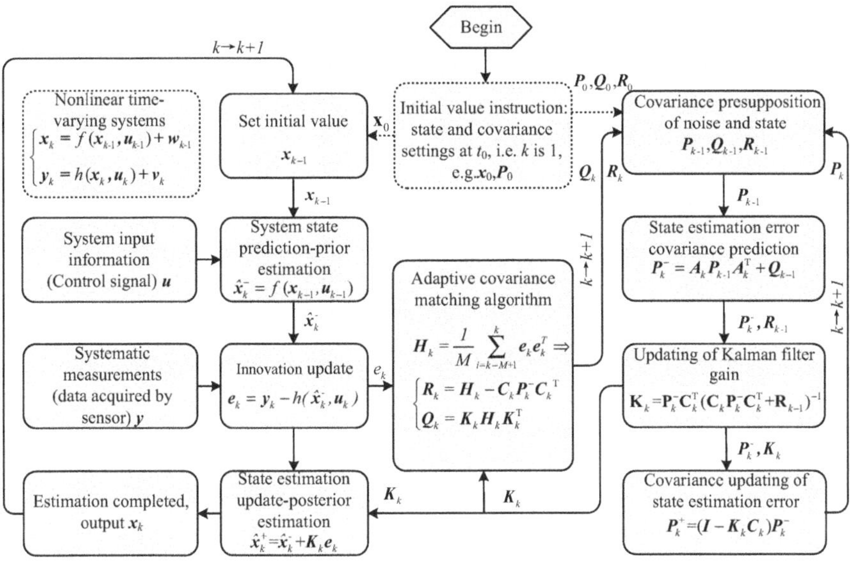

Notes: P is the covariance of state estimation error; H is a covariance function for real-time estimation of innovation based on the principle of window estimation, M denotes the size of the window; K is a Kalman gain matrix, e is innovation; $\hat{x}_k^-$ and P_k^- are a priori estimates of system state and state estimation error covariance, respectively, $\hat{x}_k^+$ and P_k^+ are a posteriori estimates of the system state and state estimation error covariance, respectively.

Fig. 4.7 Flowchart of the AEKF algorithm

(2) SOC estimation based on the AEKF

Based on the battery system state equation in Sect. 4.1.2 and the AEKF algorithm shown in Fig. 4.7, a detailed SOC estimation method is established, as shown in Table 4.2.

Among them, the elements of matrix A_k and C_k can be derived from the parameter identification method referred to in Chap. 3 of this book.

(3) Case study

Case 1: the capacity and accurate initial SOC are given

The SOC estimation method was evaluated with UDDS data of cell 2 at 25 °C. The initial SOC in Figs. 4.8 and 4.9 is set to the true SOC (90%). Figure 4.8 presents the estimation result of the terminal voltage and SOC and Fig. 4.9 shows the estimated polarization voltage.

The results reveal that the SOC estimation method can not only accurately estimate the battery voltage and SOC, but also calculate the polarization voltage of the battery. The detailed accuracy evaluation is shown in Table 4.3.

The maximum estimation error and standard deviation of SOC are less than 1%, the maximum error of terminal voltage is 38 mV, and the average error is 5.6 mV. However, the above results depend on the precise initial SOC which is difficult to

Table 4.2 Detailed calculation process of the SOC estimation method based on AEKF algorithm

The linear discretization equation of the Thevenin model can be expressed as follows: $\begin{cases} \boldsymbol{x}_k = f(\boldsymbol{x}_{k-1}, \boldsymbol{u}_{k-1}) + \boldsymbol{\omega}_{k-1} \approx \boldsymbol{A}_{k-1}\boldsymbol{x}_{k-1} + \boldsymbol{B}_{k-1}\boldsymbol{u}_{k-1} + \boldsymbol{\omega}_{k-1} \\ \boldsymbol{y}_k = h(\boldsymbol{x}_k, \boldsymbol{u}_k) + \boldsymbol{v}_k \approx \boldsymbol{C}_k\boldsymbol{x}_k + \boldsymbol{D}_k + \boldsymbol{v}_k \end{cases}$		(4.14)
where $\boldsymbol{x} = \begin{bmatrix} U_D & z \end{bmatrix}^T$, $\boldsymbol{u} = i_L$, $\boldsymbol{y} = U_t$; $\boldsymbol{A}_k \triangleq \begin{bmatrix} \exp(-\Delta t/\tau) & 0 \\ 0 & 1 \end{bmatrix}$, $\boldsymbol{B}_k \triangleq \begin{pmatrix} (1-\exp(-\Delta t/\tau))R_D \\ \eta(i_L)\Delta t/C_{max} \end{pmatrix}$ $\boldsymbol{C}_k \triangleq \begin{bmatrix} -1 & \frac{dU_{oc}(z)}{dz} \end{bmatrix}$, $\boldsymbol{D}_k \triangleq h(\boldsymbol{x}_k, \boldsymbol{u}_k) - \boldsymbol{C}_k\boldsymbol{x}_k = U_{oc,k} - U_{D,k} - R_i\boldsymbol{u}_k - \boldsymbol{C}_k\boldsymbol{x}_k$ $\frac{dU_{oc}(z)}{dz} = \alpha_1 + 2\alpha_2 z + 3\alpha_3 z^2 - \frac{\alpha_4}{z^2} + \frac{\alpha_5}{z} - \frac{\alpha_6}{1-z}$		(4.15)
• Initialization	Sets the initial values of the state observer: $\boldsymbol{x}_0, \boldsymbol{P}_0, \boldsymbol{Q}_0, \boldsymbol{R}_0$	
• Prior prediction: (calculation of the status from time $(k–1)^+$ to time $(k)^-$)	For $k = 1, 2\ldots$, complete the following priori predictions (time update), and the state and covariance are estimated from time $(k–1)^+$ to time $(k)^-$, the AEKF time update equations as follows:	
	State estimation : $\hat{\boldsymbol{x}}_k^- = f(\boldsymbol{x}_{k-1}, \boldsymbol{u}_{k-1})$	(4.16)
	Error covariance estimation $\boldsymbol{P}_k^- = \boldsymbol{A}_{k-1}\boldsymbol{P}_{k-1}\boldsymbol{A}_{k-1}^T + \boldsymbol{Q}_{k-1}$	(4.17)
• Posterior correction: (calculation of the status from time $(k)^-$ to time $(k)^+$)	This step uses the measured value $\boldsymbol{Y}_k$ to correct state estimation and covariance estimation. The estimated results are expressed by $\hat{\boldsymbol{x}}_k^+$ and $\boldsymbol{P}_k^+$, respectively. The measurement update equation of the AEKF is expressed as follows:	
	Innovation matrix : $\boldsymbol{e}_k = \boldsymbol{Y}_k - h(\hat{\boldsymbol{x}}_k^-, \boldsymbol{u}_k)$	(4.18)
	Kalman gain matrix : $\boldsymbol{K}_k = \boldsymbol{P}_k^-\boldsymbol{C}_k^T(\boldsymbol{C}_k\boldsymbol{P}_k^-\boldsymbol{C}_k^T + \boldsymbol{R}_{k-1})^{-1}$	(4.19)
	Adaptive noise covariance: $\boldsymbol{H}_k = \frac{1}{M}\sum_{i=k-M+1}^{k} \boldsymbol{e}_I\boldsymbol{e}_i^T$, $\boldsymbol{R}_k = \boldsymbol{H}_k - \boldsymbol{C}_k\boldsymbol{P}_k^-\boldsymbol{C}_k^T$, $\boldsymbol{Q}_k = \boldsymbol{K}_k\boldsymbol{H}_k\boldsymbol{K}_k^T$	(4.20)
	System state correction: $\hat{\boldsymbol{x}}_k^+ = \hat{\boldsymbol{x}}_k^- + \boldsymbol{K}_k\boldsymbol{e}_k$	(4.21)
	Error covariance correction: $\boldsymbol{P}_k^+ = (\boldsymbol{I} - \boldsymbol{K}_k\boldsymbol{C}_k)\boldsymbol{P}_k^-$	(4.22)
• Time update	The state and covariance matrix of time $(k)^+$ are used as the output to prepare the state estimation of time $(k + 1)$	

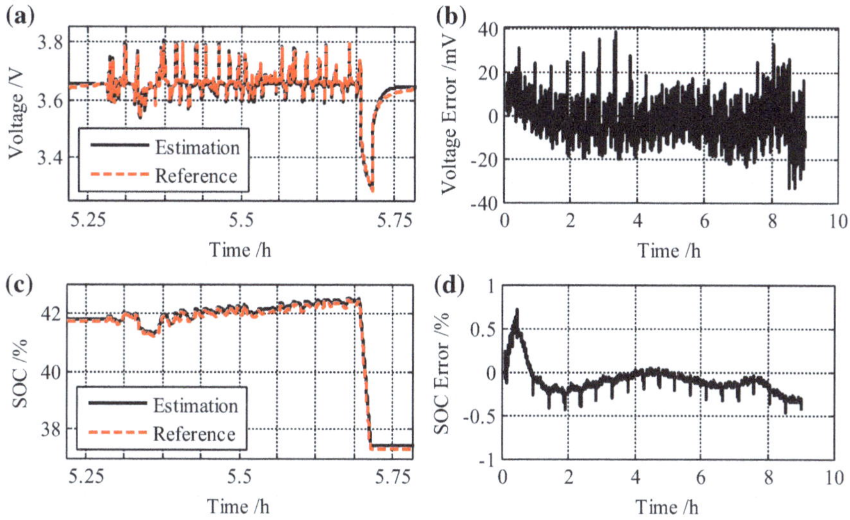

Fig. 4.8 Estimation of voltage and SOC under UDDS conditions: **a** comparison between voltage measurement and predicted value (local); **b** voltage prediction error; **c** comparison between SOC reference value and estimated value (partial); **d** SOC estimation error

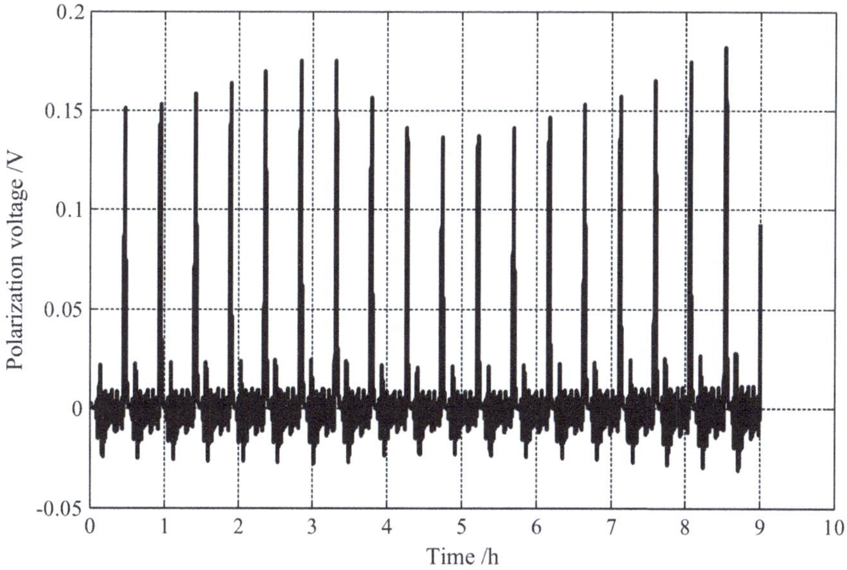

Fig. 4.9 Polarization voltage estimation results under UDDS conditions

Table 4.3 Errors statistics of the SOC and terminal voltage under UDDS condition

Targets	Absolute maximum error	Mean error	Standard deviation
SOC error	0.71%	0.16%	0.11%
Terminal voltage error	38 mV	5.6 mV	1.9 mV

be determined in practical application. Therefore, it is necessary to consider the estimation performance of the algorithm under inaccurate initial SOC values.

Case 2: the capacity and inaccurate initial SOC are given

In order to explore the convergence ability of the proposed SOC estimation method for inaccurate initial SOC, totally six groups of inaccurate initial SOC including 100, 80, 70, 60, 50, and 10% are used for analysis and discussion. Figure 4.10 shows the terminal voltage estimation results under different initial SOC values. For the convenience of analysis, only the first cycle is drawn. Figure 4.11 shows the comparison result of SOC and its error as well as the local enlarged figure.

It is revealed that with the convergence of SOC from the inaccurate initial value to the accurate value, the predicted terminal voltage also converge to the measured value very soon. It is noteworthy that this method can still correct the SOC estimation error when the battery is not working, and the correction accuracy depends on the corresponding relationship between OCV and SOC. In addition, the initial error of SOC affects the convergence rate. In fact, the deviation of initial SOC is generally less than 20%.

As shown in Fig. 4.11, when the initial SOC is 10%, the maximum estimation error is 7.2% after 60 s, and it will be reduced to 1.27% at 120 s. The estimated errors of other initial values have converged to 1% within 60 s. Therefore, the convergence speed is related to the inaccuracy degrees of the initial SOC. However, any initial

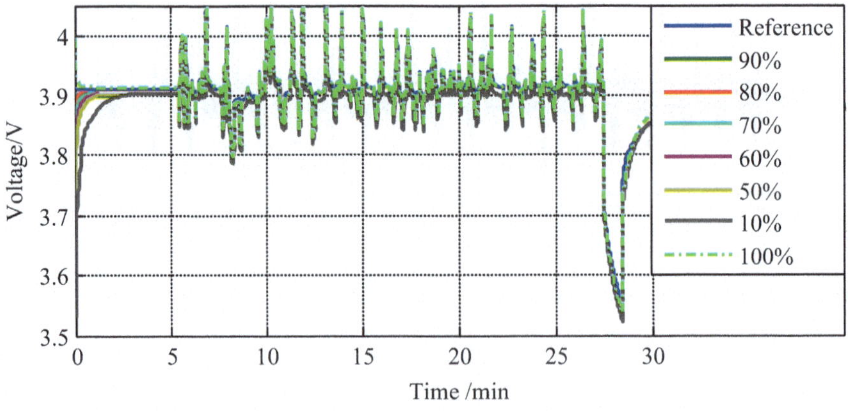

Fig. 4.10 Estimation results of terminal voltage under the inaccurate initial SOC

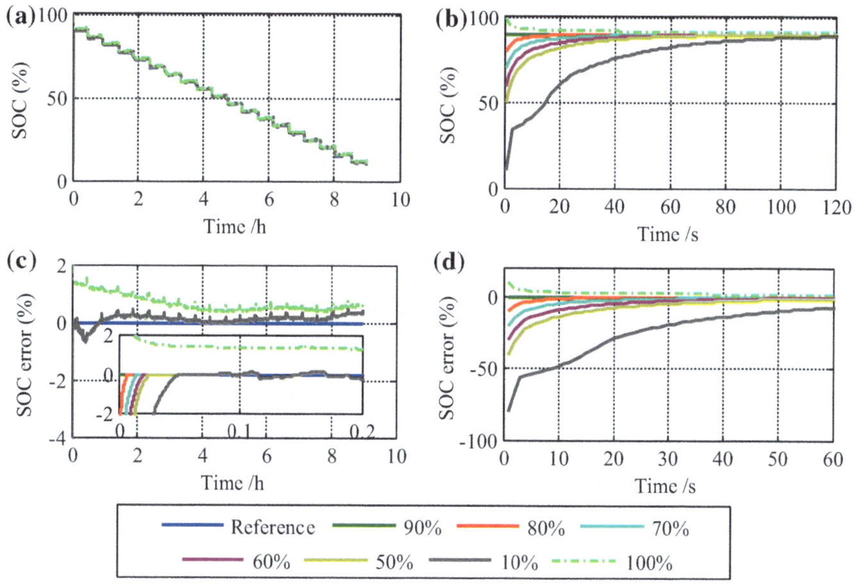

Fig. 4.11 SOC estimation results under inaccurate initial SOC values: **a** SOC estimation results; **b** SOC estimation results within 120 s; **c** SOC estimation errors; **d** SOC estimation errors within 60 s

error can be corrected accurately after a certain number of iterations. In order to systematically discuss its estimation performance, Table 4.4 illustrates the terminal voltage prediction error and the absolute value of the SOC estimation error under different SOC initial values. Only the results after the convergence of the estimation are used for the evaluation.

It is indicated that the maximum, average and standard deviations of the absolute errors of predicted voltage and estimated SOC are basically the same when the initial values of SOC are different. Therefore, the SOC estimation method based

Table 4.4 Error statistics of the voltage and SOC under different SOC initial values

Initial SOC (%)	Voltage prediction error (mV)			SOC estimation error (%)		
	Maximum error	Mean error	Standard deviation	Maximum error	Mean error	Standard deviation
100	36.45	5.29	5.11	0.981	0.279	0.279
90	38.09	5.57	4.91	0.708	0.155	0.109
80	38.09	5.57	4.91	0.708	0.155	0.109
70	38.09	5.57	4.91	0.708	0.155	0.109
60	38.09	5.57	4.90	0.708	0.155	0.109
50	38.09	5.57	4.90	0.708	0.156	0.109
10	39.92	5.59	4.91	1.270	0.158	1.111

on the AEKF algorithm shows a strong robustness ability to the inaccurate initial SOCvalues. The maximum prediction error of voltage is within 40 mV and the maximum error of SOC is less than 1.5%.

4.1.4 SOC Estimation Based on HIF Algorithm

(1) HIF algorithm

KF is an effective tool to analyze and solve the problem of system state estimation. However, as mentioned above, it is based on the assumption that the system model is accurate and the noise statistical characteristics are known. In practice, the noise statistical characteristics are difficult to obtain, and it is difficult to establish an accurate model for a phenomenon or objective.

Therefore, the model errors in the process of modeling will also affect the accuracy of estimation. In order to overcome the defects of KF algorithm and the uncertainty of model errors in modeling, the HIF algorithm was proposed. The following is a brief description of its basic theory from the perspective of game theory.

Define the following discrete state-space function:

$$\begin{cases} \boldsymbol{x}_k = \boldsymbol{A}_{k-1}\boldsymbol{x}_{k-1} + \boldsymbol{\omega}_{k-1} \\ \boldsymbol{y}_k = \boldsymbol{C}_k\boldsymbol{x}_k + \boldsymbol{v}_k \end{cases} \tag{4.23}$$

Define the following cost function:

$$J_1 = \frac{\sum_{k=0}^{N-1} \left\| \boldsymbol{x}_k - \hat{\boldsymbol{x}}_k \right\|_{\boldsymbol{S}_k}^2}{\left\| \boldsymbol{x}_0 - \hat{\boldsymbol{x}}_0 \right\|_{\boldsymbol{P}_0^{-1}}^2 + \sum_{k=0}^{N-1} \left(\|\boldsymbol{\omega}_k\|_{\boldsymbol{Q}_k^{-1}}^2 + \|\boldsymbol{v}_k\|_{\boldsymbol{R}_k^{-1}}^2 \right)} \tag{4.24}$$

where $\boldsymbol{x}_0$ and $\hat{\boldsymbol{x}}_0$ represent the initial value of the state and its initial estimation, respectively.

$\boldsymbol{P}_0, \boldsymbol{Q}_k, \boldsymbol{R}_k$, and $\boldsymbol{S}_k$ are different from the parameter matrix in the AEKF algorithm, which are the symmetric positive definite matrices defined by the designer based on the specific problem. $\boldsymbol{P}_0$ is often designed as the initial state error covariance matrix. $\boldsymbol{Q}_k$ and $\boldsymbol{R}_k$ are often designed as the state equation noise covariance matrix and the measurement equation noise covariance matrix. $\boldsymbol{S}_k$ needs to be designed based on the emphasis placed by the designer on each state quantity. For example, when we have interest in the third element of the state vector $\boldsymbol{x}_k$, we ought to make $\boldsymbol{S}_k$ (3, 3) much larger than other elements in $\boldsymbol{S}_k$.

Our goal is to find an estimate of $\boldsymbol{x}_k$ to minimize $\boldsymbol{x}_k - \hat{\boldsymbol{x}}_k$. Contrary to our goal, it is assumed that nature is our adversary (with interference), and its goal is to maximize $\boldsymbol{x}_k - \hat{\boldsymbol{x}}_k$ by finding appropriate $\boldsymbol{x}_0$, $\boldsymbol{\omega}_k$ and $\boldsymbol{v}_k$. It is apparently unfair that nature can maximize $\boldsymbol{x}_k - \hat{\boldsymbol{x}}_k$ by simply making $\boldsymbol{x}_0$, $\boldsymbol{\omega}_k$ and $\boldsymbol{v}_k$ infinite. Therefore, when the cost function is defined, we can place $\boldsymbol{x}_0$, $\boldsymbol{\omega}_k$ and $\boldsymbol{v}_k$ in the denominator. This signifies that nature must maximize $\boldsymbol{x}_k - \hat{\boldsymbol{x}}_k$ by choosing the appropriate $\boldsymbol{x}_0$, $\boldsymbol{\omega}_k$ and $\boldsymbol{v}_k$. Similarly, it is required to design an appropriate estimation strategy to minimize it.

It is not easy to directly minimize J_1, for which we choose a boundary λ to satisfy:

$$J_1 < \frac{1}{\lambda} \tag{4.25}$$

The equation is rearranged to obtain:

$$J_2 = \frac{-1}{\lambda}\left\|\boldsymbol{x}_0 - \hat{\boldsymbol{x}}_0\right\|^2_{\boldsymbol{P}_0^{-1}} + \sum_{k=0}^{N-1}\left(\left\|\boldsymbol{x}_k - \hat{\boldsymbol{x}}_k\right\|^2_{\mathbf{S}_k} - \frac{1}{\lambda}\left(\|\boldsymbol{\omega}_k\|^2_{\boldsymbol{Q}_k^{-1}} + \|\boldsymbol{v}_k\|^2_{\boldsymbol{R}_k^{-1}}\right)\right) < 0 \tag{4.26}$$

In this case, this problem becomes that when $\boldsymbol{x}_0$, $\boldsymbol{\omega}_k$ and $\boldsymbol{v}_k$ make J_2 the largest, choose appropriate $\hat{\boldsymbol{x}}_k$ to make J_2 the smallest. By solving the above problem, we can finally ascertain the recursive relationship that makes the cost function J_2 smaller than $1/\lambda$:

$$\begin{cases} \boldsymbol{K}_k = \boldsymbol{P}_k\left(\boldsymbol{I} - \lambda\boldsymbol{S}_k\boldsymbol{P}_k + \boldsymbol{C}_k^{\mathrm{T}}\boldsymbol{R}_k^{-1}\boldsymbol{C}_k\boldsymbol{P}_k\right)^{-1}\boldsymbol{C}_k^{\mathrm{T}}\boldsymbol{R}_k^{-1} \\ \hat{\boldsymbol{x}}_{k+1} = \boldsymbol{A}_k\hat{\boldsymbol{x}}_k + \boldsymbol{A}_k\boldsymbol{K}_k\left(\boldsymbol{y}_k - \boldsymbol{C}_k\hat{\boldsymbol{x}}_k\right) \\ \boldsymbol{P}_{k+1} = \boldsymbol{A}_k\boldsymbol{P}_k\left(\boldsymbol{I} - \lambda\boldsymbol{S}_k\boldsymbol{P}_k + \boldsymbol{C}_k^{\mathrm{T}}\boldsymbol{R}_k^{-1}\boldsymbol{C}_k\boldsymbol{P}_k\right)^{-1}\boldsymbol{A}_k^{\mathrm{T}} + \boldsymbol{Q}_k \end{cases} \tag{4.27}$$

where $\boldsymbol{K}_k$ represents the gain matrix and $\boldsymbol{P}_k$ denotes the matrix obtained from the selected symmetric positive definite matrix $\boldsymbol{P}_0$ based on the above recursive relationship.

(2) SOC Estimation process based on HIF

In order to facilitate calculation and application, similar to KF, the recursive relation Eq. (4.27) is often divided into two steps: time update and measurement update. Similarly, in order to distinguish the estimated results of the two-step updating, $\hat{\boldsymbol{x}}_k^-$ and $\boldsymbol{P}_k^-$ are used to represent the $\boldsymbol{x}_k$ and $\boldsymbol{P}_k$ after the time update, respectively. The $\hat{\boldsymbol{x}}_k^+$ and $\boldsymbol{P}_k^+$ are used to measure the $\boldsymbol{x}_k$ and $\boldsymbol{P}_k$ after measurement update, respectively. The detailed calculation process is shown in Table 4.5.

(3) Case study

Cell 3 with a rated capacity of 35 A h was selected in this study. The charge and discharge cutoff voltages are 4.2 and 3.0 V, respectively. The static capacity of battery is known and the experimental data were obtained at 25 °C.

The convergence ability of the algorithm with inaccurate initial SOC is analyzed directly. When the initial SOC is 80% (the accurate initial value is 100%), the predicted voltage and estimated SOC of the HIF are plotted in Fig. 4.12, respectively. The results reveal that the maximum absolute error of terminal voltage is slightly less than 0.06 V. The absolute value of SOC estimation error is shown to be within 1.5%,

Table 4.5 Detailed process of the SOC estimation based on the HIF algorithm

The linear discretization equation of Thevenin battery model can be expressed as Eq. (4.14) in Table 4.2.

• Initialization	Sets the initial value of the state observer: $\boldsymbol{x}_0, \boldsymbol{P}_0, \boldsymbol{Q}, \boldsymbol{R}, \lambda, \boldsymbol{S}$.	
• Prior prediction: (calculation of the status from time $(k\text{-}1)^+$ to time$(k)^-$)	For $k = 1, 2\ldots$, complete the following priori estimation (time update) operation, and the state and covariance are estimated from time $(k-1)^+$ to time $(k)^-$, the HIF time update equations as follows:	
	State estimation : $\hat{\boldsymbol{x}}_k^- = f(\boldsymbol{x}_{k-1}, \boldsymbol{u}_{k-1})$	(4.28)
	Prediction of $H\infty$characteristic matrix : $\boldsymbol{P}_k^- = \boldsymbol{A}_{k-1}\boldsymbol{P}_{k-1}\boldsymbol{A}_{k-1}^{\mathrm{T}} + \boldsymbol{Q}$	(4.29)
• Posterior correction: measurement update (calculation of the status from time $(k)^-$to time $(k)^+$)	This step uses the measured value $\boldsymbol{Y}_{\mathrm{k}}$ to correct the state estimation and covariance estimation. The estimated results are expressed by $\hat{\boldsymbol{x}}_k^+$ and $\boldsymbol{P}_k^+$, respectively. The measurement update equation of the HIF is expressed as follows:	
	Innovation matrix : $\boldsymbol{e}_k = \boldsymbol{Y}_k - h(\hat{\boldsymbol{x}}_k^-, \boldsymbol{u}_k)$	(4.30)
	Gain matrix : $\boldsymbol{K}_k = \boldsymbol{A}_k\boldsymbol{P}_k^-(\boldsymbol{I} - \lambda\boldsymbol{S}\boldsymbol{P}_k^- + \boldsymbol{C}_k^{\mathrm{T}}\boldsymbol{R}^{-1}\boldsymbol{C}_k\boldsymbol{P}_k^-)^{-1}\boldsymbol{C}_k^{\mathrm{T}}\boldsymbol{R}^{-1}$	(4.31)
	System state correction : $\hat{\boldsymbol{x}}_k^+ = \hat{\boldsymbol{x}}_k^- + \boldsymbol{K}_k\boldsymbol{e}_k$	(4.32)
	Correction of HIF characteristic matrix : $\boldsymbol{P}_k^+ = \boldsymbol{P}_k^-(\boldsymbol{I} - \lambda\boldsymbol{S}\boldsymbol{P}_k^- + \boldsymbol{C}_k^{\mathrm{T}}\boldsymbol{R}^{-1}\boldsymbol{C}_k\boldsymbol{P}_k^-)^{-1}$	(4.33)
• Time update	The state and covariance matrix of time $(k)^+$ are used as output to prepare the state estimation of time $(k + 1)$	

which basically meets the requirements of SOC estimation. For the convergence rate, the HIF takes about 70 s to make the SOC converge to the reference value.

4.2 SOH Estimation

The energy storage and fast charging–discharging capability of batteries decline with the battery degradation. SOH is the quantitative index to evaluate the aging degree of the batteries. The accurate estimation of SOC depends on the precise SOH value. It is not practical to estimate the SOC with the known SOH, which can only provide a reference for the SOC estimation.

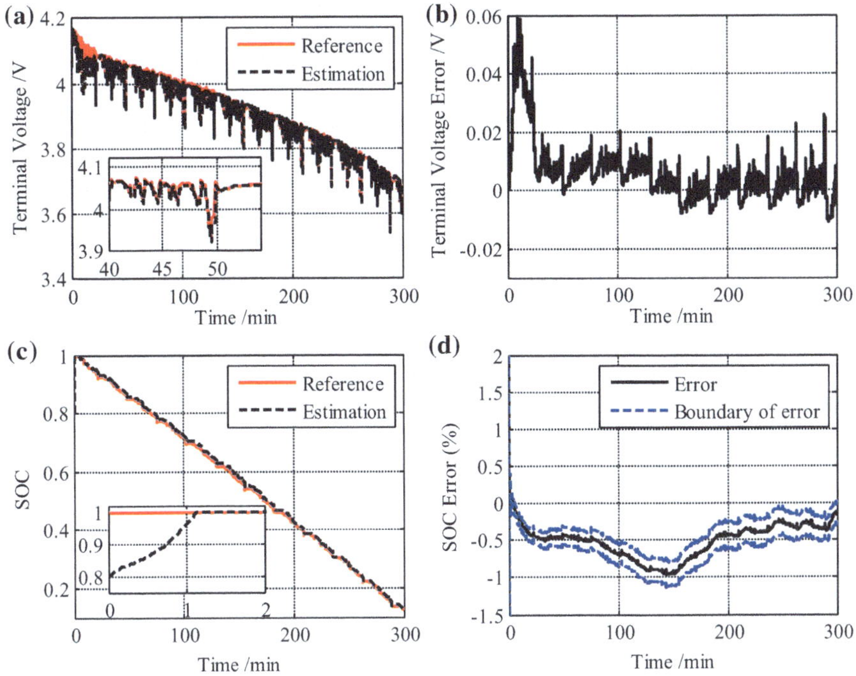

Fig. 4.12 **a** The estimation results of terminal voltage and SOC of HIF algorithm: comparison between predicted and measured values of terminal voltage; **b** the prediction error of terminal voltage; **c** the comparison between estimated and reference values of SOC; **d** the estimation error of SOC

4.2.1 Classification of SOH Estimation Methods

The most used SOH indicators are battery capacity and internal resistance, which reflect the energy capability and power capability, respectively. Generally speaking, the SOH of a new battery is considered as 100%. For BEVs with a high demand for battery capacity, it can be considered that the battery is incapable of meeting the normal demand if the battery capacity decreases to 80% of the initial capacity. For HEVs with a high demand for battery power capability, the battery needs to be replaced when the internal resistance value is twice as high as the initial value.

SOH estimation methods can be grouped into two categories [13]: experimental analysis methods and model-based methods, as shown in Fig. 4.13. By analyzing the collected experimental data on current, voltage, temperature, etc., experimental analysis methods can directly obtain some characteristic parameters reflecting the degradation degree of batteries. According to the different choice of health indicators, experimental analysis methods can be divided into direct measurement methods and indirect analysis methods. Model-based methods need to estimate the health-related

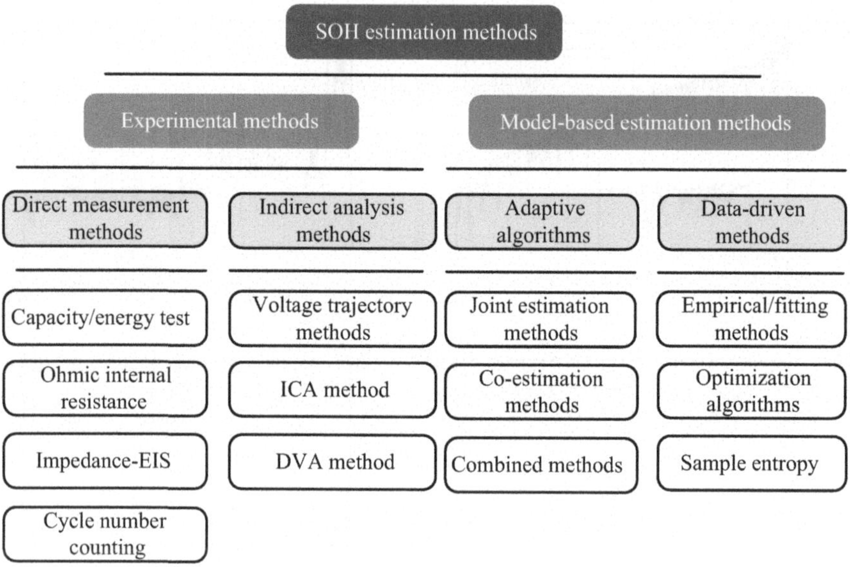

Fig. 4.13 Classification of SOH estimation methods

parameters to realize the SOH calibration. According to the different estimation algorithms, it could be categorized into adaptive algorithms and data-driven methods.

Direct measurement methods

The direct measurement methods refer to the direct measurement of some characteristic parameters to evaluate the battery health, mainly including capacity/energy measurement method, ohmic internal resistance measurement method, impedance measurement method, and cycle counting method.

(1) Capacity/energy measurement method

The capacity/energy measurement method refers to the accurate and direct measurement of the capacity or energy to determine the battery SOH. Obviously, the accurate measurement of capacity and energy requires to ensure the integrity of the charging–discharging process and the high accuracy of the data collection, which means that this method can only be used in laboratory or other relatively stable conditions.

(2) Ohmic resistance measurement method

The ohmic resistance measurement method is to evaluate SOH by measuring the ohmic resistance of the battery in real time. The calculation method is shown in Eq. (4.34). Compared to the capacity, the ohmic resistance is easier to obtain. Sudden brake or acceleration in the driving process will lead to more significant changes in current and voltage [14]. However, besides the influence of SOH and temperature, the ohmic internal resistance will also change with the SOC change. It is also significantly affected by the sampling interval of current and voltage. The measured ohmic resistance is closer to the true value with a smaller sampling interval. Also, the minimum absolute value of Δi_L ought to be limited when calculating the ohmic resistance. Otherwise, the result will vary dramatically.

$$R_i = \frac{\Delta U_t}{\Delta i_L} \tag{4.34}$$

where ΔU_t represents the pulse voltage and Δi_L indicates the pulse current.

(3) Impedance measurement method

The impedance measurement method needs to measure the impedance spectra by means of electrochemical workstation or other AC excitation equipment with similar functions [15]. Figure 2.32 shows the battery EIS under different aging conditions. It can be seen that there is a clear relationship between the EIS and the aging state of batteries. Besides, the battery response of the excitation varies with different frequency excitation. At high frequencies, the impedance is more characterized by ohmic properties, while at low frequencies, the capacitive effect is more pronounced. Therefore, after obtaining the EIS, SOH could be calibrated by extracting some characteristic parameters from EIS.

Indirect analysis methods

The indirect analysis is a typical multi-step derivation method. It does not directly calculate the capacity or internal resistance, but calibrates SOH by designing or measuring some process parameters which can reflect the change in capacity or internal resistance during the aging process these process parameters are usually named health indicators, including the SEI film resistance, capacity-OCV-SOC response surface, voltage response trajectory or charging time at constant voltage stage, incremental capacity (IC) curve or differential voltage (DV) curve, ultrasonic wave response characteristics, etc. Obviously, two or more health factors can be combined to evaluate the battery SOH.

1. Terminal voltage response directly reflects the internal reaction characteristics of the battery. Therefore, based on the controlling variable method, the terminal voltage response trajectory under specific SOC, temperature, and current excitation can be analyzed to complete the SOH calibration. Considering that the discharge conditions are complex and changeable, this method usually uses relatively stable charging process as the object of analysis [16]. At present, the most common charging method is the CCCV charging method. As shown in Fig. 2.11, it is divided into two stages: first, constant current charging to the upper cutoff voltage (CC stage), and then constant voltage charging until the minimum current threshold (CV stage). For batteries of the same material, the total charging time of this charging method is basically unchanged. However, the charging time of CV stage will increase significantly as the battery aging. Therefore, if the complete charge curve of the CV stage can be obtained, the battery SOH can be calculated accurately.

2. IC analysis (ICA) [17] and DV analysis (DVA) [18] analyze the degradation process and aging mechanism of the battery using IC curve and DV curve, respectively. IC and DV curves can be transformed from constant current charging–discharging data. The IC curve describes the relationship between dQ/dV and voltage. DV curve describes the relationship between dV/dQ and Q. These two methods will be described in detail in Sect. 4.2.4.

Adaptive algorithms

Based on the electrochemical model or equivalent circuit model, the adaptive algorithms generally identify the model parameters to complete the SOH calibration. These methods are featured with the characteristics of closed-loop control and feedback, which can adjust the estimation results adaptively with voltage. Adaptive algorithms include joint estimation method, co-estimation method, fusion estimation method, etc.

(1) Joint estimation method

The joint estimation method [19] estimates the model parameters and SOC online simultaneously, for which two or more filters/observers are generally used. The model parameters to be identified mainly include internal resistance, impedance, OCV, and so on. As the battery SOC is closely related to its capacity, the battery capacity can be determined after the relatively accurate SOC value is obtained. The battery capacity estimation method based on SOC estimation is described in detail in Sect. 4.2.2.

(2) Co-estimation method

The co-estimation method [20] realizes the simultaneous online estimation of model parameters and SOC. Compared to the joint estimation method, the capacity is listed as an additional item of model parameters. Thus, battery capacity and SOC are simultaneously estimated. From the basic framework of the algorithm, the difference between the co-estimation and joint estimation method is mainly reflected in the following two aspects:

- For the two kinds types of estimation algorithms, the usage mode of the innovation (voltage prediction error) sequence is different. Two estimators in the co-estimation method share the same innovation. But by contrast, in the joint estimation method, the innovations of the two estimators is not correlated.
- The relationship between parameter estimation and state estimation of these two kinds of estimation algorithms is different. In the co-estimation method, the state estimation and parameter estimation will interact, but there is no obvious interaction effect in the joint estimation. The detailed calculation process of the co-estimation method will be introduced in Sect. 4.3.

Data-driven based methods

The data-driven SOH estimation methods are not reliant on the accurate mathematical models to describe the aging process of the battery, but only on the historical aging data, i.e., extracting critical aging information from historical data points through specific learning algorithms.

(1) Empirical/fitting method uses the existing aging data to predict the battery life without a detailed understanding of the structure and material characteristics of the battery. Polynomial, exponential, power laws, logarithm and trigonometric functions are commonly used in empirical models and fitting models. Their computational complexity is usually small and their computational speed is fast. For example, the Arrhenius equation [21], which is not only very concise, but also accurately describes the temperature dependence of chemical reaction rate, is often used to simulate the changes of diffusion coefficient, creep rate and other thermal processes caused by temperature. The Arrhenius equation can also be used to describe the temperature-dependent aging rate of batteries, as shown in Eq. (4.35).

$$\frac{dC}{dn} = \Lambda e^{-\frac{\Delta E}{R_g T}} = \Lambda e^{-\frac{\lambda}{T}} \tag{4.35}$$

where dC/dn represents the change rate of cell capacity relative to aging cycle, Λ indicates the pre-exponential factor, R_g is defined as the general gas constant, i.e., 8.314 J/(mol K), ΔE denotes the activation energy, in units of J/mol, and T refers to the absolute temperature, in units of K. Λ and $\lambda = \Delta E/R_g$ represent two unknown parameters needing to be calibrated. Integrating Eq. (4.35) gives:

$$C_r = -\Lambda n_c e^{-\frac{\lambda}{T}} \tag{4.36}$$

where C_r indicates the capacity reduction threshold indicating the cell degradation, and n_c denotes the cell cycling life. Two different temperature points T_1 and T_2 ($T_1 > T_2$) were selected.

$$\Delta n_c = \frac{C_r}{\Lambda}\left(e^{-\frac{\lambda}{T_2}} - e^{-\frac{\lambda}{T_1}}\right) \tag{4.37}$$

where Δn_c is defined as the lifetime difference between two temperature points, which quantitatively describes the effect of temperature on the cell lifetime.

After identifying the parameters in the Arrhenius equation, SOH evaluation of the battery at variable temperature can be realized based on this equation.

(2) Sample entropy (SampEn) [22] can be used to evaluate the predictability of time series and quantify the regularity of data series. Therefore, SampEn can be used to analyze the terminal voltage data to indicate battery SOH. The calculation flow of the SampEn algorithm is shown in Table 4.6.

After obtaining a large amount of offline data, machine learning algorithms, such as support vector machine and relevance vector machine, can be directly used to learn the offline mapping relationship between SOH and SampEn. Afterward, the offline mapping relationship can be used to complete the real-time estimation of battery SOH.

Table 4.6 Calculation flow of SampEn

(a) Obtain the time series data under N sampling points, recorded as $x(1), x(2), \ldots, x(N)$;
(b) Define SampEn basic parameters: length of subseries vectors m, tolerance r;
(c) Define subseries vectors: $X_m(i) = \{x(i), x(i+1), \ldots, x(i+m-1)\}, 1 \le i \le N-m+1, i \in N$;
(d) The distance between two such vectors is defined as the maximum absolute difference of their scalar elements: $d(X_m(i), X_m(j)) = \max\limits_{k\in[0,m-1]}(\lvert X_m(i+k) - X_m(j+k)\rvert), 1 \le j \le N-m+1, j \ne i, j \in N$;
(e) For a given subseries vector $X_m(i)$, compare it with other subseries vectors $X_m(j)$. $A_i$is the number of vectors $X_m(j)$ which make $d(X_m(i), X_m(j)) \le r, j \ne i$.
(f) Define B_i^m: $B_i^m(r) = \frac{A_i}{N-m+1}$:
(g) Define $B_i^m(r)$: $B_m(r) = \frac{1}{N-m}\sum\limits_{i=1}^{N-m} B_i^m(r), 0 \le i \le N-m+1$:
(h) Increase the length of the vector to $m + 1$ and repeat the calculation steps from (c) to (g) to get $B_{m+1}(r)$;
(i) The SampEn of the time series of length N is calculated: $\mathrm{SampEn}(m, r, N) = -\ln\left[\frac{B_{m+1}(r)}{B_m(r)}\right]$

Table 4.7 summarizes the advantages and disadvantages exhibited by the various SOH estimation methods as well as their corresponding application scope.

4.2.2 Capacity Estimation Based on SOC Estimation

In Sect. 4.1, a relatively stable OCV-SOC relationship is used to estimate the SOC and impressive results are obtained. Considering different battery materials, the variation characteristics of OCV-SOC relationship with battery aging are different. For the case that OCV-SOC relationship is basically unchanged or barely changes with battery aging, the capacity estimation method based on SOC estimation in this section is often used. For the case that OCV-SOC relationship changes noticeably with battery aging, the response surface based capacity estimation method in Sect. 4.2.3 is the best choice. For the case between the two cases, a more generalized co-estimation method in Sect. 4.3 can be used.

(1) Problem description

It is difficult to measure the battery capacity directly in a running vehicle. The simplest capacity estimation method is derived from the ampere-hour integral method in Sect. 4.1.1 [23], as shown in Eq. (4.38).

$$\hat{C}_{\max} = \frac{\int_{t_0}^{t} \eta i_L(\tau)\,\mathrm{d}\tau}{z(t) - z(t_0)} = \frac{\Delta Q}{\Delta z} \tag{4.38}$$

Table 4.7 Advantages and disadvantages of various SOH estimation methods and their application scope

Method	Advantage	Disadvantage	Applicability	Convenience of operation
Capacity/energy measurement method	• Easy to realize; • Low computational cost	• Sensitive to uncertainty factors, such as temperature, working condition (current magnitude), etc.; • Long test time; • Need high data collection accuracy, especially for current	Bad	Simple
Ohmic resistance measurement	• Easy to realize; • Good real-time performance	• Sensitive to uncertainty factors, such as temperature, working condition, SOC; • High requests for data collection accuracy, sampling interval, sampling synchronization, etc.	Medium	Simple
Impedance measurement	• Reflect the aging state directly; • Consist of plenty of aging information	• Sensitive to uncertainty factors, such as temperature, working condition, SOC, test method, proficiency,; • Long test time; • Costly equipment	Bad	Medium

(continued)

Table 4.7 (continued)

Method	Advantage	Disadvantage	Applicability	Convenience of operation
Cycle counting method	• Easy to realize	• Difficult in the real application;	Bad	Medium
Voltage trajectory method	• Easy to operate	• Sensitive to uncertainty factors, such as temperature, working condition, SOC; • Need large offline data	Medium	Medium
ICA and DVA	• High accuracy; • Reflect aging mechanisms;	• High difficulty in operation; • Long test time	Medium	Hard
Joint estimation method	• High accuracy	• Require lots of debugging; • High dependent on model accuracy; • Output results fluctuate greatly	Flat OCV curve	Hard
Co-estimation method	• High accuracy	• Require lots of debugging; • High dependent on model accuracy;	Good	Hard
Empirical/fitting method	• Easy to realize	• Poor accuracy	Bad	Simple
Optimization algorithm	• High accuracy; • Stable output results	• High computation complexity	Steep OCV curve	Medium
SampEn	• High accuracy	• Need large offline data	Medium	Hard

The discretization results are as follows:

$$\hat{C}_{\max,k} = \frac{\Delta Q_k}{\Delta z_k} = \frac{\sum_{j=k-L_s+1}^{k} \eta_i i_{L,j} \Delta t}{z_k - z_{k-L_s}} (k \geq L_s) \quad (4.39)$$

where L_s denotes the length of a given time period, i.e., the number of sampling intervals.

It is revealed that the estimated value of available capacity $\hat{C}_{\max,k}$ at kth sampling time can be directly calculated after the cumulative electricity and SOC changes in the interval time are obtained. In order to ensure the accuracy and stability of the estimation results, there are still multiple challenges in the implementation process of this method, as shown in Fig. 4.14. These challenges include:

I: The accuracy of cumulative electricity directly depends on the current collection accuracy, so this method requires higher precision current sensor;

II: The accurate value of SOC is difficult to obtain under the condition of inaccurate capacity;

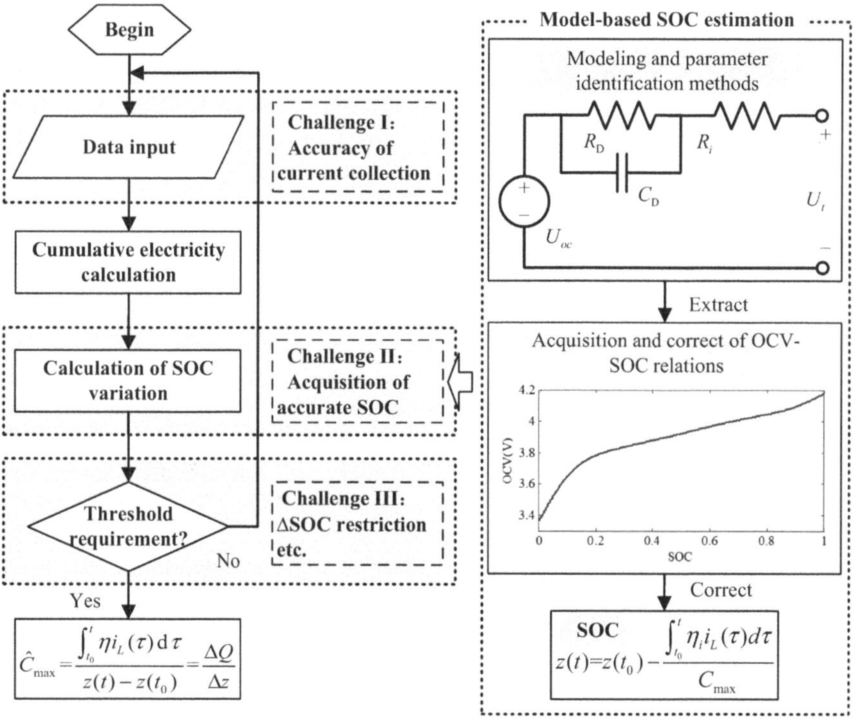

Fig. 4.14 Description of available capacity estimation method for power batteries based on SOC estimation

III: When the change of SOC in the data segment used is insignificant, capacity estimation results will fluctuate dramatically.

For challenge II, the inaccurate capacity leads to the inaccurate SOC estimate using only the Ah integral method. Therefore, the OCV-SOC relationship must be applied to continuously calibrate the battery SOC. In order to ensure the stability and accuracy of SOC correction under unknown capacity, this SOH estimation method is more suitable for OCV-SOC relationship, which is basically unchanged or has minor changes with battery aging. For the SOC estimation method, model-based SOC estimation method is often used. The accuracy of the model and the corresponding parameters ought to be improved as much as possible.

(2) Calculation flow of the algorithm

First, in view of the challenge II, an accurate battery model ought to be established (taking Thevenin model as an example), especially the accurate OCV-SOC relationship, to ensure the accurate SOC estimation. The accuracy of the OCV-SOC relationship directly determines the estimation accuracy of SOC, for which its variation with battery aging ought to be discussed in particular:

(a) When the OCV-SOC relationship is basically unchanged with battery aging, the OCV-SOC relationship is not related to battery capacity. SOC can be obtained by OCV directly;
(b) When the OCV-SOC relationship changes slightly with battery aging, single OCV-SOC curve is difficult to accurately describe the OCV-SOC relationship under different aging conditions. Thus, it is necessary to establish a capacity-OCV-SOC response surface model to calibrate the OCV-SOC curve offline at varying capacities.

When the inaccurate available capacity is obtained, despite the OCV-SOC curve obtained directly by interpolation or fitting having some errors, which will reduce the estimation accuracy of SOC, the smaller SOC error will not cause the capacity estimation result to deviate too much. Also, the varying SOC and inaccurate available capacity cannot always match the most appropriate OCV value. Only the appropriate SOC, OCV and available capacity will minimize the square sum of estimation errors, so the capacity estimation will fluctuate near the true value. Besides the OCV-SOC function, the impedance parameters of the battery will directly affect the model accuracy, and then affect the estimation accuracy of the SOC. Therefore, the online parameter identification method is often used to obtain more accurate impedance parameters of the model. The online parameter identification algorithms have been introduced in detail in Chap. 3. Here, RLS with forgetting factors is used to illustrate the implementation process of the capacity estimation method.

For challenge III, it is necessary to preset a variety of boundary conditions to ensure the accuracy and stability of the estimation, including (i) the limitation on time interval length, i.e., reserving a certain preparation time for the algorithm to accumulate sufficient charging and discharging data; (ii) the limitation on SOC variation range ΔSOC, that is, when ΔSOC is in excess of a certain threshold, the available capacity of the battery is estimated, which is because too small ΔSOC will cause dramatic fluctuations in capacity estimates; (iii) the limitation on capacity change rate, considering that the battery capacity will not change abruptly in normal use, the capacity convergence coefficient δ is used to evaluate the reliability of capacity estimation.

$$\delta = \left| \frac{\hat{C}_{\max,k} - \hat{C}_{\max,k-1}}{\hat{C}_{\max,k-1}} \right| \leq \varepsilon \tag{4.40}$$

where δ represents the convergence coefficient of capacity estimation; ε denotes the acceptable convergence range, with 2% being recommended here.

After completing the preparatory work of the above algorithm, the available capacity estimation method based on SOC estimation can be implemented. The calculation process can be divided into four steps:

(a) Data collection. The battery data including current, voltage, and temperature is sent to the controller for real-time operation of the algorithm.
(b) Parameter identification. The parameters of the Thevenin model including R_i, R_D, and C_D are identified based on online measurement data, and then used for SOC estimation.
(c) SOC estimation. Based on the collected data and identificated parameters, the AEKF-based SOC estimation algorithm mentioned in Sect. 4.1.3 is used to accurately estimate the SOC. The calculation results are used for capacity estimation.
(d) Capacity estimation. After obtaining the SOC estimation results, it should be judged whether the above three boundary conditions meet the requirements. If so, the battery capacity can be calculated directly from Eq. (4.40). Otherwise, the available capacity will not be updated.

Figure 4.15 presents the flowchart of the battery capacity estimation method based on SOC estimation, where the SOC estimation method uses the AEKF algorithm in Sect. 4.1.3.

It is noteworthy that a model-based parameters and SOC joint estimation algorithm system is introduced in this book. In this section, a joint estimation method based on RLS and AEKF algorithms is taken as an example [19]. According to the calculation rules of the algorithm, other types of algorithms can also be combined, such as RLS-UKF [24], HIF-UKF [25], which can be referred to in the Chinese national invention patent 201610802342.4.

(3) Case study

In this section, the DST data of battery 2 is used to evaluate the accuracy of the SOC and capacity joint estimation method. L_s is set to 360 s. Figure 4.16 shows the estimation results of battery 2 cell 1 under accurate initial SOC values. It is worth noting that the maximum available capacity of battery 2 cell 1 is 31.6 A h. The initial available capacity of this algorithm is set to 30 A h. As the available capacity estimation is not turned on until 360 s, the maximum available capacity within 360 s defaults to the initial value of 30 A h.

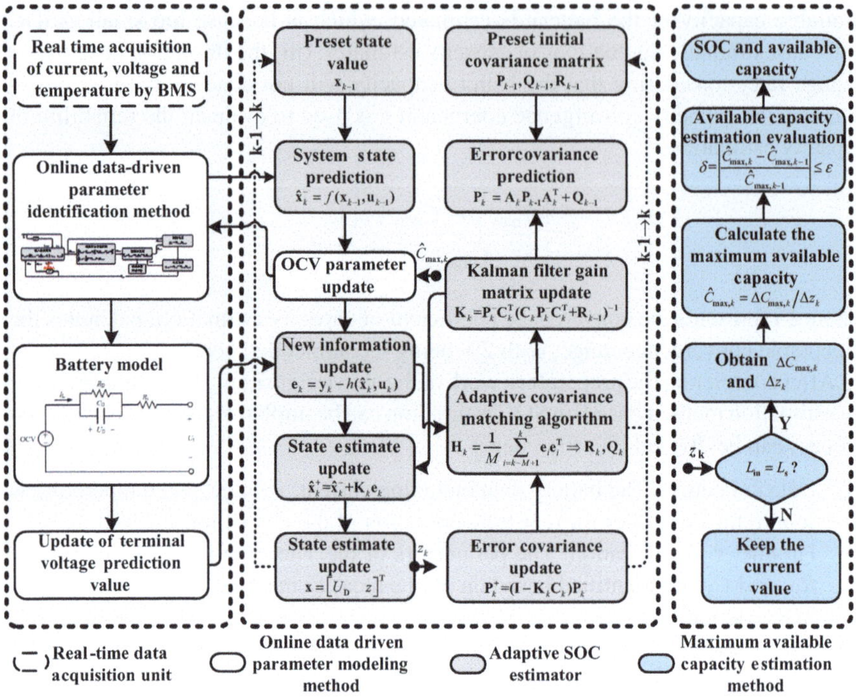

Fig. 4.15 Capacity estimation method based on SOC estimation

It is revealed that the voltage estimation error is less than 15 mV, the SOC estimation error is less than 1%, and the available capacity estimation error is less than 2 A h. When the available capacity estimator is turned on, the capacity estimator will have a jump. The effect of excessive perturbation of available capacity estimation on the stability of estimated SOC can be avoided by setting the convergence coefficient. Therefore, the estimation accuracies of the terminal voltage, SOC, available capacity under accurate initial SOC values are ideal. Nevertheless, the accurate estimation method does not need to rely on the exact initial value of the state vector. Figure 4.17 presents the estimation results under the erroneous initial SOC (the initial SOC error is 40%).

It is demonstrated that the algorithm is insensitive to the inaccurate initial values of SOC and available capacity. The initial error only affects the number of iterations when the algorithm converges. After 50 min, the maximum voltage prediction error is less than 20 mV, the maximum SOC estimation error is lower than 1%, and the maximum available capacity estimation error is less than 0.6 A h. To systematically evaluate the estimation accuracies of SOC and available capacity, the statistical results of the estimated SOC and available capacity are listed in Table 4.8. As the value of L_S is 360 s, the absolute value of estimation error after 480 s is selected for analysis, and the initial value of SOC is 60%.

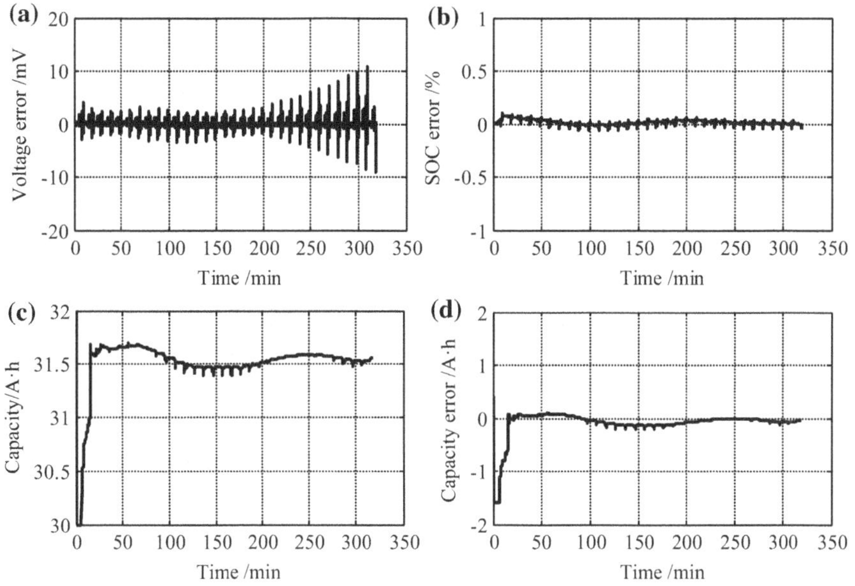

Fig. 4.16 Battery available capacity estimation based on SOC estimation (accurate initial SOC): **a** voltage estimation error; **b** SOC estimation error; **c** available capacity estimation results; **d** available capacity estimation error

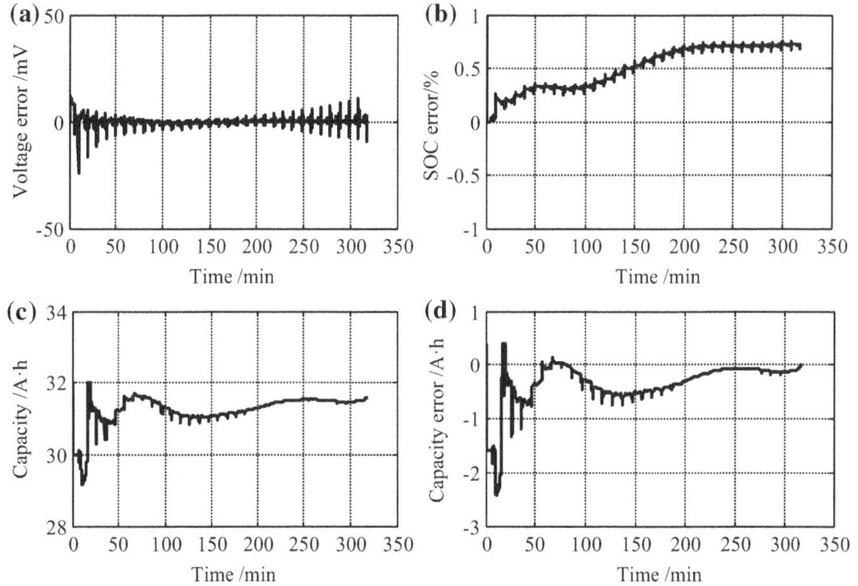

Fig. 4.17 Available capacity estimation based on SOC estimation (initial SOC value is 60%): **a** voltage estimation error; **b** SOC estimation error; **c** available capacity estimation results; **d** available capacity estimation error

Table 4.8 SOC and available capacity estimation errors based on SOC estimation (initial SOC 60%, 480 s)

Battery 2	SOC estimation error (%)			Available capacity estimation error (A h)			
	Maximum error	Average error	Standard deviation	Maximum error	Average error	Standard deviation	After convergence
Cell 01	0.753	0.513	0.191	2.431	0.334	0.367	<0.5
Cell 02	0.867	0.491	0.298	2.598	0.458	0.458	<0.6
Cell 03	0.919	0.332	0.325	3.500	0.246	0.301	<0.4

Table 4.8 shows that the SOC estimation error is less than 1%, and the available capacity estimation error after convergence is less than 0.6 A h. Considering that the rated capacity of the battery is 32A h, the estimated error of available capacity is less than 2% and less than the convergence coefficient.

4.2.3 *Available Capacity Estimation Method Based on Response Surface*

(1) Problem description

For the case that the OCV-SOC relationship changes notably with the aging process, it is difficult to correct the SOC accurately with the incorrect OCV-SOC relationship caused by the erroneous capacity. At this time, the capacity-OCV-SOC response surface can be established [26], and the OCV-SOC relationship that conforms to the current capacity could be searched for by collecting data segments. There are three main challenges in this method, as shown in Fig. 4.18, including

I: Higher requirements for the accuracy of current sensors.
II: The length of used data segments is difficult to determine. The excessively long data segments will make the calculation amount of the search algorithm increase rapidly. On the contrary, the overly short data segments will make the search algorithm easy to fall into a local optimum.
III: The search process of curves has heavy computational burden and takes a long time.

The key to the challenge III lies in the selection of an optimization algorithm, which takes the minimum terminal voltage error as the objective function. It is easy to find that the more obvious the change of OCV-SOC curve with aging, the easier it is to find the global optimum. When the OCV-SOC relations do not change significantly with the battery aging process, all capacities are optimal solutions. In other words, the capacity estimation can not be completed.

(2) Calculation flow of the algorithm

First, a curve cluster $U_{oc}(z, C_{max,i})$ can be obtained from OCV-SOC curves with different capacities:

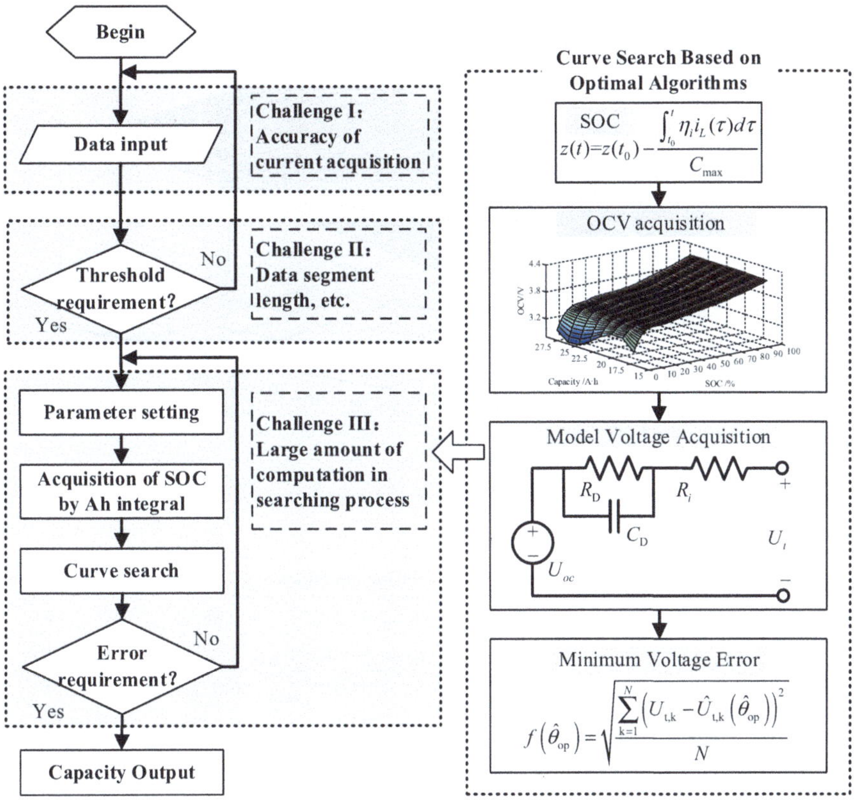

Fig. 4.18 Problem description of battery available capacity estimation method based on response surface

$$U_{oc}\left(z, C_{\max,i}\right) = \alpha_{0,i} + \alpha_{1,i} z + \alpha_{2,i} / z + \alpha_{3,i} \ln z + \alpha_{4,i} \ln(1 - z) \tag{4.41}$$

where $C_{\max,i}$ denotes the available capacity of the battery at the ith capacity point; $\alpha_{j,i}$ ($j = 0, 1, \ldots, 4$) represents the fitting coefficients of OCV-SOC curves at the ith capacity point, and $\alpha_{j,i}$ indicates functions of $C_{\max,i}$. Compared to Eqs. (4.10), (4.41) is a simplified OCV-SOC fitting equation.

Subsequently to that, the response surface of capacity-OCV-SOC can be derived from two-dimensional interpolation of the coefficients. Figure 4.19 presents a response surface constructed from OCV-SOC curves of battery 1 cell 3 at different capacity points.

The relationship between the fitting coefficients and available capacity is shown in Table 4.9, so that the unique OCV-SOC curve can be determined by available capacity.

In the model parameter identification process, the change in model parameters is reflected by the difference of terminal voltage. Using optimization methods (such as GA) to identify model parameters, different combinations of model parameters

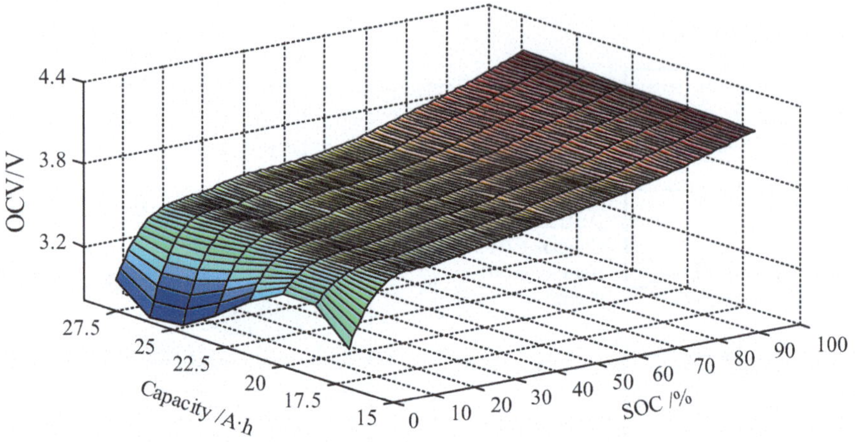

Fig. 4.19 Capacity-OCV-SOC response surface

Table 4.9 OCV fitting coefficients at different capacities

Coefficients	Fitting equations
α_0	$\alpha_0 = 0.0212C_{\max}^2 - 0.8828C_{\max} + 11.8447$
α_1	$\alpha_1 = -0.0231C_{\max}^2 + 0.9595C_{\max} - 8.6399$
α_2	$\alpha_2 = 0.0019C_{\max}^2 - 0.0773C_{\max} + 0.7259$
α_3	$\alpha_3 = 0.0151C_{\max}^2 - 0.6285C_{\max} + 5.9196$
α_4	$\alpha_4 = 0.000487C_{\max}^2 - 0.0224C_{\max} + 0.1579$

correspond to different terminal voltage output results. The set of parameters that minimize the root mean square (RMS) error of terminal voltage is considered to be the optimal model parameter identification results. Taking the Thevenin model as an example, the terminal voltage is expressed as Eq. (4.42).

$$U_t = U_{oc}(z, C_{\max}) - U_D - i_L R_i \tag{4.42}$$

In Eq. (4.42), the capacity-OCV-SOC response surface has been introduced into the battery model. Thus, the difference between different available capacity values is reflected in the OCV-SOC curve, and will eventually be reflected in the terminal voltage output values. Therefore, the available capacity of the battery needs to be identified along with other model parameters.

It is noted that $U_{oc}(z, C_{\max})$ in Eq. (4.42) is also related to the battery SOC. The available capacity can determine the OCV-SOC relationship, but the initial SOC remains unknown. The OCV is only available when both are known. Therefore, the initial value of SOC ought to be included in the parameters vector as well to be identified. The SOC at each sampling time in the working condition can be obtained by the ampere-hour integral method.

The estimation of available capacity can be transformed into searching for the best OCV-SOC relationship from the capacity-OCV-SOC response surface. The parameters to be identified for offline estimation of the battery available capacity can be expressed as follows:

$$\hat{\theta}_{\text{op}} = \left[z_{\text{ini}} \ R_i \ R_{\text{D}} \ C_{\text{D}} \ C_{\max} \right] \tag{4.43}$$

Taking the RMS error of the terminal voltage as the fitness function shown in Eq. (4.44), the appropriate optimization algorithm is selected to identify the parameter vector and obtain the battery capacity.

$$\begin{cases} \min\left\{ f\left(\hat{\theta}_{\text{op}}\right)\right\} \\ f\left(\hat{\theta}_{\text{op}}\right) = \sqrt{\dfrac{\sum_{k=1}^{N}\left(U_{\text{t,k}} - \hat{U}_{\text{t,k}}\left(\hat{\theta}_{\text{op}}\right)\right)^2}{N}} \end{cases} \tag{4.44}$$

(3) Case study

The capacity-OCV-SOC response surface is obtained by using the cycling aging test of battery1 cell 3. The accuracy of available capacity estimation based on response surface method is evaluated by the DST profile of battery1 cell 3 under 25 °C. The available capacity of this fresh battery is 27.31 A h. The terminal voltage and current data in the range of 3,000–13,800 s are selected so as to identify model parameters. The identification results are shown in Table 4.10.

At the 3,000 s, the real SOC of the battery is 82.7%, the identification result is 84.3%, and the estimation error is 1.6%. The capacity estimate differs from the real value of available capacity by 0.18 A h, and the relative error is 0.66%. It is revealed that the identified initial SOC and available capacity can well estimate the real state of the battery.

In addition, under the temperature of 25 °C, the DST data at different capacity points are selected to evaluate the estimation method during the aging process. The estimated results of available capacity, initial SOC and the estimation errors are presented in Tables 4.11 and 4.12. Figure 4.20 shows the relative error of the estimated results in a statistical histogram.

It can be seen from Fig. 4.20 that the relative errors of available capacity and initial SOC estimation can be guaranteed within 4% at five capacity points. Thus, the available capacity can be effectively identified using the optimization method based on response surface.

Table 4.10 Parameter identification results

Identification results	Model parameters				
	z_{ini} (%)	R_{o} (mΩ)	R_{DI} (mΩ)	C_{DI} (F)	$C_{\max}$ (A h)
	84.3	2.1	2.3	22800	27.13

Table 4.11 The estimated available capacity and error with DST loading profiles at different capacity points

Capacity points	Real value (A h)	Estimate value (A h)	Absolute error (A h)
1	27.31	27.13	0.18
2	25.83	25.84	0.01
3	25.28	24.45	0.83
4	24.86	24.13	0.73
5	23.82	23.19	0.62

Table 4.12 Initial SOC estimation and error under DST loading profiles at different capacity points

Capacity points	Real value (%)	Estimate value (%)	Absolute error (%)
1	82.7	84.3	1.6
2	81.6	84.6	3.0
3	81.3	83.3	2.0
4	80.8	83.4	2.6
5	80.2	83.4	3.1

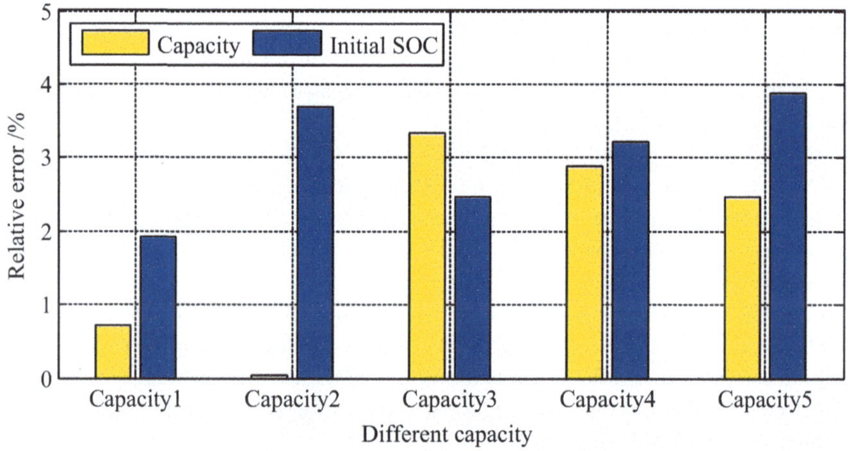

Fig. 4.20 Relative errors of available capacity and initial SOC estimation at different capacity points

4.2.4 SOH Estimation Method Based on ICA/DVA

(1) Incremental capacity analysis

As a nondestructive method, the incremental capacity analysis (ICA) is conducted to study the degradation process and aging mechanism by analyzing the subtle changes in IC curves during the battery aging process. It has become an effective method to study the aging mechanism of batteries and estimate the SOH in the offline environment [27].

Compared with the original curve before the transformation, the most obvious characteristic of IC curve lies in IC peak. Each IC peak represents an electrochemical process occurring inside the battery, which has its unique shape, height, and position. Therefore, any change to the position and shape of the IC peaks represents the battery aging. Taking battery 4 as an example, the relationships among the IC curve, electrochemical reaction and aging process in the battery are analyzed. Figure 4.21 shows the voltage-capacity curves and corresponding IC curves derived from charging–discharging tests at C/20 and C/3 under 25 °C. The IC peaks in the IC curve correspond to the phase transformation processes in the electrode of the battery. In the graphite negative electrode (NE), the lithium intercalation process converts C to LiC_6, which needs a minimum of five different phase transformation processes, recorded as ⑤–①. In the positive electrode (PE), two-phase transformation processes of $LiMn_2O_4$ material at 4.1 and 4.0 V are involved and recorded as I and II. The corresponding phase transformation processes of positive and negative materials are shown in Table 4.13.

Due to the different phase transformation processes in NE, the IC peaks corresponding to the phase transformation II in PE are divided into four peaks, as shown in Fig. 4.21b (solid line). Similarly, IC peaks corresponding to the phase transformation

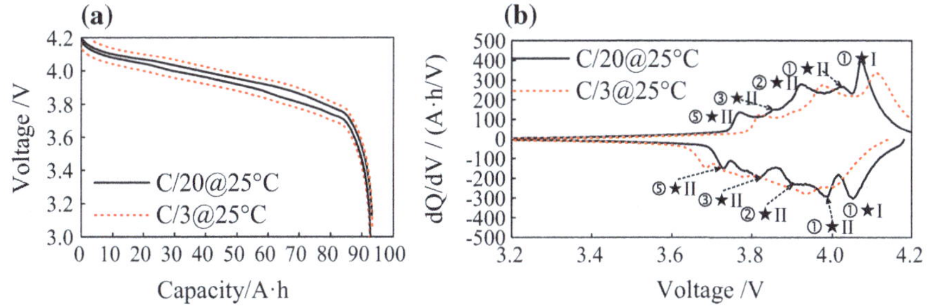

Fig. 4.21 **a** Voltage-capacity curve; **b** Charge–discharge IC curves

Table 4.13 Phase structure changes in positive and negative materials

Type	Number	Phase transformation processes
PE	I	$Li_{0.5}Mn_2O_4 \leftrightarrow \lambda Mn_2O_4$
	II	$LiMn_2O_4 \leftrightarrow Li_{0.5}Mn_2O_4$
NE	①	$LiC_{12} \leftrightarrow LiC_6$
	②	$LiC_{18} \leftrightarrow LiC_{12}$
	③	$LiC_{36} \leftrightarrow LiC_2 \leftrightarrow LiC_{18}$
	④	$LiC_{72} \leftrightarrow LiC_{36}$
	⑤	$C_6 \leftrightarrow LiC_{72}$

processes in NE can be divided into different peaks by different phase transformation processes in PE. For the sake of simplicity, each IC peak is numbered. The IC peak numbered as ①★I denotes the phase transformation ① in the NE convoluting with the phase transformation I in the PE. It can be discovered that the IC curve has higher sensitivity than the traditional charge–discharge curve, which ensures that any behavioral changes in the aging process of batteries can be monitored.

In order to track SOH and ensure a more accurate understanding of the aging mechanism, IC curves are always driven by very low current, such as C/25. Nevertheless, in practical applications, it is difficult to complete the relatively complete charging and discharging process with a such small current rate. Therefore, IC curves based on a larger current (C/3) have been gradually used to analyze the behavior of batteries. Figure 4.21b shows that there are only three distinct IC peaks in the IC curve (dotted line) of C/3 charging process, noted as ①★I, ②★II, and ⑤★II. Therefore, the IC curve derived from a larger current can also provide aging information for batteries. The area under each peak represents the capacity change during the associated reaction process.

Figure 4.22 presents the comparison of IC curves of batteries cycled under different temperatures. The aging temperatures include 40 and 10 °C, and the DOD is 50%. The IC curves are derived from the charging curves of capacity experiments (25 °C and C/3 rate) under the corresponding aging cycles. IC peaks of the battery show the relationship between the voltage response and the battery aging behavior. It is found that with the increase in the cycle number, IC peaks usually move regularly to the lower right. Thus, the SOH of the battery can be indicated by the change of IC peaks. Furthermore, the SOH value of the battery can also be derived from the relationship between the cycle number and IC peaks.

(2) Differential voltage analysis

The differential voltage analysis (DVA) method can also be used to analyze the degradation process and aging mechanism of batteries including the loss of active material in both electrodes, loss of lithium ions and so on. The characteristic of the DVA method is that there are always clear and identifiable peaks shown on the DV curve. Figure 4.23 shows the voltage-capacity curve and the corresponding DV curve of the battery at 25 °C and C/3 rate.

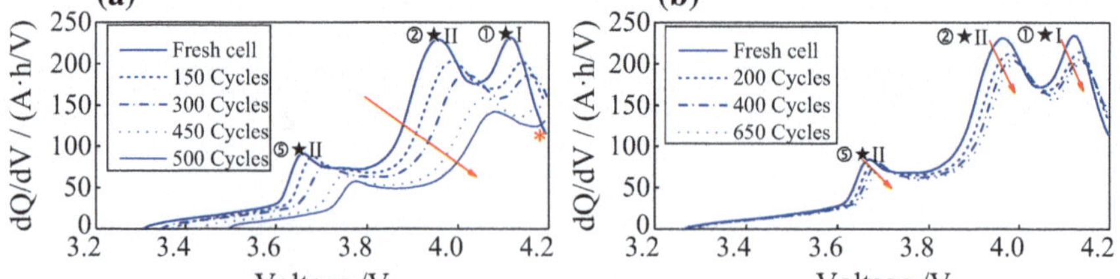

Fig. 4.22 Evolution of IC curves indifferent thermal cycling: **a** 40 °C, 50% DOD; **b** 10 °C, 50% DOD

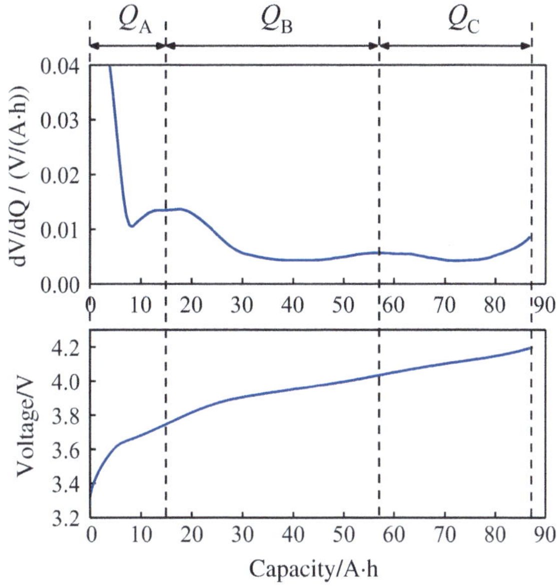

Fig. 4.23 Voltage-capacity curves and corresponding DV curves of batteries at 25 °C and C/3 rate

In the DV curve, the IC peaks describing the phase transition phenomena become valleys while the DV peaks represent non-stoichiometry in the single-phase regions (solid solution). The DV curve is split into three regions by the DV peak, each region corresponds to the capacity of phase transformation involved in the reaction. The charging capacity is divided into Q_A, Q_B, and Q_C according to the above three regions, which are associated with the areas under the peaks ⑤★II, ②★II, and ①★I in the IC curve, respectively. Similarly, the aging mechanism of batteries can be analyzed by comparing the characteristics of DV curves. The comparison of DV curves for the cells cycled at 40, 10 °C and 50% DOD at different cycle number are plotted in Fig. 4.24.

The DVA method is easier to quantify the capacity loss corresponding to the three peaks in IC curve [28]. Figure 4.25 shows the evolution of Q_A, Q_B, and Q_C

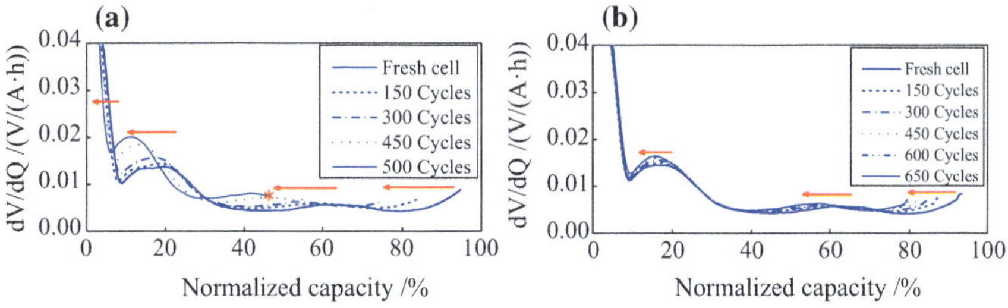

Fig. 4.24 Evolution of DV curves in different thermal cycling: **a** 40 °C, 50% DOD; **b** 10 °C, 50% DOD

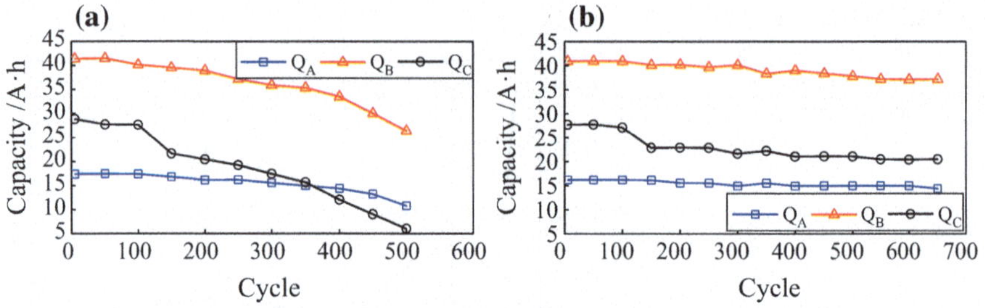

Fig. 4.25 Evolution of Q_A, Q_B, and Q_C in different thermal cycling: **a** 40 °C, 50% DOD; **b** 10 °C, 50% DOD

with the cycle number at different temperatures. The aging temperatures include 40 and 10 °C, and the DOD is 50%.

The results indicate that Q_A, Q_B, and Q_C decrease significantly with the increase in cycle number. The decline rates of Q_A and Q_B are almost the same, suggesting that the capacity loss of Q_A and Q_B is caused by the loss of active material in NE. Q_C decreases at a higher rate, indicating that in addition to the loss of active material, lithium ions are also lost. Obviously, the mapping relationship between Q_B and battery aging is suitable for SOH estimation.

4.3 Multi-timescale Co-Estimation of SOC and SOH

The coupling relationship between SOC and SOH makes the estimation algorithms of SOC and SOH interdependent. For accurate SOC estimation in the complex and variable driving conditions, it is necessary to build a battery model reflecting the correlation between SOC and SOH, and design an algorithm for the co-estimation of SOC and SOH. In the co-estimation framework of battery SOC and SOH, the capacity value used in SOC estimation is dynamic. As compared to the static capacity used in Sect. 4.1, the dynamic capacity could be adaptively adjusted with the change of battery operating conditions, as a result of which a better estimation performance can be achieved.

4.3.1 Problem Description

The model parameters, including OCV, impedance parameters and capacity, which directly reflect the battery SOH, can be directly regarded as the SOH mapping parameter value. At this time, the co-estimation of SOC and SOH can be transformed into the co-estimation of state (SOC) and parameters (SOH). To solve this problem, two estimators with mutual input, namely co-estimation algorithm, can be applied to estimate the state and parameters of the battery separately. Two estimators share the same innovation to ensure the stability and convergence of the algorithm.

The battery parameters (capacity, internal resistance, etc.) have slow time-varying characteristics, the states of the battery (SOC, SOP, etc.) have fast time-varying characteristics [29]. If all the parameters and states are estimated at the same timescale, it will require a huge calculation cost of BMS. Also, too frequent parameter update will make the parameters fluctuate sharply and even affect the accuracy of state estimation. Therefore, the parameters of the battery can be estimated on a macro-timescale, including available capacity, polarization resistance, and ohmic resistance. The state of the battery can be estimated in micro-timescale. A multi-timescale-based method for battery parameters and state estimation can be established.

Here, a multi-timescale system with implicit states $\boldsymbol{\chi}$ and parameters $\boldsymbol{\theta}$ is described by the nonlinear discrete system shown in Eq. (4.45):

$$\begin{cases} \boldsymbol{\chi}_{k,l+1} = \boldsymbol{F}(\boldsymbol{\chi}_{k,l}, \boldsymbol{\theta}_k, \boldsymbol{u}_{k,l}) + \boldsymbol{\omega}_{k,l}, \boldsymbol{\theta}_{k+1} = \boldsymbol{\theta}_k + \boldsymbol{\rho}_k \\ \boldsymbol{Y}_{k,l} = \mathbf{G}(\boldsymbol{\chi}_{k,l}, \boldsymbol{\theta}_k, \boldsymbol{u}_{k,l}) + \boldsymbol{v}_{k,l} \end{cases} \tag{4.45}$$

where $\boldsymbol{\chi}_{k,l}$ is the state of the system at time $t_{k,l} = t_{k,0} + l \times \Delta t$ $(1 \leq l \leq L_z)$. Two timescales k and l describe the macro- and micro-timescales, respectively. L_z represents the scale conversion limit, that is, a macro-timescale is equal to L_z micro-timescale. $\boldsymbol{u}_{k,l}$ is the input information of the system (input matrix) at time $t_{k,l}$. $\boldsymbol{Y}_{k,l}$ indicates the observation matrix of the system (measurement matrix) at time $t_{k,l}$.$\boldsymbol{\omega}_{k,l}$ and $\boldsymbol{\rho}_k$ represent the white noise of system state and parameter, the covariance of which are $\boldsymbol{Q}^{\chi}_{k,l}$ and $\boldsymbol{Q}^{\theta}_{k}$. $\boldsymbol{v}^{l}_{k}$ denotes the measurement white noise, the covariance of which is $\boldsymbol{R}_{k,l}$. The state of the system is estimated by micro-timescale, and the parameters of the system are estimated by macro-timescale. For the system parameters, the macroscale values remain unchanged from 0 to L_z−1, i.e., $\boldsymbol{\theta}_k = \boldsymbol{\theta}_{k,0:L_z-1}$.

Based on the established multi-timescale system, two estimators are used to estimate the system state and parameters step by step. One estimator is responsible for model parameter estimation and the other is responsible for SOC estimation. The two estimators work alternately for mutual input, establishing the basic framework of SOC-SOH co-estimation for batteries in multi-timescale. Figure 4.26 presents the schematic diagram of the multi-timescale dual filtering algorithms, including parameter prior estimation (parameter time update), state prior estimation (state time update), state posterior estimation (state measurement update), and parameter posterior estimation (parameter measurement update). The AEKF algorithm and HIF algorithm introduced in Sect. 4.2 are applied to this framework, forming the multi-timescale AEKF (MAEKF) algorithm [30] and multi-timescale HIF (MHIF) algorithm [31], respectively.

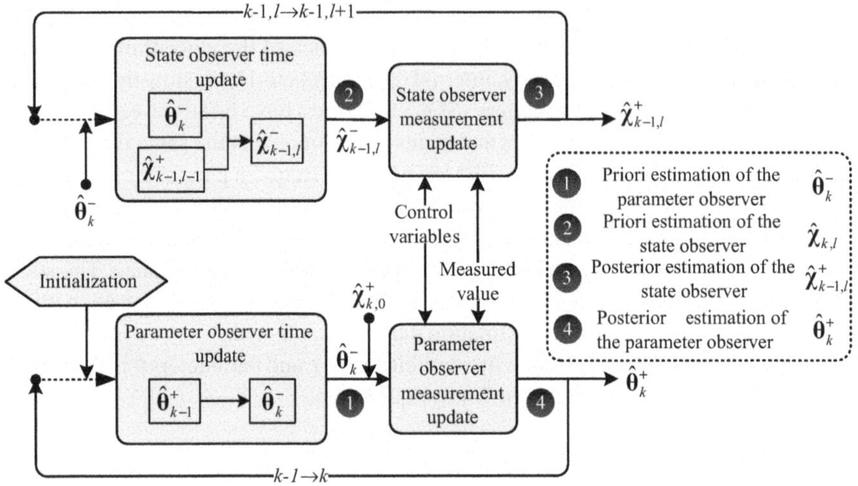

Fig. 4.26 Schematic diagram of multi-timescale filtering algorithms

4.3.2 Co-Estimation Method Based on MAEKF

(1) Calculation flow of the algorithm

Based on the flowchart of the multi-timescale filtering algorithm described in Fig. 4.26, $AEKF_\chi$ is used as the state observer with micro-timescale, $AEKF_\theta$ is used as the parameter observer with macro-timescale, and two-layer AEKF algorithm is used to perform the multi-timescale estimation of system parameters and states. The detailed calculation steps are summarized as follows:

Initialization: set the initial parameters of the parameter observer $AEKF_\theta$ and the state observer $AEKF_\chi$, respectively.

$$\boldsymbol{\theta}_0,\ \boldsymbol{P}_0^{\theta},\ \boldsymbol{Q}_0^{\theta},\ \boldsymbol{R}_0,\ \boldsymbol{\chi}_{0,0},\ \boldsymbol{P}_{0,0}^{\chi},\ \boldsymbol{Q}_{0,0}^{\chi},\ \boldsymbol{R}_{0,0} \tag{4.46}$$

where $\boldsymbol{\theta}_0$ represents the initial parameter of $AEKF_\theta$, $\boldsymbol{P}_0^{\theta}$ indicates the initial value of the parameter estimation error covariance matrix, and $\boldsymbol{Q}_0^{\theta}$ denotes the initial value of the system noise covariance matrix. $\boldsymbol{\chi}_{0,0}$ indicates the initial system state. $\boldsymbol{P}_{0,0}^{\chi}$ is defined as the initial state estimation error covariance matrix. $\boldsymbol{Q}_{0,0}^{\chi}$ refers to the initial system noise covariance matrix for $AEKF_\chi$. $\boldsymbol{R}_0$ and $\boldsymbol{R}_{0,0}$ represent the covariance of the measurement noises, and satisfy $\boldsymbol{R}_k = \boldsymbol{R}_{k,0:Lz-1}$, for which $\boldsymbol{R}_0 = \boldsymbol{R}_{0,0}$. When the estimation begins, values of time (0) are converted to that of time (k–1), and values of time (0, 0) are converted to that of time (k–1, l–1).

For macro-timescale series, $k = 1, 2, \ldots$ (from time $(k-1)^+$ to time $(k)^-$).

Step 1: Time update of parameter observer $AEKF_\theta$ based on macro-timescale (prior estimation)

$$\hat{\boldsymbol{\theta}}_k^- = \hat{\boldsymbol{\theta}}_{k-1}, \boldsymbol{P}_k^{\theta,-} = \boldsymbol{P}_{k-1}^{\theta} + \boldsymbol{Q}_{k-1}^{\theta} \tag{4.47}$$

For micro-timescale series, (from time $(k–1, l–1)^+$ to time $(k–1, L_z)^+$).

Step 2: Time update of state observer AEKF_χ based on macro-timescale (prior estimation)

$$\hat{\boldsymbol{\chi}}_{k-1,l}^- = \boldsymbol{F}(\hat{\boldsymbol{\chi}}_{k-1,l-1}, \hat{\boldsymbol{\theta}}_k^-, \boldsymbol{u}_{k-1,l-1}), \boldsymbol{P}_{k-1,l}^{\chi,-} = \boldsymbol{A}_{k-1,l-1}\boldsymbol{P}_{k-1,l-1}^{\chi}\boldsymbol{A}_{k-1,l-1}^{\mathrm{T}} + \boldsymbol{Q}_{k-1,l-1}^{\chi} \tag{4.48}$$

Step 3: Measurement update (posterior estimation) of state observer AEKF_χ based on micro-timescale

$$\text{Innovation matrix}: \boldsymbol{e}_{k-1,l} = \boldsymbol{Y}_{k-1,l} - \boldsymbol{G}(\hat{\boldsymbol{\chi}}_{k-1,l}^-, \hat{\boldsymbol{\theta}}_k^-, \boldsymbol{u}_{k-1,l}) \tag{4.49}$$

$$\text{Kalman gain matrix}: \boldsymbol{K}_{k-1,l}^{\chi} = \boldsymbol{P}_{k-1,l}^{\chi,-}(\boldsymbol{C}_{k-1,l}^{\chi})^{\mathrm{T}}\left(\boldsymbol{C}_{k-1,l}^{\chi}\boldsymbol{P}_{k-1,l}^{\chi,-}(\boldsymbol{C}_{k-1,l}^{\chi})^{\mathrm{T}} + \boldsymbol{R}_{k-1,l-1}\right)^{-1} \tag{4.50}$$

Adaptive covariance matching (estimation error window function of voltage):

$$\boldsymbol{H}_{k-1,l}^{\chi} = \frac{1}{M_\chi}\sum_{i=l-M_\chi+1}^{l} \boldsymbol{e}_{k-1,i}\boldsymbol{e}_{k-1,i}^{\mathrm{T}} \tag{4.51}$$

Noise covariance update:

$$\boldsymbol{R}_{k-1,l} = \boldsymbol{H}_{k-1,l}^{\chi} - \boldsymbol{C}_{k-1,l}^{\chi}\boldsymbol{P}_{k-1,l}^{\chi,-}(\boldsymbol{C}_{k-1,l}^{\chi})^{\mathrm{T}}, \boldsymbol{Q}_{k-1,l}^{\chi} = \boldsymbol{K}_{k-1,l}^{\chi}\boldsymbol{H}_{k-1,l}^{\chi}(\boldsymbol{K}_{k-1,l}^{\chi})^{\mathrm{T}} \tag{4.52}$$

State estimate measurement update:

$$\hat{\boldsymbol{\chi}}_{k-1,l}^+ = \hat{\boldsymbol{\chi}}_{k-1,l}^- + \boldsymbol{K}_{k-1,l}^{\chi}\left[\boldsymbol{Y}_{k-1,l} - \boldsymbol{G}(\hat{\boldsymbol{\chi}}_{k-1,l}^-, \hat{\boldsymbol{\theta}}_k^-, \boldsymbol{u}_{k-1,l})\right] \tag{4.53}$$

Error covariance update of state estimation:

$$\boldsymbol{P}_{k-1,l}^{\chi,+} = (\boldsymbol{I} - \boldsymbol{K}_{k-1,l}^{\chi}\boldsymbol{C}_{k-1,l}^{\chi})\boldsymbol{P}_{k-1,l}^{\chi,-} \tag{4.54}$$

For time series calculation (For $l = 1{:}\ L_z$) and timescale transform (when $l = L_z$):

$$\hat{\boldsymbol{\chi}}_{k,0}^+ = \hat{\boldsymbol{\chi}}_{k-1,L_z}^+, \boldsymbol{P}_{k,0}^{\chi,+} = \boldsymbol{P}_{k-1,L_z}^{\chi,+}, \boldsymbol{Y}_{k,0} = \boldsymbol{Y}_{k-1,L_z}, \boldsymbol{u}_{k,0} = \boldsymbol{u}_{k-1,L_z} \tag{4.55}$$

So far, the micro-timescale cycle calculation under a macro-timescale is completed, and the next step is to return to the macro-timescale to update the measurement of parameter estimation (posterior estimation).

Step 4: Measurement update (posterior estimation) of state observer AEKF$_\theta$ based on macro-timescale

Innovation matrix innovation of parameter estimation:

$$\boldsymbol{e}_k^{\theta} = \boldsymbol{Y}_{k,0} - \boldsymbol{G}\left(\hat{\boldsymbol{\chi}}_{k,0}^{+}, \hat{\boldsymbol{\theta}}_k^{-}, \boldsymbol{u}_{k,0}\right) \tag{4.56}$$

Kalman gain matrix:

$$\boldsymbol{K}_k^{\theta} = \boldsymbol{P}_k^{\theta,-}\left(\boldsymbol{C}_k^{\theta}\right)^{\mathrm{T}}\left(\boldsymbol{C}_k^{\theta}\boldsymbol{P}_k^{\theta,-}\left(\boldsymbol{C}_k^{\theta}\right)^{\mathrm{T}} + \boldsymbol{R}_{k-1}\right)^{-1} \tag{4.57}$$

Adaptive covariance matching (estimation error window function of voltage):

$$\boldsymbol{H}_k^{\theta} = \frac{1}{M_\theta}\sum_{i=k-M_\theta+1}^{k} \boldsymbol{e}_i^{\theta}\left(\boldsymbol{e}_i^{\theta}\right)^{\mathrm{T}} \tag{4.58}$$

Noise covariance update:

$$\boldsymbol{R}_k = \boldsymbol{H}_k^{\theta} - \boldsymbol{C}_k^{\theta}\boldsymbol{P}_k^{\theta,-}\left(\boldsymbol{C}_k^{\theta}\right)^{\mathrm{T}}, \; \boldsymbol{Q}_k^{\theta} = \boldsymbol{K}_k^{\theta}\boldsymbol{H}_k^{\theta}\left(\boldsymbol{K}_k^{\theta}\right)^{\mathrm{T}} \tag{4.59}$$

Parameter estimate update:

$$\hat{\boldsymbol{\theta}}_k^{+} = \hat{\boldsymbol{\theta}}_k^{-} + \boldsymbol{K}_k^{\theta}\mathbf{e}_k^{\theta} \tag{4.60}$$

Error covariance update of parameter estimation:

$$\boldsymbol{P}_k^{\theta,+} = \left(\boldsymbol{I} - \boldsymbol{K}_k^{\theta}\boldsymbol{C}_k^{\theta}\right)\boldsymbol{P}_k^{\theta,-} \tag{4.61}$$

where

$$\boldsymbol{A}_{k-1,l-1} = \frac{\partial \boldsymbol{F}(\boldsymbol{\chi}, \hat{\boldsymbol{\theta}}_k^{-}, \boldsymbol{u}_{k-1,l-1})}{\partial \boldsymbol{\chi}}\bigg|_{\boldsymbol{\chi}=\hat{\boldsymbol{\chi}}_{k-1,l-1}}, \; \boldsymbol{C}_{k-1,l}^{\chi} = \frac{\partial \boldsymbol{G}(\boldsymbol{\chi}, \hat{\boldsymbol{\theta}}_k^{-}, \boldsymbol{u}_{k-1,l})}{\partial \boldsymbol{\chi}}\bigg|_{\boldsymbol{\chi}=\hat{\boldsymbol{\chi}}_{k-1,l}^{-}},$$

$$\boldsymbol{C}_k^{\theta} = \frac{\partial \boldsymbol{G}(\hat{\boldsymbol{\chi}}_{k,0}, \boldsymbol{\theta}, \boldsymbol{u}_{k,0})}{\partial \boldsymbol{\theta}}\bigg|_{\boldsymbol{\theta}=\hat{\boldsymbol{\theta}}_k^{-}} \tag{4.62}$$

The calculation of $\boldsymbol{C}_k^{\theta}$ is as follows:

$$\boldsymbol{C}_k^{\theta} = \frac{\partial \boldsymbol{G}(\hat{\boldsymbol{\chi}}_{k,0}, \boldsymbol{\theta}, \boldsymbol{u}_{k,0})}{\partial \boldsymbol{\theta}}\bigg|_{\boldsymbol{\theta}=\hat{\boldsymbol{\theta}}_k^{-}} = \frac{\partial \boldsymbol{G}(\hat{\boldsymbol{\chi}}_{k,0}, \hat{\boldsymbol{\theta}}_k^{-}, \boldsymbol{u}_{k,0})}{\partial \hat{\boldsymbol{\theta}}_k^{-}} + \frac{\partial \boldsymbol{G}(\hat{\boldsymbol{\chi}}_{k,0}, \hat{\boldsymbol{\theta}}_k^{-}, \boldsymbol{u}_{k,0})}{\partial \hat{\boldsymbol{\chi}}_{k,0}}\frac{d\hat{\boldsymbol{\chi}}_{k,0}}{d\hat{\boldsymbol{\theta}}_k^{-}} \tag{4.63}$$

Combining $\hat{\chi}_{k,0} = \hat{\chi}^{+}_{k-1,L_z}$ and Eq. (4.53) gives:

$$\frac{d\hat{\chi}_{k,0}}{d\hat{\theta}_k^-} = \frac{d\hat{\chi}^{+}_{k-1,L_z}}{d\hat{\theta}_k^-} = \frac{d}{d\hat{\theta}_k^-}\left(\hat{\chi}^{-}_{k-1,L_z-1}+K^{\chi}_{k-1,L_z-1}\left(Y_{k-1,L_z-1} - G\left(\hat{\chi}^{-}_{k-1,L_z-1}, \hat{\theta}_k^-, u_{k-1,L_z-1}\right)\right)\right) \tag{4.64}$$

$$\frac{d}{d\hat{\theta}_k^-}\left(K^{\chi}_{k-1,L_z-1}Y_{k-1,L_z-1}\right)=Y_{k-1,L_z-1}\frac{\partial K^{\chi}_{k-1,L_z-1}}{\partial\hat{\theta}_k^-} \tag{4.65}$$

$$\begin{aligned}&\frac{d}{d\hat{\theta}_k^-}\left(K^{\chi}_{k-1,L_z-1}G\left(\hat{\chi}^{-}_{k-1,L_z-1}, \hat{\theta}_k^-, u_{k-1,L_z-1}\right)\right)\\ &= K^{\chi}_{k-1,L_z-1}\frac{dG\left(\hat{\chi}^{-}_{k-1,L_z-1}, \hat{\theta}_k^-, u_{k-1,L_z-1}\right)}{d\hat{\theta}_k^-} + \frac{\partial K^{\chi}_{k-1,L_z-1}}{\partial\hat{\theta}_k^-}G\left(\hat{\chi}^{-}_{k-1,L_z-1}, \hat{\theta}_k^-, u_{k-1,L_z-1}\right)\end{aligned} \tag{4.66}$$

So far, the multi-timescale estimation of parameters and states at the kth sampling time is completed. The state estimation at time $(k + 1)$ is prepared. Set $\hat{\chi}_{k,0} = \hat{\chi}^{+}_{k,0}$, $\hat{\theta}_k = \hat{\theta}_k^{+}$.

Based on the calculation flow of the MAEKF algorithm mentioned above, in view of the slow time-varying characteristics shown by the battery parameters and the fast time-varying characteristics of the battery state, the micro-timescale and macro-timescale are used to estimate the battery parameters and state, respectively. The state equation is transformed into a nonlinear discrete system as shown in Eq. (4.67):

$$\begin{bmatrix} U^{\mathrm{D}}_{k,l} \\ z_{k,l} \end{bmatrix} = \begin{bmatrix} \exp\left(-\frac{\Delta t}{R_{\mathrm{D}}C_{\mathrm{D}}}\right) & 0 \\ 0 & 1 \end{bmatrix}\begin{bmatrix} U^{\mathrm{D}}_{k,l-1} \\ z_{k,l-1} \end{bmatrix}+\begin{bmatrix} \left(1-\exp\left(-\frac{\Delta t}{R_{\mathrm{D}}C_{\mathrm{D}}}\right)\right)R_{\mathrm{D}} \\ -\frac{\eta_i\Delta t}{C_{\max}} \end{bmatrix}i_{k,l} \tag{4.67}$$

where $U^{\mathrm{D}}_{k,l}$ and $z_{k,l}$ represent the polarization voltage and SOC of the battery at $t_{k,1}$. $i_{k,l}$ is defined as the current of the battery at $t_{k,1}$.

Then the system equation of the battery can be described as

$$\begin{cases} \chi_{k,l+1} = F(\chi_{k,l}, \theta_k, u_{k,l}) = \begin{bmatrix} \exp\left(-\frac{\Delta t}{R_{\mathrm{D}}C_{\mathrm{D}}}\right) & 0 \\ 0 & 1 \end{bmatrix}\chi_{k,l} + \begin{bmatrix} \left(1-\exp\left(-\frac{\Delta t}{R_{\mathrm{D}}C_{\mathrm{D}}}\right)\right)R_{\mathrm{D}} \\ -\frac{\eta_i\Delta t}{C_{\max}} \end{bmatrix}u_{k,l+1} \\ Y_{k,l} = \mathbf{G}(\chi_{k,l}, \theta_k, u_{k,l}) = g(z_{k,l}, C_{\max}) - U^{\mathrm{D}}_{k,l} - R_{\mathrm{o}}u_{k,l} \end{cases} \tag{4.68}$$

where the state matrix, parameter matrix, input matrix, and output matrix of the system are as follows:

$$\begin{cases} \chi_{k,l} = \left[\, U^{\mathrm{D}}_{k,l} \; z_{k,l} \,\right]^{\mathrm{T}} \\ \theta_k = [\, R_{\mathrm{o}} \; R_{\mathrm{D}} \; C_{\mathrm{D}} \; C_{\max} \,]^{\mathrm{T}}_k \\ u_{k,l}=i_{k,l} \\ Y_{k,l} = U^{\mathrm{t}}_{k,l} \end{cases} \tag{4.69}$$

where $U_{k,l}^{\mathrm{t}}$ denotes the terminal voltage of the battery at $t_{\mathrm{k,l}}$.

Based on the calculation flow of the algorithm, the implementation flowchart of the multi-timescale estimation algorithm for battery parameters and states is established. The state matrix and observation matrix of the system are

$$\boldsymbol{A}_{k-1,l-1} = \begin{bmatrix} \exp\left(-\frac{\Delta t}{R_{\mathrm{D}}C_{\mathrm{D}}}\right) & 0 \\ 0 & 1 \end{bmatrix} \tag{4.70}$$

$$\boldsymbol{C}_{k-1,l}^{\chi} = \begin{bmatrix} -1 & \frac{\partial g_{k-1,l}(z, C_{\max})}{\partial z_{k-1,l}^{-}} \end{bmatrix} \tag{4.71}$$

$$C_k^{\theta} = \begin{bmatrix} -i_{k,0} & 0 & 0 & \frac{\partial g(z, C_{\max})}{\partial z_{k,0}} \frac{\partial z_{k,0}}{\partial \hat{C}_{max,k}^{-}} \end{bmatrix} + C_{k,0}^{\widehat{\chi}} \left(\frac{\partial F\left(\widehat{\chi}_{k-1,L_z-1}, \widehat{\theta}_k^{-}, u_{k-1,L_z-1}\right)}{\partial \widehat{\theta}_k^{-}} + \frac{\partial F\left(\widehat{\chi}_{k-1,L_z-1}, \widehat{\theta}_k^{-}, u_{k-1,L_z-1}\right)}{\partial \widehat{\chi}_{k-1,L_z-1}} \frac{d\widehat{\chi}_{k-1,L_z-1}}{d\widehat{\theta}_k^{-}} \right) \tag{4.72}$$

According to the OCV-SOC relationship, we obtain:

$$\frac{\partial g_{k-1,l}(z, C_{\max})}{\partial z_{k,l}} = \alpha_1 + 2\alpha_2 z_{k,l} + 3\alpha_3 \left(z_{k,l}\right)^2 - \alpha_4 / \left(z_{k,l}\right)^2 + \alpha_5 / z_{k,l} - \alpha_6 / \left(1 - z_{k,l}\right) \tag{4.73}$$

In Eq. (4.72):

$$\frac{\partial z_{k,0}}{\partial \hat{C}_{\max,k}^{-}} = \frac{\eta_i i_{k,l-1} \Delta t}{\left(\hat{C}_{\max,k}^{-}\right)^2} \tag{4.74}$$

$$\frac{\partial \boldsymbol{F}\left(\hat{\boldsymbol{\chi}}_{k-1,L_z-1}, \hat{\boldsymbol{\theta}}_k^{-}, \boldsymbol{u}_{k-1,L_z-1}\right)}{\partial \hat{\boldsymbol{\theta}}_k^{-}} = \begin{bmatrix} 0 & \sigma_1 & \sigma_2 & 0 \\ 0 & 0 & 0 & \sigma_3 \end{bmatrix} \tag{4.75}$$

where σ_1, σ_2, σ_3 are

$$\begin{cases} \sigma_{1,k} = \frac{\hat{U}_{k-1,L_z-1}^{\mathrm{D}} \Delta t}{R_{\mathrm{D}}^2 C_{\mathrm{D}}} \exp\left(-\frac{\Delta t}{R_{\mathrm{D}}C_{\mathrm{D}}}\right) - \frac{i_{k-1,L_z-1} \Delta t}{R_{\mathrm{D}}C_{\mathrm{D}}} \exp\left(-\frac{\Delta t}{R_{\mathrm{D}}C_{\mathrm{D}}}\right) - \\ \quad i_{k-1,L_z-1}\left(\exp\left(-\frac{\Delta t}{R_{\mathrm{D}}C_{\mathrm{D}}}\right) - 1\right) \\ \sigma_{2,k} = \frac{\hat{U}_{k-1,L_z-1}^{\mathrm{D}} \Delta t}{R_{\mathrm{D}} C_{\mathrm{D}}^2} \exp\left(-\frac{\Delta t}{R_{\mathrm{D}}C_{\mathrm{D}}}\right) - \frac{i_{k-1,L_z-1} \Delta t}{C_{\mathrm{D}}^2} \exp\left(-\frac{\Delta t}{R_{\mathrm{D}}C_{\mathrm{D}}}\right) \\ \sigma_{3,k} = \eta_i i_{k-1,L_z-1} \Delta t \Big/ \left(\hat{C}_{max,k}^{-}\right)^2 \end{cases} \tag{4.76}$$

Figure 4.27 shows a multi-timescale estimation flowchart of battery parameters and states based on MAEKF algorithm. When the current or power excitation is

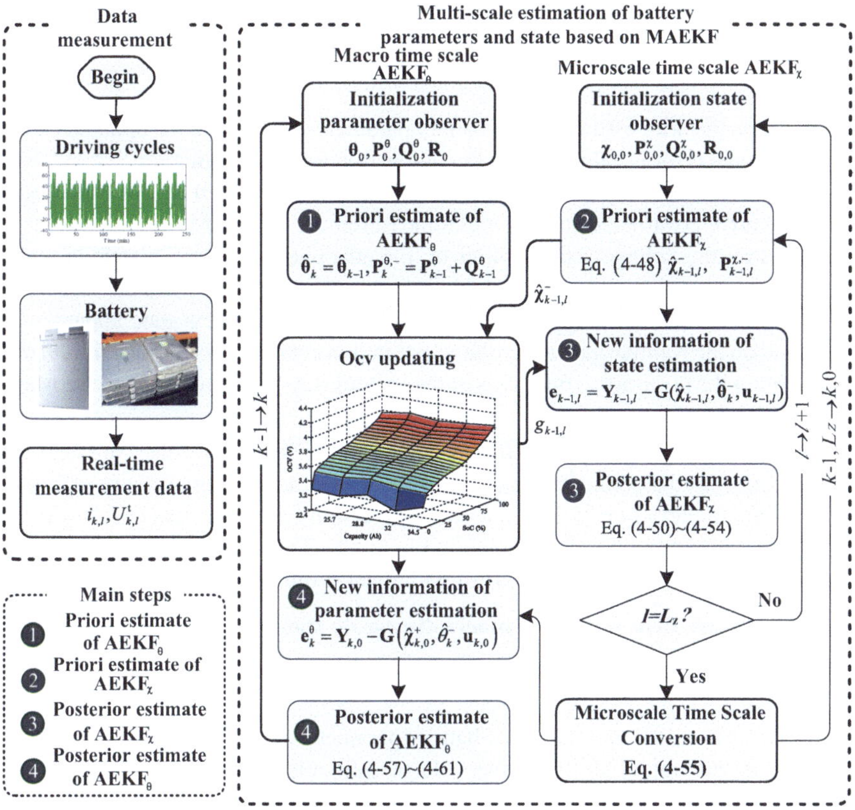

Fig. 4.27 Multi-timescale estimation of battery parameters and states based on MAEKF algorithm

loaded into the battery, the data collection system will measure the current $i_{k,l}$ and voltage $U_{k,l}^{t}$ in real time before transmitting them into the multi-timescale calculation system based on MAEKF algorithm. The specific process of battery parameters and state estimation can be divided into four steps:

Step 1: Time update of macro-timescale parameter observer (prior estimation) At the beginning of each macro-timescale parameter estimation, the parameter observer AEKF$_{\theta}$ needs to perform time update operation and calculate the priori parameter estimate $\hat{\boldsymbol{\theta}}_{k}^{-}$ and error covariance $\boldsymbol{P}_{k}^{\theta,-}$.

Step 2: Time update of micro-timescale state observer (prior estimation) After the prior estimation of AEKF$_{\theta}$, the time update of the state observer AEKF$_{\chi}$ is started based on the parameter prior estimation $\hat{\boldsymbol{\theta}}_{k}^{-}$ and the initial value of the state observer. The prior estimation of the system state $\hat{\boldsymbol{\chi}}_{k-1,l}^{-}$ and its error covariance $\boldsymbol{P}_{k-1,l}^{\chi,-}$ are obtained. Based on the prior estimated results of available capacity $\hat{C}_{\max,k}^{-}$ and state $\hat{z}_{k-1,l}^{-}$, then the battery OCV $g_{k-1,l}\left(\hat{z}_{k-1,l}^{-}, \hat{C}_{\max,k}^{-}\right)$ is updated.

Step 3: Measurement update of micro-timescale state observer (posterior estimation)

This step consists of five main calculation substeps:

① Based on the prior estimations of battery parameters, states and the updated OCV, the voltage error is calculated as the innovation for state estimation.
② The Kalman gain $\boldsymbol{K}_{k-1,l}^{\chi}$ is calculated in real time based on the prior estimation of the error covariance and the initial measurement noise covariance.
③ Based on the voltage estimation error, the adaptive covariance matching technique is applied to update the system noise covariance $\boldsymbol{R}_{k-1,l}$ and the measurement noise covariance $\boldsymbol{Q}_{k-1,l}^{\chi}$, which are intended for the next state update.
④ Based on the Kalman gain and voltage estimation error, the system state is revised to obtain the posterior estimations of the system state $\hat{\boldsymbol{\chi}}_{k-1,l}^{+}$ and the error covariance $\boldsymbol{P}_{k-1,l}^{\chi,+}$.
⑤ Judge the micro-timescale l. If this value fails to reach the timescale conversion limit L_z, the prediction and correction of the micro-timescale state estimate continue to carry out. If the timescale conversion limit L_z is reached, the scale conversion is carried out, as shown in Eq. (4.55). Then the system state is transferred to the parameter observer for parameter estimation correction.

Step 4: Measurement update of macro-timescale state observer (posterior estimation)

This step includes four main calculation parts:

① Based on the prior estimation of battery parameters and system state, the OCV value is calibrated, and then the voltage estimation error $\mathbf{e}_k^{\theta}$ is calculated as the innovation for parameter correction. It is noteworthy that $\boldsymbol{e}_k^{\theta} = \boldsymbol{e}_{k-1,L_z}$ at this time. In MAEKF, parameter observer and state observer share the same innovation matrix, namely, voltage estimation error. The algorithm has a coupling structure and is closed loop to ensure the stability of the final state value.
② Based on the prior estimation of parameter estimation error covariance and the initial measurement noise covariance, the parameter estimation Kalman gain $\boldsymbol{K}_k^{\theta}$ is updated.
③ Based on the voltage estimation error, the adaptive covariance matching technique is applied to update the system and the measurement noise covariances, and the sum is obtained for the next parameter updation.
④ Based on the Kalman gain and voltage estimation error, the system parameter is revised to derive the posterior estimation results of the system parameters $\hat{\boldsymbol{\theta}}_k^{+}$ and their error covariance $\boldsymbol{P}_k^{\theta,+}$. Up to now, the multi-timescale estimation of battery parameters and states is completed.

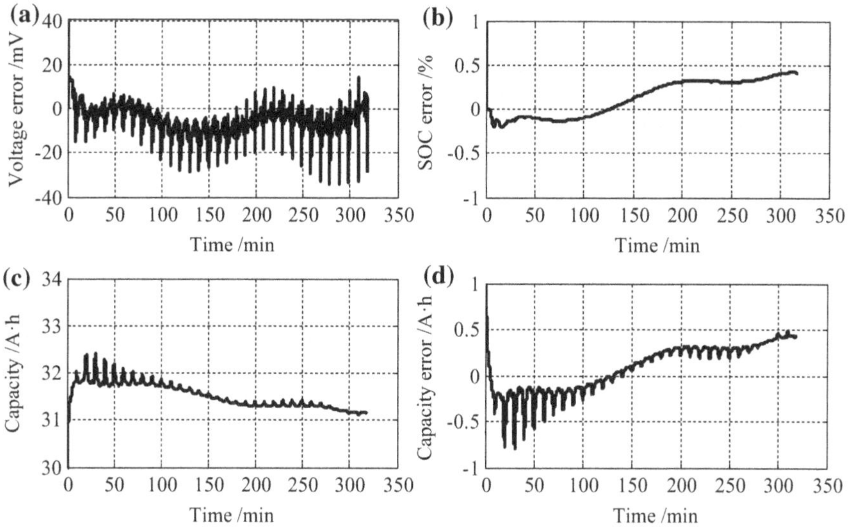

Fig. 4.28 Battery available capacity estimation results based on multi-timescale algorithm (L_z = 1 s, initial SOC value is 60%): **a** voltage estimation error; **b** SOC estimation error; **c** available capacity estimation results; **d** available capacity estimation error

(2) Case study

Case 1: Evaluation of estimation effect on different timescales

Figure 4.28 presents the estimation results for battery 2 cell 2 under inaccurate initial SOC and available capacity. It is notable that the timescale L_z is 1 s in this study, for which the multi-timescale algorithm is transformed into a single timescale algorithm.

It is revealed that the terminal voltage, SOC and available capacity of the battery can converge rapidly. The maximum estimation error of terminal voltage after convergence is lower than 35 mV. The maximum estimation error of SOC is less than 1%. The maximum estimation error of available capacity is lower than 1 A h. As demonstrated by the results, the MAEKF algorithm can guarantee the accurate estimation of SOC and available capacity. Even under the large initial error of SOC and available capacity, this method is still capable of obtaining accurate parameter and state estimation. It is noteworthy that, although the method based on a single timescale can obtain accurate estimates, it remains vulnerable to uncertain conditions. When the working current of the battery is significant, the voltage and available capacity estimates fluctuate massively. There are noticeable peaks in Fig. 4.28a, c, when the DST working condition varies from large current excitation to static state. In addition, owing to the large number of matrix calculations, the average computation time consumed by the algorithm is about 4.709 s. It should be noted that the processor of the simulation computer used here is Intel Core i5-3320 CPU @ 2.6 GHz, RAM is 8 GB, and the MATLAB version is 2012b.

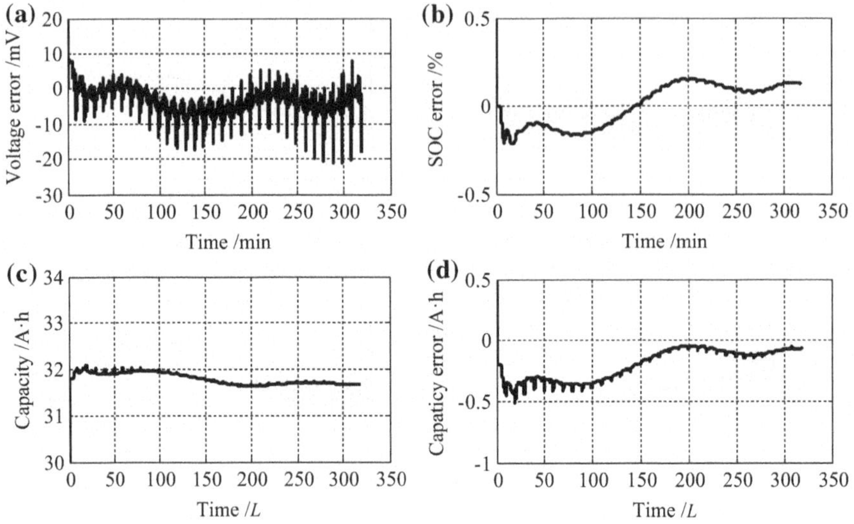

Fig. 4.29 Battery available capacity estimation results based on multi-timescale algorithm ($L_z =$ 60 s, initial SOC value is 60%): **a** voltage estimation error; **b** SOC estimation error; **c** available capacity estimation results; **d** available capacity estimation error

To discuss the estimation performance of MAEKF in multi-timescales, a comparison is drawn in the computational capabilities of different timescales. The limit value for scale conversion L_z in Fig. 4.29 is 60 s.

It is indicated that when using multi-timescales, the terminal voltage, SOC and available capacity of the battery can also converge rapidly with erroneous initial available capacity and SOC value. The estimation error of terminal voltage after convergence could be effectively limited to less than 25 mV. The estimation error of SOC is less than 0.5% and the estimation error of available capacity is less than 0.5 A h. More importantly, the estimation results of available capacity do not fluctuate due to the large current or power excitation. The estimation results are more stable and can converge to the reference value quickly. On the other hand, under the same computer configuration, the average computing time consumed by the algorithm is 2.512 s.

The results of different timescales show that (1) The MAEKF algorithm displays a better correction ability for inaccurate initial available capacity and SOC values, that is, the robustness of the algorithm is superior; (2) As compared to the single timescale counterpart, the multi-timescale method can obtain more accurate and stable results. In addition, the calculation time is reduced from 4.709 s to 2.512 s, which reduces the calculation cost by 47%.

It is worth noting that the above results do not suggest that the larger L_z brings better results. By comparing and analyzing the estimation results of different timescales, it is concluded that for the battery used in this section, the ideal range of L_z is between 60 and 150 s. When L_z is too small, capacity estimates are prone to fluctuate. The

Table 4.14 Battery SOC and available capacity estimation errors based on SOC estimation value (initial SOC is 60%, $L_Z = 60$ s)

Battery 2	Estimation error of SOC (%)			Estimation error of available capacity (A h)			
	Maximum error	Average value	Standard deviation	Maximum error	Average value	Standard deviation	After stability
Cell 1	0.424	0.215	0.120	0.678	0.055	0.086	<0.2
Cell 2	0.392	0.199	0.108	0.597	0.049	0.069	<0.2
Cell 3	0.673	0.756	0.204	0.843	0.061	0.071	<0.3

reason is that the available capacity of the battery changes slowly, and the battery capacity will not have a significant impact on the estimation error of the terminal voltage in a short time. Battery capacity is also vulnerable to the impact of operating conditions. When L_z is too large, the convergence speed of capacity estimation is too slow. The reason is that when the SOC changes significantly, the parameters of the battery model will change, which affects the convergence of capacity estimation.

Case 2: Evaluation of the estimation effect of indifferent aging conditions
Based on the above analysis, $L_Z = 60$ s is selected to evaluate the estimation accuracy of MAEKF algorithm in different aging states. As revealed by the results listed in Table 4.14, the available capacity of batteries and the initial value of SOC are both 30 A h and 60%, and the absolute value of estimation error after 10 min (10 L_Z) convergence is selected for statistical analysis.

It is indicated that the SOC estimation error is less than 1%, and the capacity estimation error is less than 0.3 A h (0.94%).

Therefore, the multi-timescale estimation method based on MAEKF algorithm is capable of accurately estimating the available capacity and SOC of the battery. For different aging degrees, inaccurate available capacity and initial value of SOC, the estimation error of SOC and capacity after convergence is less than 1%. This method not only guarantees the estimation accuracy, but also has an excellent robustness to uncertain current excitation, aging state, available capacity and initial value of SOC.

4.3.3 Co-Estimation Method Based on MHIF

(1) Calculation flow of the algorithm

Based on the principle flowchart of the multi-timescale filtering algorithm described in Fig. 4.26, HIF_χ is applied as the state observer with micro-timescale, HIF_θ is applied as the parameter observer with macro-timescale, and the two-layer HIF algorithm is applied to conduct the multi-timescale estimation of the system parameters and states. The detailed calculation steps are summarized as follows:

Initialization of algorithm: The initial parameters of parameter observer HIF_θ and state observer HIF_χ are set, respectively.

$$\boldsymbol{\theta}_0,\ \lambda^\theta,\ \boldsymbol{S}^\theta,\ \boldsymbol{P}_0^\theta,\ \boldsymbol{Q}^\theta,\ \boldsymbol{R}^\theta,\ \boldsymbol{\chi}_{0,0},\ \lambda^\chi,\ \boldsymbol{S}^\chi,\ \boldsymbol{P}_{0,0}^\chi,\ \boldsymbol{Q}^\chi,\ \boldsymbol{R}^\chi \tag{4.77}$$

where $\boldsymbol{\theta}_0$ represents the initial value of HIF_θ, λ^θ indicates the performance boundary of HIF_θ, $\boldsymbol{S}^\theta$, $\boldsymbol{P}_0^\theta$, $\boldsymbol{Q}^\theta$ and $\boldsymbol{R}^\theta$ denote the customized matrices of HIF_θ, $\boldsymbol{\chi}_{0,0}$ refers to the initial parameter of HIF_χ, and λ^χ indicates the performance boundary of HIF_χ. $\boldsymbol{S}^\chi$, $\boldsymbol{P}_{0,0}^\chi$, $\boldsymbol{Q}^\chi$ and $\boldsymbol{R}^\chi$ represent the customized matrices of HIF_χ. Section 4.1.4 describes the common design method of HIF customized matrices in detail. It is worth noting that the MHIF introduced here does not involve the noise covariance matching algorithm, for which the corresponding matrix is also transformed from a variable to a constant.

For macro-timescale series, $k = 1, 2$, (from time $(k-1)^+$ to time $(k)^-$).

Step 1: Time update equations for parameters based on macro-timescale (prior estimation)

$$\hat{\boldsymbol{\theta}}_k^- = \hat{\boldsymbol{\theta}}_{k-1},\ \boldsymbol{P}_k^{\theta,-} = \boldsymbol{P}_{k-1}^\theta + \boldsymbol{Q}_{k-1}^\theta \tag{4.78}$$

For micro-timescale series, $l = 1, 2 \cdots\cdots, L_z$ (from time $(k-1, l-1)^+$ to time $(k-1, L_z)^-$).

Step 2: Time update equations for state based on micro-timescale (prior estimation)

$$\hat{\boldsymbol{\chi}}_{k-1,l}^- = \boldsymbol{F}(\hat{\boldsymbol{\chi}}_{k-1,l-1}, \hat{\boldsymbol{\theta}}_k^-, \boldsymbol{u}_{k-1,l-1}),\ \boldsymbol{P}_{k-1,l}^{\chi,-} = \boldsymbol{A}_{k-1,l-1}\boldsymbol{P}_{k-1,l-1}^\chi \boldsymbol{A}_{k-1,l-1}^{\mathrm{T}} + \boldsymbol{Q}^\chi \tag{4.79}$$

Step 3: Measurement update for state based on micro-timescale (posterior estimation)

Update innovation matrix:

$$\boldsymbol{e}_{k-1,l} = \boldsymbol{Y}_{k-1,l} - \boldsymbol{G}(\hat{\boldsymbol{\chi}}_{k-1,l}^-, \hat{\boldsymbol{\theta}}_k^-, \boldsymbol{u}_{k-1,l}) \tag{4.80}$$

Gain matrix of $H\infty$:

$$\boldsymbol{K}_{k-1,l}^\chi = \boldsymbol{A}_{k-1,l}\boldsymbol{P}_{k-1,l}^{\chi,-}\left(\boldsymbol{I} - \lambda^\chi \boldsymbol{S}^\chi \boldsymbol{P}_{k-1,l}^{\chi,-} + \left(\boldsymbol{C}_{k-1,l}^\chi\right)^{\mathrm{T}}(\boldsymbol{R}^\chi)^{-1}\boldsymbol{C}_{k-1,l}^\chi \boldsymbol{P}_{k-1,l}^{\chi,-}\right)^{-1}\left(\boldsymbol{C}_{k-1,l}^\chi\right)^{\mathrm{T}}(\boldsymbol{R}^\chi)^{-1} \tag{4.81}$$

Correction of state estimate:

$$\hat{\boldsymbol{\chi}}_{k-1,l}^+ = \hat{\boldsymbol{\chi}}_{k-1,l}^- + \boldsymbol{K}_{k-1,l}^\chi\left[\boldsymbol{Y}_{k-1,l} - \boldsymbol{G}(\hat{\boldsymbol{\chi}}_{k-1,l}^-, \hat{\boldsymbol{\theta}}_k^-, \boldsymbol{u}_{k-1,l})\right] \tag{4.82}$$

Update of H_∞ characteristic matrix:

$$\boldsymbol{P}_{k-1,l}^{\chi,+} = \boldsymbol{P}_{k-1,l}^{\chi,-}\left(\boldsymbol{I} - \lambda^{\chi}\boldsymbol{S}^{\chi}\boldsymbol{P}_{k-1,l}^{\chi,-} + \left(\boldsymbol{C}_{k-1,l}^{\chi}\right)^{\mathrm{T}}\left(\boldsymbol{R}^{\chi}\right)^{-1}\boldsymbol{C}_{k-1,l}^{\chi}\boldsymbol{P}_{k-1,l}^{\chi,-}\right)^{-1} \tag{4.83}$$

For time series calculation (For $l = 1{:}\, L_z$) and timescale transform (when $l = L_z$):

$$\hat{\boldsymbol{\chi}}_{k,0}^{+} = \hat{\boldsymbol{\chi}}_{k-1,L_z}^{+},\ \boldsymbol{P}_{k,0}^{\chi,+} = \boldsymbol{P}_{k-1,L_z}^{\chi,+},\ \boldsymbol{Y}_{k,0} = \boldsymbol{Y}_{k-1,L_z},\ \boldsymbol{u}_{k,0} = \boldsymbol{u}_{k-1,L_z} \tag{4.84}$$

So far, the micro-timescale cycle calculation under a macro-timescale is completed, and the next step is to return to the macro-timescale to update the measurement equations for parameters (posterior estimation).

Step 4: Measurement update of HIF_θ based on macro-timescale (posterior estimation)

Update innovation matrix:

$$\boldsymbol{e}_k^{\theta} = \boldsymbol{Y}_{k,0} - \boldsymbol{G}\left(\hat{\boldsymbol{\chi}}_{k,0}^{+}, \hat{\theta}_k^{-}, \boldsymbol{u}_{k,0}\right) \tag{4.85}$$

Gain matrix of H∞:

$$\boldsymbol{K}_k^{\theta} = \boldsymbol{P}_k^{\theta,-}\left(\boldsymbol{I} - \lambda^{\theta}\boldsymbol{S}^{\theta}\boldsymbol{P}_k^{\theta,-} + \left(\boldsymbol{C}_k^{\theta}\right)^{\mathrm{T}}\left(\boldsymbol{R}^{\theta}\right)^{-1}\boldsymbol{C}_k^{\theta}\boldsymbol{P}_k^{\theta,-}\right)^{-1}\left(\boldsymbol{C}_k^{\theta}\right)^{\mathrm{T}}\left(\boldsymbol{R}^{\theta}\right)^{-1} \tag{4.86}$$

Correction of parameter estimate:

$$\hat{\boldsymbol{\theta}}_k^{+} = \hat{\boldsymbol{\theta}}_k^{-} + \boldsymbol{K}_k^{\theta}\boldsymbol{e}_k^{\theta} \tag{4.87}$$

Update of H_∞ characteristic matrix:

$$\boldsymbol{P}_k^{\theta,+} = \boldsymbol{P}_k^{\theta,-}\left(\boldsymbol{I} - \lambda^{\theta}\boldsymbol{S}^{\theta}\boldsymbol{P}_k^{\theta,-} + \left(\boldsymbol{C}_k^{\theta}\right)^{\mathrm{T}}\left(\boldsymbol{R}^{\theta}\right)^{-1}\boldsymbol{C}_k^{\theta}\boldsymbol{P}_k^{\theta,-}\right)^{-1} \tag{4.88}$$

The multi-timescale estimation of parameters and state at kth time is completed. The state filter is deduced from time $(k)^+$ to time $(k+1)^-$, and the state estimation at $(k+1)$ is prepared. Set $\hat{\boldsymbol{\chi}}_{k,0} = \hat{\boldsymbol{\chi}}_{k,0}^{+}, \hat{\boldsymbol{\theta}}_k = \hat{\boldsymbol{\theta}}_k^{+}$.

Figure 4.30 presents a flowchart of the multi-timescale estimation method for battery parameters with state based on the MHIF algorithm. The specific process can be referred to in Sect. 4.3.2 or the Chinese invention patent 201610675853.4.

(2) Case study

Based on the DST data of battery 5 at room temperature, the co-estimation of SOC and available capacity using the MHIF algorithm was evaluated. The available capacity of the battery at room temperature is 2.096 A h and the initial SOC error is 20%. The initial available capacity is 1.5 A h. Figure 4.31a–f show the predicted terminal voltages and their errors, estimated SOCs and their errors, available capacity

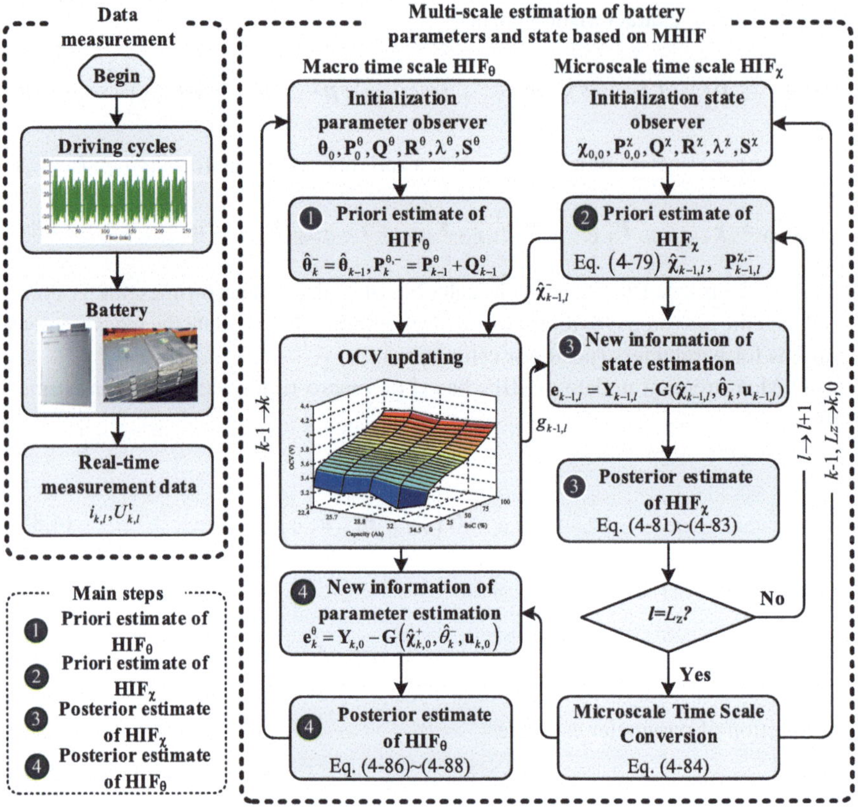

Fig. 4.30 Multi-timescale estimation of parameters and state based on MHIF algorithms

estimation results, and their errors, respectively, where MHIF represents the MHIF algorithm estimates and L_z is set to 60 s.

Figure 4.31a, b show that there is a significant error in the estimation of terminal voltage when the algorithm is turned on, which is due to the inaccurate initial SOC and capacity setting. With the continuous self-adaptive correction of the algorithm, the estimation of terminal voltage gradually converges to the measured value, and the absolute error of terminal voltage after stabilization basically remains below 30 mV. Figure 4.31c, d reveal that the estimation of SOC of the algorithm can not only converge to the reference value quickly, but also maintain the absolute error within 1% after stabilization. Figure 4.31e, f shows that the capacity estimation of the algorithm can also converge to the reference value quickly, with the absolute error being constantly lower than 1% after stabilization.

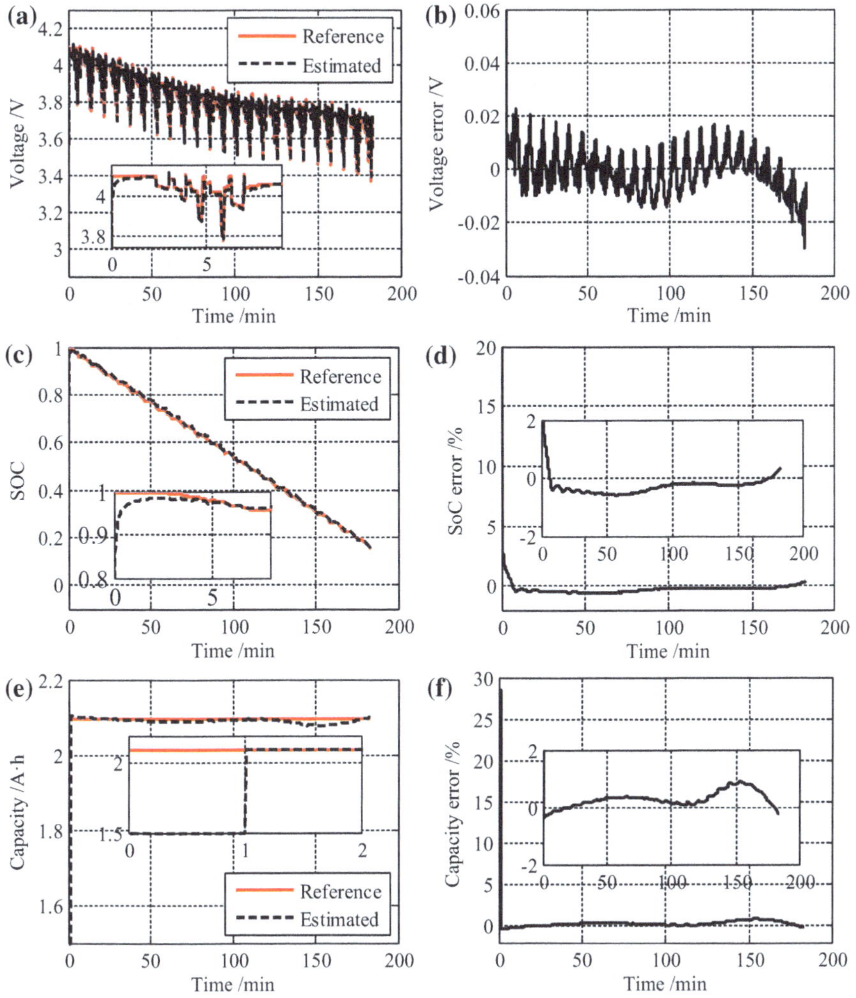

Fig. 4.31 Estimation results of MHIF algorithm under DST working conditions: **a** comparison of terminal voltage estimates with measurements, **b** error of terminal voltage estimates, **c** comparison of SOC estimates with reference values, **d** error of SOC estimates, **e** comparison of capacity estimates with reference values, **f** error of capacity estimates

4.4 Conclusion

Due to the complex electrochemical reactions inside the lithium-ion battery in EVs, the battery has strong nonlinearity and time-varying characteristics, which poses a massive challenge for the BMS to obtain the current status of the battery. On the other hand, the complex and changeable external environment and operating conditions increase the uncertainty of battery performance degradation and further increase the difficulty in states estimation. Therefore, the state estimation of lithium-ion battery for vehicles have always been technical bottlenecks for the industry and the academic fields.

An accurate SOC estimation can objectively reflect the driving ability of batteries and effectively prevent the battery from abuse conditions. This chapter summarizes and classifies the commonly used SOC estimation methods, in addition to introducing procedures of the algorithms. As the equivalent circuit model can balance the relationship between estimation accuracy and algorithm complexity, the SOC estimation method based on the equivalent circuit model is an effective estimation method. The results show that the SOC estimation absolute errors of AEKF and HIF are less than 1.5% under known static capacity. At the same time, battery degradation is inevitable with time. Therefore, SOC estimation is often inseparable from accurate dynamic capacity (SOH) real-time updating and calibration. This chapter introduces and classifies the commonly used SOH methods, and describes their procedures in detail. Finally, considering the coupling relationship between SOC, SOH and the different change timescales of states and parameters, the multi-timescale co-estimation algorithm is emphatically described. The results reveal that this method can quickly and steadily realize the accurate estimation of SOC and capacity. The SOC estimation error of MAEKF and MHIF is less than 1% with unknown capacity, and the capacity estimation error is less than 1%.

References

1. Xiong R, Cao J, Yu Q, He H, Sun F (2018) Critical review on the battery state of charge estimation methods for electric vehicles. IEEE Access 6:1832–1843
2. Andre D, Meiler M, Steiner K (2011) Characterization of high-power lithium-ion batteries by electrochemical impedance spectroscopy. II: Model. J Power Sour 196:5349–5356
3. He H, Xiong R, Guo H (2012) Online estimation of model parameters and state-of-charge of LiFePO4 batteries in electric vehicles. Appl Energy 89:413–420
4. Zhang Y, Song W, Lin S, Feng Z (2014) A novel model of the initial state of charge estimation for $LiFePO_4$ batteries. J Power Sour 248:1028–1033
5. Plett GL (2004) Extended Kalman filtering for battery management systems of LiPB-based HEV battery packs Part 1. Backgr. J Power Sour 134:252–261
6. Xiong R, He H (2013) Evaluation on state of charge estimation of batteries with adaptive extended Kalman filter by experiment approach. IEEE Trans Veh Technol 62:108–117
7. Zhang Y, Xiong R (2017) Lithium-ion battery pack state of charge and state of energy estimation algorithms using a hardware-in-the-loop validation. IEEE Trans Power Electron 32:4421–4431
8. Kang LW, Zhao X, Ma J (2014) A new neural network model for the state-of-charge estimation in the battery degradation process. Appl Energy 121:20–27
9. He W, Williard N, Chen C, Pecht M (2014) State of charge estimation for Li-ion batteries using neural network modeling and unscented Kalman filter-based error cancellation. Int J Electr Power Energy Syst 62:783–791
10. Chemali E, Kollmeyer PJ, Preindl M (2018) Long short-term memory networks for accurate state-of-charge estimation of Li-ion batteries. IEEE Trans Ind Electron 65:6730–6739
11. Lin C, Mu H, Xiong R, Shen W (2016) A novel multi-model probability battery state of charge estimation approach for electric vehicles using H-infinity algorithm. Appl Energy 166:76–83

12. Plett GL (2004) Extended Kalman filtering for battery management systems of LiPB-based HEV battery packs Part 2. Model. identif. J Power Sour 134:262–276
13. Xiong R, Li L, Tian J (2018) Towards a smarter battery management system: A critical review on battery state of health monitoring methods. J Power Sour 405:18–29
14. Waag W, Käbitz S, Sauer DU (2013) Experimental investigation of the lithium-ion battery impedance characteristic at various conditions and aging states and its influence on the application. Appl Energy 102:885–897
15. Xiong R, Tian J, Mu H, Wang C (2017) A systematic model-based degradation behavior recognition and health monitoring method for lithium-ion batteries. Appl Energy 207:372–383
16. Xiong R, Zhang Y, Wang J, He H, Peng S, Pecht M (2019) Lithium-ion battery health prognosis based on a real battery management system used in electric vehicles. IEEE Trans Veh Technol 68(5):4110–4121
17. Tian J, Xiong R (2019) Fractional-order model-based incremental capacity analysis for degradation state recognition of Lithium-ion batteries. IEEE Trans Ind Electron 66:1576–1584
18. Ma Z, Wang Z, Xiong R, Jiang J (2018) A mechanism identification model based state-of-health diagnosis of lithium-ion batteries for energy storage applications. J Clean Prod 193:379–390
19. Xiong R, Sun F, Gong X, Gao C (2014) A data-driven based adaptive state of charge estimator of lithium-ion polymer battery used in electric vehicles. Appl Energy 113:1421–1433
20. Plett GL (2006) Sigma-point Kalman filtering for battery management systems of LiPB-based HEV battery packs Part 2 : Simultaneous state and parameter estimation 161:1369–1384
21. Lam L, Bauer P (2013) Practical capacity fading model for Li-ion battery cells in electric vehicles. IEEE Trans Power Electron 28:5910–5918
22. Hu X, Jiang J (2016) Battery health prognosis for electric vehicles using sample entropy and sparse bayesian predictive modeling. IEEE Trans Ind Electron 63:2645–2656
23. Xiong R, Yu Q, Lin C, Shen WX (2019) A sensor fault diagnosis method for a Lithium-ion battery pack in electric vehicles. IEEE Trans Power Electron 34(10):9709–9718
24. Wang J, Xiong R, Li L, Fang Y (2018) A comparative analysis and validation for double-filters-based state of charge estimators using battery-in-the-loop approach. Appl Energy 229:648–659
25. Yu Q, Xiong R, Lin C (2017) Lithium-ion battery parameters and state-of-charge joint estimation based on H-infinity and unscented Kalman filters. IEEE Trans Veh Technol 66:8693–8701
26. Yang R, Xiong R, He H (2017) A novel method on estimating the degradation and state of charge of lithium-ion batteries used for electrical vehicles. Appl Energy 207:336–345
27. Ma Z, Jiang J, Shi W (2015) Investigation of path dependence in commercial lithium-ion cells for pure electric bus applications: Aging mechanism identification. J Power Sour 274:29–40
28. Ma Z, Yang R, Wang Z (2019) A novel data-model fusion state-of-health estimation approach for lithium-ion batteries. Appl Energy 237:836–847
29. Hu C, Youn BD, Chung J (2012) A multiscale framework with extended Kalman filter for lithium-ion battery SOC and capacity estimation. Appl Energy 92:694–704
30. Xiong R, Sun F, Chen Z, He H (2014) A data-driven multi-scale extended Kalman filtering based parameter and state estimation approach of lithium-ion polymer battery in electric vehicles. Appl Energy 113:463–476
31. Chen C, Xiong R, Shen W (2018) A Lithium-ion battery-in-the-loop approach to test and validate multiscale dual H-infinity filters for state-of-charge and capacity estimation. IEEE Trans Power Electron 33:332–342

Chapter 5
State Estimation of Battery System

A battery system mainly consists of battery modules, a BMS, and a battery pack case. A battery cell has maximum available capacity and SOC, the estimation of which has clear reference values and evaluation methods. While there is no uniform definition of the capacity of the battery pack, and the available capacity values vary under different research purposes and balancing strategies. On account of the uncertainty differences between voltage, available capacity, SOC, and internal resistance of each cell in application, the widely used "large battery cell model" and "barrel effect" have obvious limitations in practical applications. In addition, the uncertainty differences change as the aging state, application environment and operating temperature, leading to uncertainty changes of the available pack capacity and poor state estimation results. Aiming at the three technical difficulties which are "unrefined model, inaccurate state estimation, and poor system management" caused by "broom effect" in the pack, this chapter discusses basic theory and practical applications including the battery grouping, uncertainty modeling, state estimation, peak power capability prediction, etc.

5.1 Battery Grouping Analysis

The connection modes of the battery pack include series, parallel, and hybrid connection to meet different power demands. Battery cells in series and in parallel can satisfy the demand of high operation voltage and large capacity, respectively. The hybrid mode including series and parallel connections of cells can take the dual demand of high voltage and high capacity of new energy vehicles into consideration at the same time. However, there are certain differences in the cell capacity and internal resistance during manufacture affected by manufacturing materials, process parameters, etc. Such inconsistencies can be further deteriorated in service due to the differences in service conditions, operation conditions, connection modes of battery pack, and management methods, resulting in an obvious "broom effect" at the end of

R. Xiong, *Battery Management Algorithm for Electric Vehicles*,
https://doi.org/10.1007/978-981-15-0248-4_5

charge and discharge, thus the charging and discharging performance of the battery system is further affected.

5.1.1 "Broom Effect" of Battery Pack

The strong polarization and nonlinearity at the end of charge and discharge would cause the "broom effect" of the battery pack. As is shown in Fig. 5.1, the cell inconsistency experiences a significant increase at the end of charge and discharge. The consistency performs well before the broom neck, and deteriorates greatly after exceeding the cutoff threshold of broom neck.

The inconsistency of battery cells restricts the battery pack's energy/capacity utilization ratio to a great extent and also accelerates the battery aging. To solve this problem, the determination of cutoff threshold value and control rules of charging/discharging to broom neck is proposed by Sun Fengchun's team from Beijing Institute of Technology. The distribution rules of battery cells and systems inconsistency in the same battery manufacturing batch are determined through experiments, especially the mapping relationship between voltage change and capacity at the end of charge/discharge. And the next is to establish rules to determinate the broom neck and improve the charge–discharge control strategy, which enhances the consistency and energy utilization ratio of the battery system and thus reduces the risk of overcharge or over-discharge of battery cells.

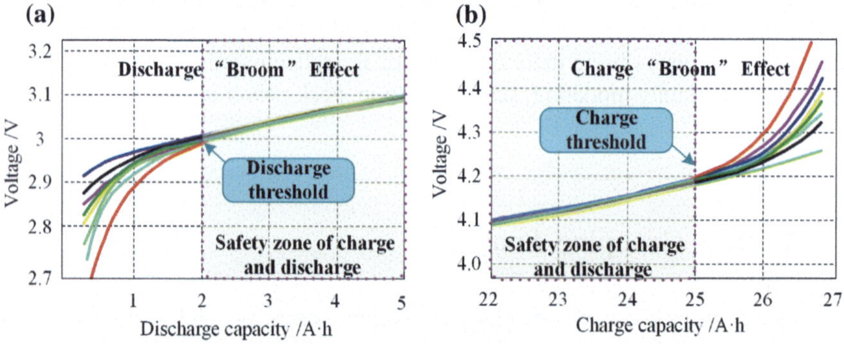

Fig. 5.1 The "broom effect" of battery pack: **a** "broom effect" curve at the end of discharge for different cells; **b** "broom effect" curve at the end of charge for different cells

5.1.2 Battery Pack in Series and Parallel

Battery cells connection of series and parallel are the most basic module units in a battery pack, and different connection modes may cause different inconsistency spreading trends. Figure 5.2a is the topology diagram of a "4-series" battery pack composed of four cells connected in series, while Fig. 5.2b is the topology diagram of "2-parallel" battery pack composed of two cells connected in parallel. The influences of cell inconsistency on the performance of battery packs mainly demonstrated in the following two aspects:

Capacity inconsistency
The current through each cell is the same during the charge and discharge process in series-connected battery packs. Cells with less capacity are prone to be charged or discharged first, and battery pack is required to stop charging or discharging to prevent safety hazards like overcharge or over-discharge. It can be seen that the cell with minimum capacity restricts the charge–discharge capacity of the pack in series, which is the right reason for employing the "barrel effect" to analyze the pack performance. It should be pointed out that it is the cell with minimum residual capacity that determines the pack discharge capacity other than the cell with the minimum capacity. The residual capacity is the product of the current SOC and the maximum available capacity. Therefore, the "barrel effect" is not suitable for the charge–discharge behavior analysis of battery pack in series. On the other hand, the available capacity of the parallel-connected battery pack is the sum of available capacity of each cell. When the battery pack continues to discharge, the cell with the lowest OCV may be charged; when the pack continues to charge, the cell with the highest OCV may be discharged. Therefore, the charge–discharge capacity of the battery pack connected in parallel is not limited by the cell with minimum capacity, and the charge and discharge behavior does not conform to the "barrel effect" either [1].

Internal resistance inconsistency
For a series-connected battery pack with similar capacity of cells, the cell with large internal resistance will first reach the cutoff voltage when charging. If the battery pack continues to be charged in this case, it would bring some safety risks caused by overcharging to the cells with large resistance; while stopping charging could make the cells with small resistance fail to be fully charged. The cells with large resistance

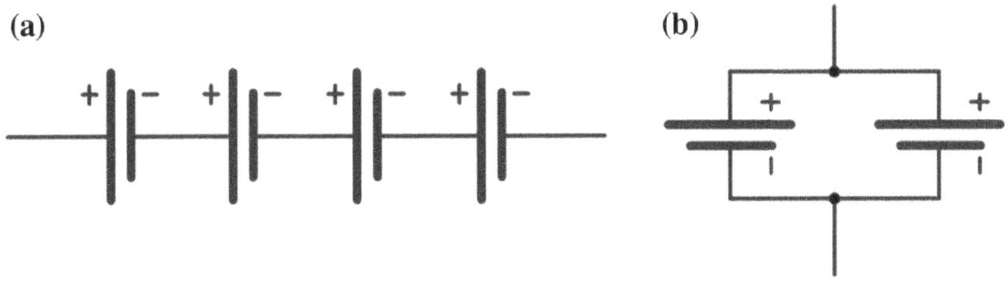

Fig. 5.2 Topology diagram of series and parallel pack: **a** 4-series pack; **b** 2-parallel pack

are prone to produce more heat, and experience a rapid rise in the internal temperature when discharged at a high rate. When the battery pack in parallel is charged or discharged, the terminal voltage of each cell remains equal, whereas the current of each branch varies caused by the resistance inconsistency, which further leads to a different charge or discharge electricity and aging trajectory. So the inconsistency of internal resistance not only directly affects the voltage and heat dissipation of the series-connected battery pack to a great extent, but also affects the long-term performance decline path of the parallel-connected battery pack.

5.1.3 Performance Analysis of Typical Hybrid Connected Battery Pack

Both parallel–series and series–parallel are the common connected modes for the battery pack. For example, the battery pack in Tesla Model S consists of 96 battery modules in series and each module contains 74 cylindrical cells of 3.1A h in parallel, while the battery pack used in Nissan's Leaf contains two parallel battery modules and in each module there are 96 cells of 33A h are connected in series. Different connection modes can affect the performances of battery packs. Certain basis for determining the connection mode of battery pack is provided by analyzing the performances of the battery pack of different connection modes. To simplify the analysis process, simplified connection modes shown in Fig. 5.3 are adopted in this section and 01–08 represent the battery cell. The hybrid connected modes of the battery pack in this chapter are defined as

① nPmS: A battery pack consists of m modules in series and in each module, there are n cells in parallel.
② mSnP: A battery pack consists of n modules in parallel and in each module, there are m cells in series.

Reliability analysis

The concept of component reliability is used to analyze the reliability of battery packs. It is assumed that the reliability of the normal operation of the cell is r $(0 < r < 1)$ for two different battery packs grouping modes, and the reliability is independent of each other. The probabilities of the normal operation of 2P4S and 4S2P are denoted as P_a and P_b, respectively, then:

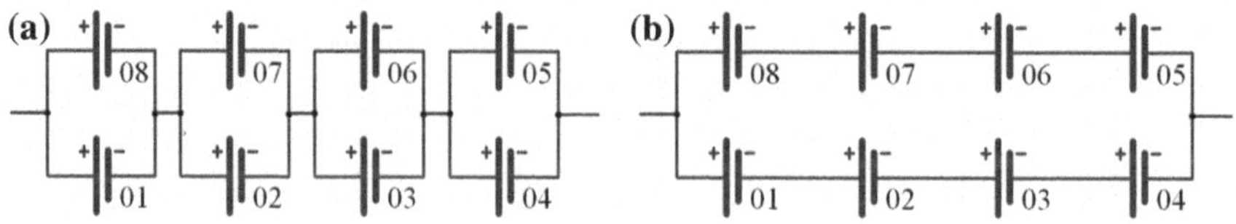

Fig. 5.3 Topology diagram of series and parallel pack: **a** 2P4S; **b** 4S2P

$$P_a = \left[1 - (1 - r)^2\right]^4 \tag{5.1}$$

$$P_b = 1 - \left(1 - r^4\right)^2 \tag{5.2}$$

When r is 0.8, the values of P_a and P_b are 0.8493 and 0.6514, respectively; when r is 0.9, the values of P_a and P_b change to 0.9606 and 0.8817, respectively. No matter what r is, it can be obtained that P_a outnumbers P_b, that is, the reliability of the battery system in parallel–series is superior to that of the battery system in series–parallel. It should be pointed out that the actual conclusions may be different from those based on the reliabilities of components when the short-circuit fault, the disconnection fault and costs of different battery management methods are considered in the actual situation.

Economy and safety analysis

In the parallel–series mode of the battery pack, modules in series are independent of each other. The self-balancing of cells can restrain the inconsistency of capacity and internal resistance among cells to a certain extent. The terminal voltages of cells are equal in the parallel module, and only one voltage acquisition sensor is needed, leading to a low cost. If the current through each cell in the parallel module needs to be analyzed, a large number of shunts or hall sensors are demanded measuring the currents of cells, which will significantly increase the system cost. However, it is not required to measure the branch currents of cells for the conventional parallel–series system. Self-balancing takes place in series branches of the series–parallel pack, which has low balancing efficiency. The number of BMC acquisition boards is large to obtain the voltage of each cell, thus causing a high cost.

The parallel module in the battery pack in parallel–series is directly connected by battery cells in parallel. The battery pack has high safety because the influence of each battery cell is small even if there is a certain deviation in the terminal voltage of battery cells. However, there is plenty of battery cells in each series branch in the series–parallel battery pack. It is necessary to ensure the terminal voltage consistency in each branch. Otherwise, even if the terminal voltage difference between cells is small, a large voltage deviation will occur between the series branches after connection for a battery pack, even causing safety accidents in severe cases.

When a cell is disconnected in the series–parallel battery pack, it will cause that the entire series branch of the cell disconnects from the battery pack and a larger charge/discharge current flows through the other cells and their branches. On the contrary, when the same case happens in the parallel–series battery pack, it only reflects on the other cells in the same parallel module while others work normally. Therefore, the battery pack with cells of small capacity and large number of cells in the parallel module has less influence, while the pack with cells of large capacity and small number of cells in the parallel module has large influence. When one battery cell is short-circuited in the series–parallel battery pack, the charge/discharge current of the branch where the short-circuited cell is located at will increase to stabilize the input/output power. This branch may be charged by the normal branch. However, for the parallel–series battery pack, the other cells of the module in which the short-circuited cell is located will become short-circuiting as well, and the other modules will suffer a heavier power load, also easily causing some safety hazards. To sum up, the parallel–series battery pack is safer than series–parallel battery pack in term of the disconnection fault, while the series–parallel battery pack has higher safety for the short-circuit fault. The disconnection fault is related to the connection methods and manufacturing level, while the short-circuit fault mainly comes from the abnormal operation modes such as external impact, connection failure, etc. At current battery manufacturing processes and grouping levels, the probability of both faults occurring on the battery is small.

Capacity analysis

Battery available capacity includes available charge capacity and available discharge capacity. The former refers to the capacity that the battery can be charged from the current state by CCCV to the cutoff voltage. The latter, namely the remaining capacity in the current state, refers to the capacity discharged from the current state to the cutoff voltage. The available capacity of the series battery pack is equal to the sum of the minimum available charge capacity and the minimum available discharge capacity among cells, while that of the parallel battery pack is the sum of the available capacities of all cells. Therefore, the calculation of available capacity depends on the capacities of all cells and SOC values for packs with different connection modes. Eight cells in different aging stages which are numbered cell 01–08 with the same initial SOC of 50% are selected and constituted 2P4S or 4S2P to analyze the capacity characteristics in different connection modes. The available capacities of the cells at 25 °C are shown in Fig. 5.4.

For a battery pack with the configuration of 2P4S or 4S2P, the number of connection modes up to 40,320 (i.e., 8!). The capacities of the two battery packs and their proportions in all permutations are shown in Fig. 5.5, in which the proportions of the battery pack capacity is the proportions of the number of the permutations with the same capacity to the overall number of permutations. The capacities of battery packs and their distribution characteristics in overall arrangements can be determined by means of numerical simulation. The results indicate that the battery pack capacities of 2P4S are more dispersed than that of 4S2P. The maximum capacity of 4S2P reaches 4.9240 A h, accounting for only 2.9%, while the proportion of 2P4S with the capacity more than 4.9240 A h is 17.2%. So the probability that the capacity of 2P4S is larger than the maximum capacity of 4S2P is 17.2%. Moreover, since the expected battery pack capacity of 4S2P is 4.8346 A h while the value of 2P4S is 4.7415 A h, 2P4S has better capacity performance. It should be noted that both 2P4S and 4S2P can reach their maximum capacities when the 8 cells are arranged according to Fig. 5.3a, b, respectively. And four parallel modules in 2P4S have similar capacities and the capacity difference between the two series branches in 4S2P is larger than that in 2P4S.

Capacities of the two battery packs are expressed as $C_{2P4S(0)}$ and $C_{4S2P(0)}$, respectively. Table 5.1 shows the comparison results of the two capacities with the same

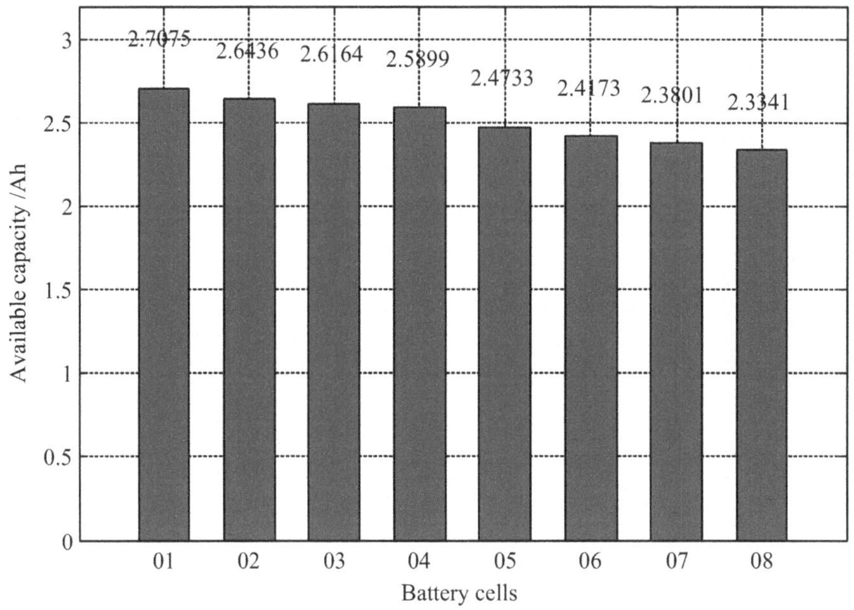

Fig. 5.4 Available capacities of battery cells at 25 °C

arrangement. The relationship $C_{2P4S(0)} \geq C_{4S2P(0)}$ can be observed and the proportion of $C_{2P4S(0)} > C_{4S2P(0)}$ accounts for 75%.

Charge/discharge electricity analysis

The electricity of the battery pack directly affects the driving mileage of EVs. Different connection modes for battery pack will lead to the differences in charge/discharge electricity. The charge/discharge electricity of the two battery packs shown in Fig. 5.3 are tested at an initial SOC of 50%. The testing steps are as follows:

Step 1: After resting for 2 min, the battery pack is discharged at a constant-current rate of 0.5C. Once the terminal voltage of any cell drops to 3.2 V, it is allowed for the pack to rest for 1 h.

Step 2: The battery pack is charged at a constant rate of 0.5C until the terminal voltage of any cell reaches 4.2 V.

The charge/discharge electricity of the battery packs are shown in Fig. 5.6. It can be seen that the difference between 2P4S and 4S2P up to 0.15 A h.

The conclusions based on numerical simulation and experiment analysis are as follows. The parallel–series battery pack is superior in the reliability, economy, and safety. It has better capacity performance no matter how the cells are arranged. In addition, it also has a larger charge/discharge electricity under the same conditions. It should be noted that although only two NMC lithium-ion battery packs of 2P4S and 4S2P are analyzed in this section, the same conclusions have also been verified

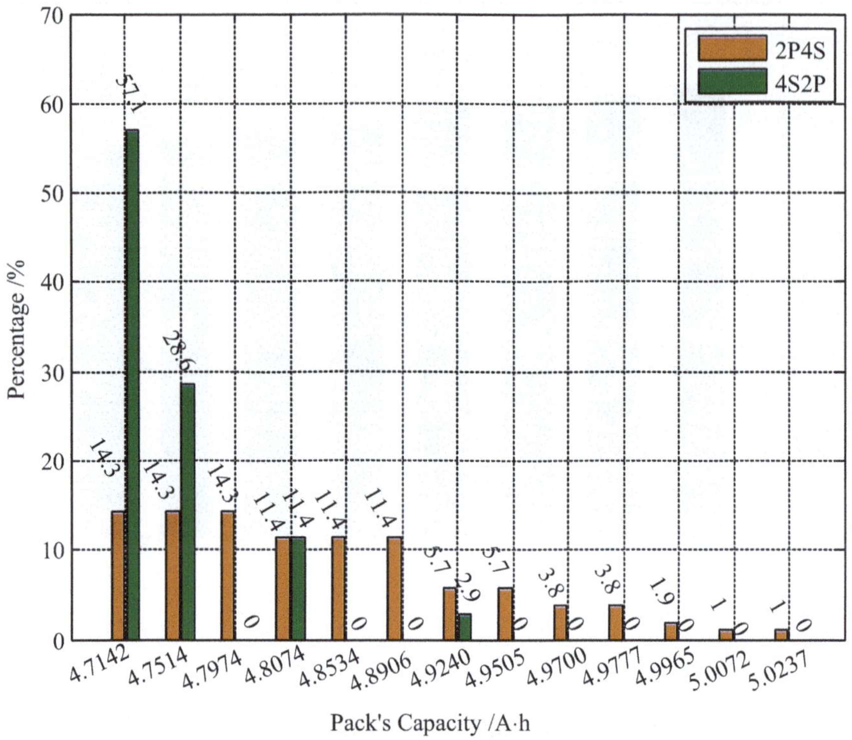

Fig. 5.5 The group capacities of the two battery packs and their proportion

Table 5.1 Comparison between group capacities of the two battery packs

	$C_{2P4S(0)} > C_{4S2P(0)}$	$C_{2P4S(0)} = C_{4S2P(0)}$	$C_{2P4S(0)} < C_{4S2P(0)}$	Summation
Number	30,240	10,080	0	40,320
Proportion (%)	75	25	0	100

in other battery types and packs with different number of cells through numerical simulation analysis.

5.2 State Estimation for Battery Pack

The performances of battery packs are highly coupled with their inconsistency, the inconsistent analysis of battery packs is a prerequisite for accurate modeling and state estimation. In view of the inconsistency of the battery pack, this section detailed introduces the screening method of the battery cells. The battery pack model and SOC

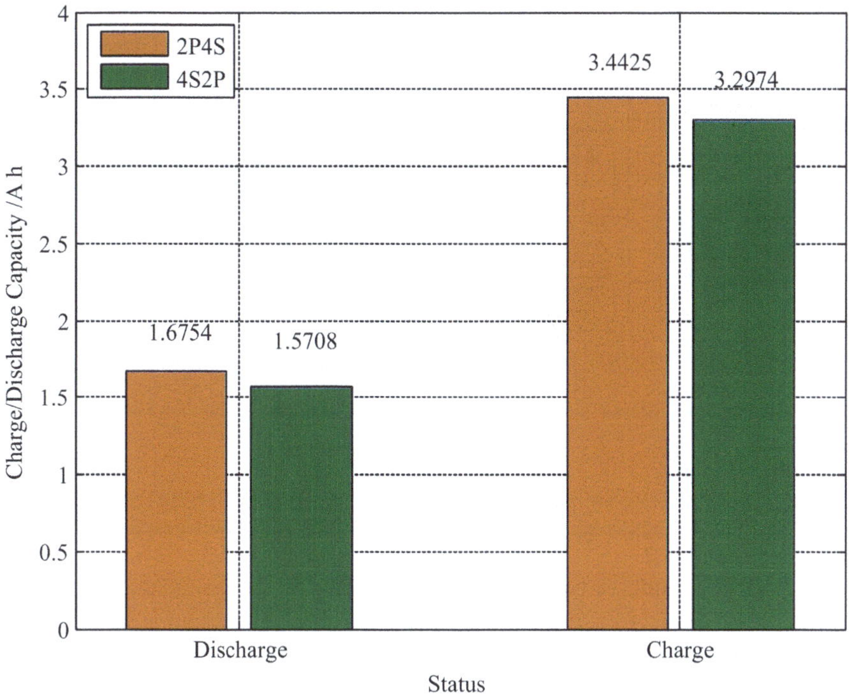

Fig. 5.6 The charge/discharge electricity of the two kinds of battery packs

estimation are completed based on the quantitative analysis on the inconsistency of the battery pack. The "average model", namely the "large cell model", is used in the rough modeling of the entire battery pack, and the difference model is utilized to quantify the inconsistency of every single cell. Then the battery pack SOC estimation is realized by means of the "average model" and the "difference model". Note that:

① Since inconsistency is hard to avoid, it is difficult to accurately estimate the state of the battery pack by simply adopting the "large cell model".

② The "large cell model" can be replaced by the selected model, which can characterize the battery cells. In this case, the "difference model" is quantified by the mapping the relationship between the voltage uncertainty deviation of the characterized cells and the residual cells and their descriptive parameters (SOC, charge/discharge rate, capacity, etc.)

5.2.1 The Inconsistency Analysis of Battery Pack

Owing to the difference of initial performance parameters and external service conditions, the inconsistency of the battery pack is difficult to avoid and results in the

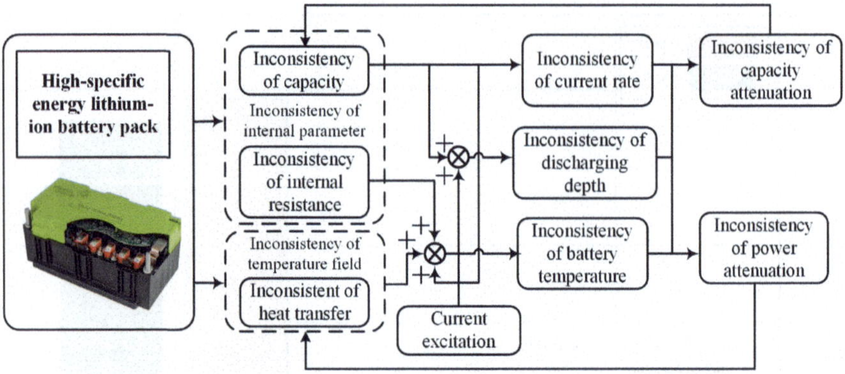

Fig. 5.7 Internal and external factors affecting battery consistency and their interactions

differences in battery characteristics such as current, voltage, temperature. These differences further aggravate the inconsistency of cells [2]. Their interaction effects are shown in Fig. 5.7.

The analysis results of battery aging mechanism and the degradation rate show that the key factors of battery performance attenuation are operating temperature, operating voltage, depth of discharge, current rate, etc. For each cell in a battery pack, the actual charge and discharge current, temperature rising, and operating voltage vary greatly under the same current condition due to the uncertainty differences of the available capacities and internal resistance. The inconsistency of temperature rising of each cell could be easily deteriorated by the unreasonable optimization of the temperature field of the battery system. The interaction of the above factors in the actual operation process would further give rise to the inconsistency of the attenuation rates of each cell. It is presented as the disaccord decline rates of the available capacity, the internal resistance, and power, forming the positive feedback effect of inconsistent amplification.

Initial inconsistencies such as the available capacity and internal resistance of cell have a great influence on battery characteristics, and will aggravate secondary inconsistencies in application. Therefore, in order to improve the efficiency and reliability of battery system in electric vehicles, it is necessary to screen strictly the initial inconsistencies when battery cells are connected to battery pack. After that, the secondary inconsistencies of batteries should be quantitatively evaluated and monitored in use.

5.2.2 *Battery Screening Methods*

The battery screening process is a process of eliminating a very small number of different battery cells or classifying them according to different cluster centers according to the inconsistency information reflected by the battery cells on certain parameters

Table 5.2 Analysis and comparison of four battery screening methods

Screening methods	Advantages	Disadvantages
Single parameter sorting method	Simple, low data amount, and high efficiency	Single information, and high limitation to the application environment
Multi-parameter sorting method	Comprehensive parameter information and mature data processing means	Failure in reflecting the change of dynamic characteristics, and demand of multiple measurements to obtain the screening quantity
Curve feature sorting method	Comprehensive response information	Large amount of curve recognition data, complex clustering, and large workload
Electrochemical impedance spectrum sorting method	Strong physical significance	High requirements on the testing equipment, harsh conditions, and poor batch testing operability

or characteristic curves. Commonly employed screening methods can be divided into single parameter sorting method, multi-parameter sorting method, curve feature sorting method, and electrochemical impedance spectrum sorting method according to different screening parameters or characteristic curve information. The characteristics of the four methods are listed in Table 5.2 [3].

As a mature data processing method, the multi-parameter sorting method could perform correlation calculation between the parameter variables for the multi-parameter variable information presented by plenty of batteries, thereby reducing the data at the time of screening. However, this method must rely on static parameter information, ignoring the dynamic change process. The curve feature sorting method utilizes a standard charge–discharge test curve to screen the battery cells. With a lot of battery characteristic information contained on the charge–discharge curve, it is reasonable to sample the voltage characteristic curve at an equal distance and extract the voltage difference at the corresponding point of the curve, after which the cell classification is implemented on basis of the set threshold, and finally the screening is completed. High-frequency sampling is demanded preventing missing some key information. The huge number of batteries could easily generate massive statistical data, bringing a decrease of sorting efficiency.

Therefore, the combination of battery curve features and multi-parameter information would manage to transform the curve recognition into data processing of multi-screening variables. Then, the differences of batteries could be classified comprehensively and effectively using the data processing method in multi-parameter screening, thus a new battery screening method called dynamic parameters-based cell screening method is proposed [3].

Dynamic parameters-based battery screening method

According to the cell voltage and current characteristic information and the ECM, the electrochemical reaction process is analyzed in-depth to extract the key information and construct the characteristic vector. Combined with the data processing method in multi-parameter sorting, the data extraction process is further simplified to obtain the characteristics that can represent the battery category and conduct cell screening. Based on the above idea, the battery charge–discharge curves are adopted to extract the characteristic quantities representing the key information of the battery and conduct correlation analysis. With the obtained uncorrelated new characteristic quantities, the batteries are clustered and the classification results are evaluated under various operation conditions, current and SOC conditions. The specific steps are as follows:

(1) Extraction of key feature information

The battery characteristic quantities are extracted using the CCCV curve. A complete cycle of charge and discharge of the standard capacity experiment of battery 6-cell 01 at 25 °C is shown in Fig. 5.8. Point A represents the discharge cutoff time of the last cycle, and point B represents the discharge cutoff time of the current cycle.

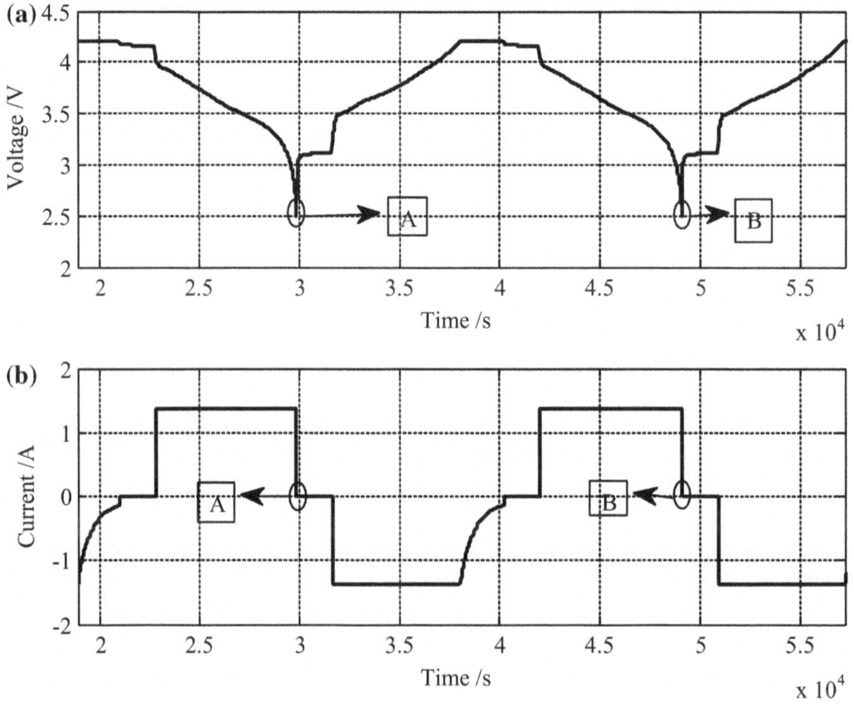

Fig. 5.8 The standard charge–discharge voltage and current curves of battery 6-cell 01 (one cycle contained): **a** voltage curve; **b** current curve

The following analysis is according to the voltage and current curves in Fig. 5.8. After discharge current of the last cycle at point A, the terminal voltage experiences an upward trend in which the rapid rises is mainly caused by the discharge ohmic resistance while the slowly rises mainly reflects the battery polarization effect. The mapping relationship between charging capacity and aging state can be established based on the charging curve in a specific voltage range.

The process of the characteristic curves can be analyzed from the perspective of ECM parameters, variables related to the ohmic resistance, polarization characteristics, aging state, and charge–discharge capacity can be extracted. Based on the analysis results of battery charge–discharge curves, a comprehensive characterization of the corresponding characteristics of the battery is obtained using five curve characteristic quantities from F_1 to F_5:

F_1: The voltage rise within 1 s after discharge to the cutoff voltage (characteristic of the ohmic internal resistance of discharge);
F_2: The voltage rise within 1–100 s after discharge to the cutoff voltage (characteristic of the dynamic polarization of battery);
F_3: The voltage rise in 1 s at the beginning of the constant-current charging phase (characteristic of the ohmic resistance of charge);
F_4: The ratio of the charge amount in the constant-current charging phase to that in the constant voltage charging phase (dual representation of the charging capacity and the degree of battery aging);
F_5: Count the discharge capacity of the entire constant-current phase.

It should be noted that voltage rapid recovery component caused by the discharge ohmic internal resistance of the previous stage is apparently higher than the voltage transient response caused by the charge ohmic internal resistance at the beginning of charging after the discharge is cut off and the current in the constant-current phase is equal, which indicates that the discharge ohmic resistance at the end is higher than the charge ohmic resistance, both of which should be discussed separately when extracting feature quantities. The more severe the aging status of the battery, the easier it is to reach the cutoff voltage at the end of the CC phase. Since the electric quantity charged in the constant-current phase is to be reduced consequently, a corresponding change of the ratio between the electric quantity charged in the CC phase and that in the CV phase is to take place, which could be employed as the characteristic value to quantify battery aging status.

Using the above five characteristic quantities, the ohmic effect, polarization characteristics, charge–discharge capacity, and battery aging state are able to be fully characterized. As these five feature quantities are directly extracted from the curve, the battery screening process is simplified and the screening efficiency is improved.

(2) Cluster analysis

Cluster is to divide a certain number of things that have the same characteristics into several categories according to certain criteria, which is commonly used for data sorting and screening. When it comes to battery screening, it is necessary to classify them quantitatively to comprehensively evaluate the consistency of the five characteristic quantities. The extracted characteristic quantities of each cell are converted into each dimension of the geometric space. Every cell is a point in the geometric space, and then calculates the distance between points. The expression is

$$d = \sum_{m} (x_i^m - x_{i+1}^m)^2 \tag{5.3}$$

where m donates the dimension, and d donates the space distance between the ith cell and $i + 1$th cell.

Due to a certain coupling relationship among the above five characteristic quantities, the fewer new factors can be extracted to explain original characteristic quantities by factor analysis. Whether the characterization is reasonable is judged by the degree of interpretation of the new factors. If the interpretation ratio exceeds 90%, the new factors can be used as the dimension of the cluster.

All characteristic quantities of cells are imported into the Statistical Product and Service Solutions software (SPSS) to generate a correlation matrix among them for factor analysis. It should be noted that the matrix characteristics must be checked before the factor analysis, that is, the corresponding probability of the Bartlett sphericity statistics should be less than the given significant level. The factor matrix and the scoring coefficient matrix are eventually obtained to describe the relationship between characteristic quantities and new factors.

(3) Battery grouping test

To evaluate the effectiveness of the screening methods, all battery clusters need to be tested under specific operation conditions. If the consistency of all samples is high, the voltage response of the battery clusters has high consistency, which illustrates a close distance a among the response vectors, the average distance of the voltage response vector of battery clusters is calculated to evaluate the consistency of the battery clusters. The value of a is obtained by

$$a = \frac{\sum_{i=1}^{k-1} \sum_{m=i+1}^{k} norm(\gamma_i - \gamma_m)}{C_k^2} \tag{5.4}$$

where k represents the total number of clusters, and γ_i represents the corresponding voltage vector.

The dimensions of the voltage response vector of all samples must be the same in statistics, which means the same number of current excitation points. In addition, since the distance refers to the mean value of the distance between every two batteries, the size of the battery clusters would not affect the evaluation of consistency. If the number of samples of the battery clusters to be screened changes, the consistency of the battery clusters can still be evaluated.

Case study

(1) Experiments

The experimental object is the battery 6 and the experimental temperature is 10 °C. In order to maintain a sufficient sample size, 16 cells were used as the initial samples, numbered as cell 09–24. Different test conditions such as DST and HPPC were used to evaluate the effectiveness of this method. The process of test is as follows:

(a) Charge/discharge test is used to obtain the charge/discharge curve of the battery clusters;
(b) DST is used to extract the voltage responses of each cell. The average distance *a* of the voltage response vector among the battery clusters is calculated with Eq. (5.4), and then the effectiveness of the screening methods is evaluated.
(c) HPPC test is used to obtain the voltage responses of each cell at three SOC points (90, 50, and 20%) covering high, medium, and low SOC stages under pulse current. For the current pulse at 20% SOC point, two C rates (0.5C and 1.0C) should be calculated.

(2) Extraction of characteristic quantities

According to the above screening methods, the extraction results of characteristic quantities in the standard charge/discharge voltage characteristic curves of 16 battery cells are shown in Table 5.3.

It can be seen that the characteristic quantities vary greatly for different batteries, which reflects the degree of inconsistency between battery clusters.

(3) Factor analysis and cluster

The curve characters F_1–F_5 were extracted to form the database of battery inconsistency to be screened, and factor analysis was performed by SPSS. Bartlett test was performed using the correlation matrix between characteristic quantities at first. Because the significant level calculated from the experimental data is less than 0.001, factor analysis can be carried out. Then the characteristic quantities are converted to fewer factors.

The total variance of the characteristic quantities interpretation was calculated after the factor analysis by SPSS. The new factors T_1–T_6 were obtained by principal component analysis shown in Table 5.4, which is the interpretation of the characteristic quantities F_1–F_5 of cell 09. T_1–T_3 can be selected as the screening variables because the cumulative interpretation of them on F_1–F_5 accounts for more than 90%, and then the new factors of all battery clusters were calculated.

The factor matrix between the new factors T_1–T_3 and the original characteristic quantities are shown in Table 5.5, where the numbers represent how to characterize the original characteristic quantities based on the new factors, such as $F_1 = 0.98T_1 - 0.098T_2 - 0.092T_3$, and the value reflects the characterization degree.

The score coefficient matrix is shown in Table 5.6. According to this matrix, the new factors are calculated using the original characteristic quantities. For example,

Table 5.3 Extracted characteristic quantities F_1–F_5 in the charge/discharge curves of 16 battery cells

Numbers	F_1	F_2	F_3	F_4	F_5
Cell 09	0.5968	0.0223	0.1937	5.9898	2.4407
Cell 10	0.5968	0.0239	0.1928	5.7952	2.4437
Cell 11	0.5980	0.0189	0.1913	5.6116	2.4390
Cell 12	0.6200	0.0198	0.1931	5.8457	2.4483
Cell 13	0.6020	0.0220	0.1875	5.9784	2.4557
Cell 14	0.6092	0.0205	0.1910	6.1898	2.4510
Cell 15	0.6300	0.0186	0.1950	5.9550	2.4653
Cell 16	0.6048	0.0193	0.1925	5.8352	2.4303
Cell 17	0.5075	0.0514	0.1894	5.3751	2.4420
Cell 18	0.5131	0.0515	0.1897	5.4049	2.4213
Cell 19	0.5180	0.0511	0.1906	5.5718	2.4393
Cell 20	0.5118	0.0571	0.1844	5.7935	2.4403
Cell 21	0.5140	0.0508	0.1866	5.8852	2.4177
Cell 22	0.5137	0.0546	0.1922	5.3453	2.4540
Cell 23	0.5137	0.0471	0.1854	6.0164	2.4220
Cell 24	0.5038	0.0558	0.1857	5.6622	2.4253

Table 5.4 Interpretation ratio of new factors to the original characteristics

Number of new factors	Interpretation ratio (%)	Cumulative interpretation ratio (%)
T_1	42.045	42.045
T_2	39.734	81.779
T_3	8.583	90.362
T_4	5.448	95.810
T_5	4.039	99.849
T_6	0.151	100.000

Table 5.5 Factor matrix

Initial characteristics	T_1	T_2	T_3
F_1	0.980	−0.098	−0.092
F_2	−0.952	0.145	0.192
F_3	0.759	0.525	−0.298
F_4	0.544	−0.800	0.146
F_5	0.695	0.391	0.604

Table 5.6 Score coefficient matrix of factors

Initial characteristics	T_1	T_2	T_3
F_1	0.299	0.201	−0.045
F_2	−0.387	−0.206	0.220
F_3	0.663	−0.376	−0.174
F_4	−0.277	0.748	0.018
F_5	−0.380	−0.030	1.174

the expression of screening variable T_1 is $T_1 = 0.299F_1 - 0.387F_2 + 0.663F_3 - 0.277F_4 - 0.380F_5$. Similarly, the corresponding screening variables T_1–T_3 can be obtained by the characteristic quantities F_1–F_5 of 16 cells. T_1–T_3 can be determined as the cluster dimensions owing that the interpretation ratio of the original five characteristic quantities accounts for more than 90%. In this way, all sample cells can be mapped to points in the cluster space.

According to the three dimensions of T_1–T_3, K-means cluster method is used to classify the 16 battery cells. The flowchart of the K-means cluster method is illustrated in Fig. 5.9. K samples are randomly selected as the initial cluster centers, and then the distance between the remaining samples and each cluster center is calculated to distribute each sample to the nearest cluster center. Once all the samples are allocated, each center of cluster is recalculated based on the existing samples in the category to minimize the sum of squared errors locally. It does not terminate until the termination condition is satisfied which is no (or minimum number of) samples are reassigned to the corresponding categories or no (or minimum) changes in cluster centers.

(4) Evaluation of screening results

All the screened samples are evaluated under DST and HPPC at different SOC points. The value of a is calculated and the screening effectiveness is evaluated.

Table 5.7 shows that the screening results have achieved good effects in a wide SOC range and various current rates, especially in the case of high SOC and high current rates. Consequently, this method impressively facilitates the consistency of battery clusters.

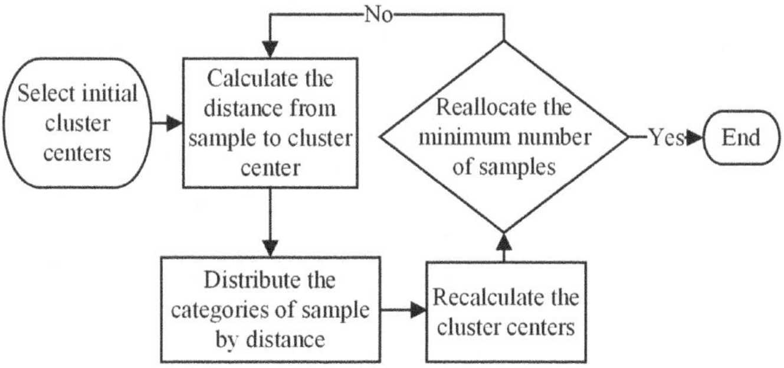

Fig. 5.9 Flowchart of K-means cluster method

Table 5.7 Comparison of calculation results of *a* before and after screening

Samples	Current pulse excitation of DST	HPPC (90%SOC, current pulse of 0.5C)	HPPC (50%SOC, current pulse of 0.5C)	HPPC (20%SOC, current pulse of 0.5C)	HPPC (20%SOC, current pulse of 1.0C)
a (V) (before screening)	0.0263	0.0243	0.0185	0.0118	0.0164
a (V) (after screening)	0.0231	0.0167	0.0118	0.0109	0.0139

5.2.3 *Quantitative Methods of Inconsistency*

When it comes to the vehicle operation application, the inconsistency of the battery pack is required to be evaluated. The results are important to the battery life analysis, safety management, state estimation, balancing management and after-sales maintenance. The difference model of battery pack is an effective way to realize the real-time calculation of the pack inconsistency. Based on that, the inconsistency of battery pack can be quantified in real time and the SOC of battery pack can be estimated.

The accurate SOC estimation of battery packs depends on the available capacity of battery cells and their SOC. If cells in pack with large differences in available capacity and SOC are known, the calculation efficiency of state estimation would be significantly improved. Thus, the cell difference analysis of available capacity and SOC is the foundation for high-accuracy estimation of SOC of the battery pack. The difference of the maximum available capacity is directly related to the SOC difference, and there is a monotonous mapping relationship between the SOC and OCV. Therefore, the inconsistency of OCV between cells can approximately represent the capacity inconsistency within a certain range, that is, the inconsistency of OCV obtained by the difference model is able to identify the capacity inconsistency.

Modeling theory

Regarding a certain cell in pack as a reference cell, the difference model mainly considers the difference in internal resistance and OCV between the reference cell and others as shown in Fig. 5.10. In this figure, ΔE_i, ΔR_i, and ΔU_i represent the OCV, internal resistance, terminal voltage difference between the cell numbered *i* and the reference cell, respectively.

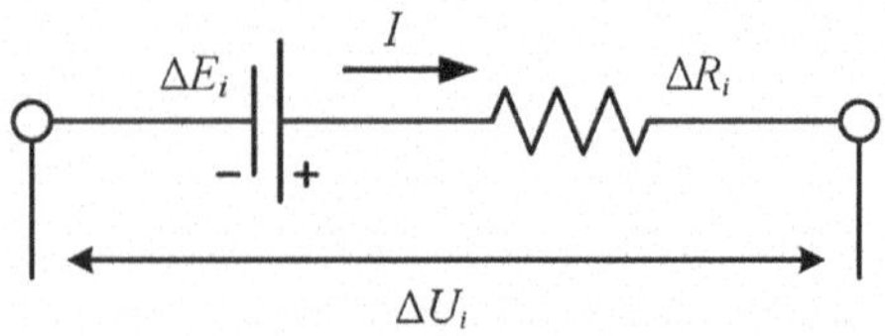

Fig. 5.10 Battery difference model

The discrete state equation of the model is

$$\Delta U_i(k\Delta t) = \Delta E_i - \Delta R_i I(k\Delta t) \tag{5.5}$$

The standard expression of the above equation is

$$\boldsymbol{u}_i = \boldsymbol{A}_i \boldsymbol{x}_i \tag{5.6}$$

where

$$\boldsymbol{u}_i = \begin{bmatrix} \Delta U(k\Delta t) \\ \cdots \\ \Delta U((k+n-1)\Delta t) \end{bmatrix}, \boldsymbol{A}_i = \begin{bmatrix} 1 & -I(k\Delta t) \\ \cdots & \cdots \\ 1 & -I((k+n-1)\Delta t) \end{bmatrix}, \boldsymbol{x}_i = \begin{bmatrix} \Delta E_i \\ \Delta R_i \end{bmatrix}$$

Since the cell inconsistency changes gradually, which makes it feasible to identify its OCV difference by using data within a time interval. The length of the time interval is known as window length. In fact, a tiny change is observed in the cell SOC within a short window and the aging state can be ignored, while the OCV of cells is affected by the above two factors (SOC and aging state). So the OCV of cells changes very little, which can be ignored as well. Therefore, the OCV difference obtained during this period can be taken as the OCV difference at any time within the window.

As for a long-time operation condition, a moving window is used to identify the OCV difference. The small magnitude of OCV difference makes it easy to be disturbed, bringing large errors whose high-frequency components can be eliminated by a low-pass filter. In order to overcome the biased estimation caused by the measurement errors of both the coefficient matrix and the output matrix, the total least squares (TLS) method is used in each window length to achieve an unbiased estimation of parameters.

Case study

The experiment was conducted on the battery 1 with the configuration of 7S1P, which was numbered as cell 03–09, respectively. To evaluate the effectiveness of this method, cells with different aging states were selected and connected. The calibration capacity of each cell is shown in Table 5.8. The seven cells can be divided into three

Table 5.8 Cell calibration capacity of 7S1P battery pack at 25 °C

Number	Charge capacity (A h)	Discharge capacity (A h)
Cell 03	28.14	28.08
Cell 04	28.48	28.44
Cell 05	27.12	27.11
Cell 06	24.47	24.45
Cell 07	27.25	27.23
Cell 08	28.30	28.24
Cell 09	28.24	28.19

groups according to the aging state: the first group includes cell 03, cell 04, cell 08, and cell 09 which are totally new. Cell 05 and cell 07 compose the second group whose capacities have declined about 5%. The last group is cell 06 whose capacity has been reduced to almost 85%.

The inconsistency of seven cells was quantified under UDDS. The OCV differences between the reference cell (cell 03) and other 6 cells were identified by the TLS method. The filtered results are shown in Fig. 5.11. Owing to the slow time-varying characteristics of OCV differences between cells, low-pass filtering manages to eliminate the high-frequency noise and remove the peaks and rags on the result curves without changing to the results, which is helpful to analyze the change trend of the curves.

It can be seen from Fig. 5.11 that the curves are divided into three categories.

(a) The first category consists of OCV difference curves of cell 04, cell 08, and cell 09, these three curves have almost no fluctuations and the value is always close to zero over the entire operation range, indicating that cell 04, cell 08, and cell 09 possess a high degree of consistency and a tiny difference from cell 03 as well.
(b) The second category consists of OCV difference curves of cell 05 and cell 07 which are highly coincided overall, but a certain difference from OCV of the reference cell exists in the initial stage. As the discharge continues, the OCV difference exhibits a small amplitude fluctuation and the OCV difference gradually decreases. The reason is that the amount of electricity released by all series-connected cells is equal during the same time. However, the maximum available capacity is different due to the different aging states, leading to the SOC of different cells change inconsistently.

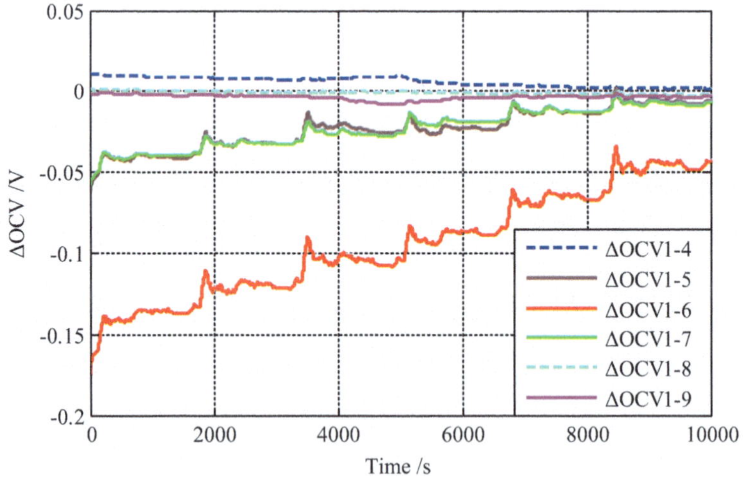

Fig. 5.11 Result curves of OCV difference

(c) The third category only contains cell 06, the OCV difference of which is greatly different at the initial stage and fluctuates obviously with the discharge, indicating the severe inconsistency with the reference cell.

The results show that the OCV difference is consistent with the capacity inconsistency at the initial stage of the operation condition. Therefore, the capacity inconsistency of pack can be quantified based on the difference model. The difference of maximum available capacity approximately obtained based on OCV difference is ultimately applied to the SOC estimation of the series-connected battery pack.

5.2.4 Modeling of Battery Pack

The modeling method of the battery pack is different from that of the cell. According to the inconsistency of cells in battery pack, the model of battery pack is classified into average model and selected cell model [4]. The former can be regarded as a battery cell with large capacity based on the battery pack of excellent consistency. Owing to the poor consistency, the latter needs a characteristic screening of cells based on inconsistent quantitative indexes to acquire the cell which is able to characterize the battery pack performance. Based on the theory of ECM in Chap. 3, the selected cell model can be built.

Average model of battery pack

The terminal voltage of the series-connected battery pack is the sum of the terminal voltage of each cell. There is an assumption that the structure of battery pack is nS1P where each cell is modeled by Thevenin model. Thus, the pack model is equivalent to the superposition of each Thevenin model. As can be seen in Fig. 5.12, the model includes the OCV $U_{\mathrm{oc},i}$, the ohmic internal resistance $R_{i,i}$, and the RC network $C_{\mathrm{D},i}$ and $R_{\mathrm{D},i}$ mapping polarization effect, where $i = 1\ldots n$. Those model parameters can be simplified to the total OCV $U_{\mathrm{oc,N}}$, total ohmic internal resistance $R_{\mathrm{i,N}}$, RC network $C_{\mathrm{D,N}}$, and $R_{\mathrm{D,N}}$ representing the battery pack polarization, and the total terminal voltage $U_{\mathrm{t,N}}$ according to principle of equivalent circuit simplification.

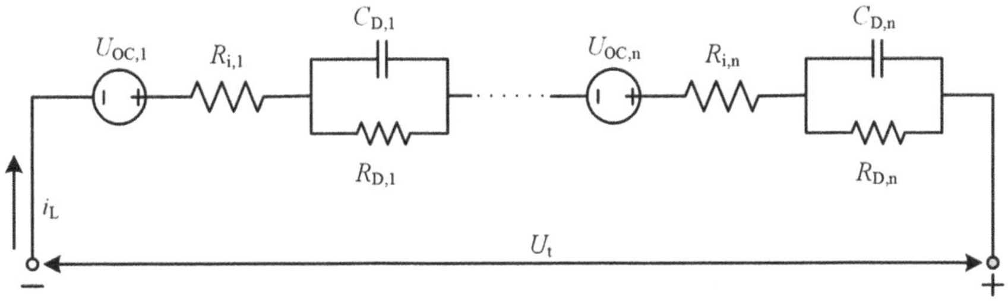

Fig. 5.12 Lumped parameter model of battery pack

The corresponding assumptions and simplified processes are shown as follows:

$$U_{OC,N} = \sum_{i=1}^{n} U_{OC,i} = NU_{OC} \tag{5.7}$$

$$R_{i,N} = \sum_{i=1}^{n} R_{i,i} = NR_{i} \tag{5.8}$$

$$R_{D,N} = \sum_{i=1}^{n} R_{D,i} = NR_{D} \tag{5.9}$$

$$C_{D,N} = C_{D,i} = C_{D} \tag{5.10}$$

$$U_{t,N} = U_{OC,N} - U_{D,N} - U_{R,N} = \sum_{i=1}^{n} U_{t,i} = NU_{t} \tag{5.11}$$

In view of the above-simplified analysis, the external characteristics of the battery pack can be represented by Thevenin model. The parameters identification in the average model can be obtained based on the current and mean terminal voltage of each cell. The estimated terminal voltage of battery pack is the product of the terminal voltage of the average model and the number of cells in series. However, the physical meaning of parameters is different from Thevenin model of cell, as shown in Fig. 5.13.

The discrete parameter identification method of this model can refer to Sect. 3.2 in Chap. 3. In this model, i_L donates the total current flowing through the bus bar of the battery pack, terminal voltage U_t donates the mean value of all cell voltages, and U_{oc} donates the mean value of all cell OCVs in the battery pack.

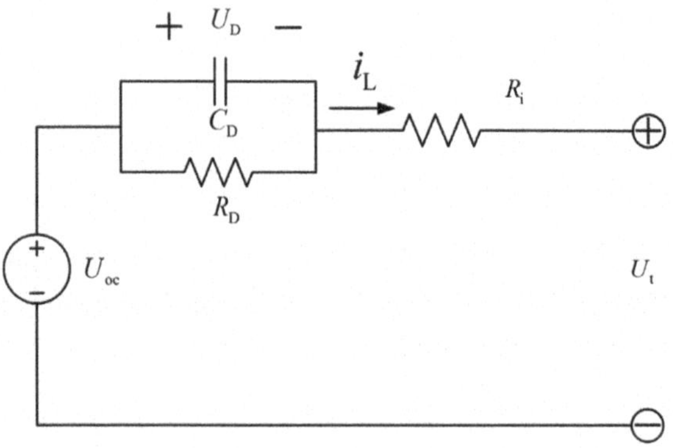

Fig. 5.13 Average model of battery pack

Selected cell model of the battery pack

The selected cell can represent the characteristics of the battery pack to some extent, it is essential to select a representative cell from pack using the inconsistency quantization method and describe the characteristic cell model in a mathematical way [5–7]. This section will concentrate on the strategy of screening the selected cell rather than the modeling method which has already been discussed in detail in Sect. 3.2 of Chap. 3.

As significant indexes to quantify the inconsistency of battery pack, the OCV difference and internal resistance difference of battery cells are selected for screening. The voltage and SOC of the battery pack are presented by certain cells in a specific category. The screening principle of the selected cell is shown in Fig. 5.14.

The screening method described in Fig. 5.14 is divided into two steps including the screening of OCV and ohmic internal resistance. The involved screening quantities are the identified value based on the difference model introduced in Sect. 5.2.3 rather than the experimental results. Different cell categories can be eventually obtained through screening, that is, the OCV and ohmic internal resistance of the selected cell are closest to the average of cells in this category.

The specific steps of the screening of the selected cells are as follows.

(1) The first step of screening is carried out based on OCV. Calculate the average voltage of the battery pack and apply the Thevenin model to identify the parameters of the average voltage model. The OCV difference is identified based on the difference between mean voltage and cell voltage. The cells are classified according to the category of the cells in pack to complete the first step.
(2) The second step of screening is carried out based on the ohmic resistance. The resistance difference reflected in the volatility of terminal voltage can realize the rescreening of cells. Any cell is randomly selected from each category of battery, remove the selected cells in the first step, and then all the selected cells of battery packs are screened. The comprehensive characterization of pack inconsistency, aging and dynamic characteristics. It should be noted that the screening algorithm is suitable for not only the charging conditions, but also the real-time operation conditions.

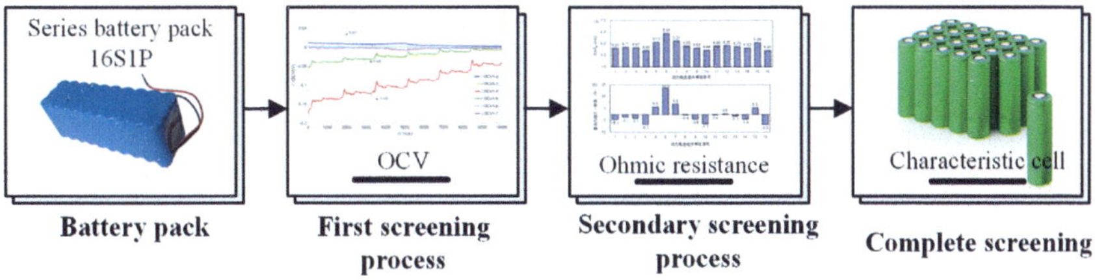

Fig. 5.14 Principle of characteristic cell screening in battery pack

5.2.5 State Estimation of Battery Pack Based on Selected Cells

SOC of the battery pack can be defined as the ratio of the maximum capacity that a battery pack can be discharged to its maximum available capacity under certain conditions. Regardless of balancing control, the SOC of battery pack in series and parallel are defined, respectively, as follows [8].

The SOC of battery pack in series
For battery pack in series, the maximum available capacity is defined as the sum of the maximum discharge capacity and the maximum charge capacity of the battery pack under certain conditions, which is

$$\begin{cases} C^{\mathrm{s}}_{\mathrm{pack-chg}} = \min\{C_{\mathrm{a1}}(1 - z(1)), C_{\mathrm{a2}}(1 - z(2)), \ldots, C_{an_{\mathrm{s}}}(1 - z(n_{\mathrm{s}}))\} \\ C^{\mathrm{s}}_{\mathrm{pack-dis}} = \min\{C_{\mathrm{a1}}z(1), C_{\mathrm{a2}}z(2), \ldots, C_{an_{\mathrm{s}}}z(n_s)\} \end{cases} \tag{5.12}$$

$$\begin{aligned} C^{\mathrm{s}}_{\mathrm{pack}} = {} & \min\{C_{\mathrm{a1}}(1 - z(1)), C_{\mathrm{a2}}(1 - z(2)), \ldots, C_{an_{\mathrm{s}}}(1 - z(n_{\mathrm{s}}))\} \\ & + \min\{C_{\mathrm{a1}}z(1), C_{\mathrm{a2}}z(2), \ldots, C_{an_{\mathrm{s}}}z(n_{\mathrm{s}})\} \end{aligned} \tag{5.13}$$

where $C^{\mathrm{s}}_{\mathrm{pack-chg}}$ is the maximum charge capacity of pack in series, $C^{\mathrm{s}}_{\mathrm{pack-dis}}$ is maximum discharge capacity, $C^{\mathrm{s}}_{\mathrm{pack}}$ is the maximum available capacity, $C_{\mathrm{a}i}$ is the maximum available capacity of cell i, and $z(i)$ is the SOC of cell i in which $i = 1 \ldots n_{\mathrm{s}}$.

The battery pack SOC can be defined as

$$\begin{aligned} SOC^{\mathrm{s}}_{\mathrm{pack}} & = \frac{C^{\mathrm{s}}_{\mathrm{pack-dis}}}{C^{\mathrm{s}}_{\mathrm{pack}}} \\ & = \frac{\min\{C_{\mathrm{a1}}z(1), C_{\mathrm{a2}}z(2), \ldots, C_{an_{\mathrm{s}}}z(n_{\mathrm{s}})\}}{\begin{array}{l}\min\{C_{\mathrm{a1}}(1 - z(1)), C_{\mathrm{a2}}(1 - z(2)), \ldots, C_{an_{\mathrm{s}}}(1 - z(n_{\mathrm{s}}))\} \\ + \min\{C_{\mathrm{a1}}z(1), C_{\mathrm{a2}}z(2), \ldots, C_{an_{\mathrm{s}}}z(n_{\mathrm{s}})\}\end{array}} \end{aligned} \tag{5.14}$$

where $SOC^{\mathrm{s}}_{\mathrm{pack}}$ is the SOC of battery pack in series.

The SOC of battery pack in parallel
Self-balancing in the parallel battery pack can fully charge each cell. Thus the maximum available capacity of battery pack in parallel can be defined as:

$$C^{\mathrm{p}}_{\mathrm{pack}} = \sum_{i=1}^{N_{\mathrm{p}}} C_{\mathrm{a}i} \tag{5.15}$$

where $C^{\mathrm{p}}_{\mathrm{pack}}$ donates the maximum available capacity of the battery pack in parallel and N_{p} is the number of parallel cells.

The SOC of the battery pack in parallel can be defined as

$$SOC_{\text{pack}}^{\text{p}} = \frac{C_{\text{pack-dis}}^{\text{p}}}{C_{\text{pack}}^{\text{p}}} = \frac{\sum_{i=1}^{N_{\text{p}}} C_{\text{a}i} z(i)}{\sum_{i=1}^{N_{\text{p}}} C_{\text{a}i}} \tag{5.16}$$

where $SOC_{\text{pack}}^{\text{p}}$ donates the SOC of the battery pack in parallel and $C_{\text{pack-dis}}^{\text{p}}$ donates the remaining capacity of the battery pack in parallel.

It should be noted that for the series-connected cells in parallel branch, the maximum available capacity and SOC of each branch can be obtained using Eqs. (5.13) and (5.14), and then the capacity and SOC of battery pack in parallel can be calculated as a "single cell".

Taking the battery pack in series as an example, the SOC estimation method and the case study are introduced as follows:

Algorithm framework

As can be seen from Eq. (5.14), the SOC of the battery pack in series depends on the capacity and SOC of all cells. As ignoring those cells with large differences, it is difficult to accurately obtain the SOC of battery pack when selected cells with good consistency of capacity and SOC are used. The accuracy of SOC estimation of battery pack in series is mainly affected by the cells with obvious inconsistency. Therefore, this section will give a detailed method for estimating the SOC of a series pack based on the selected cells [8]. The flowchart of this method is shown in Fig. 5.15.

(a) Data collection: load the driving cycle into the battery pack, and collect the real-time current and voltage of each cell in series, where $U_{\text{t},k}$ donates the terminal voltage of the kth cell.
(b) Screening of the selected cells: difference identification refers to the identification of OCV and ohmic differences based on the average model using the real-time collected current and voltage data as output. A low-pass filter method is adopted to filter the noise component of the experiment to acquire smooth difference curve cluster. The method in Sect. 5.2.4 is used to complete the screening of selected cells [9, 10].
(c) Online estimate parameters of selected cells: using the data-driven method to update the parameters of the selected cell model in real time whose specific process of can refer to the relevant parameter identification algorithms in Chap. 3.
(d) Online estimate the SOC of selected cells: The SOC estimation can be realized based on EKF whose detailed process can refer to Chap. 4.
(e) Calculation of battery pack SOC: the SOC estimation results of all selected cells are substituted into the Eq. 5.14 to obtain the SOC of the battery pack in series.

Case study

The UDDS and NEDC are used to evaluate the SOC estimation method of 7S1P battery pack. In order to verify the robustness and accuracy, the algorithm was evaluated in combination with UDDS discharge condition and NEDC charge condition with an initial SOC error of 40%. The initial SOC on UDDS and NEDC were set to

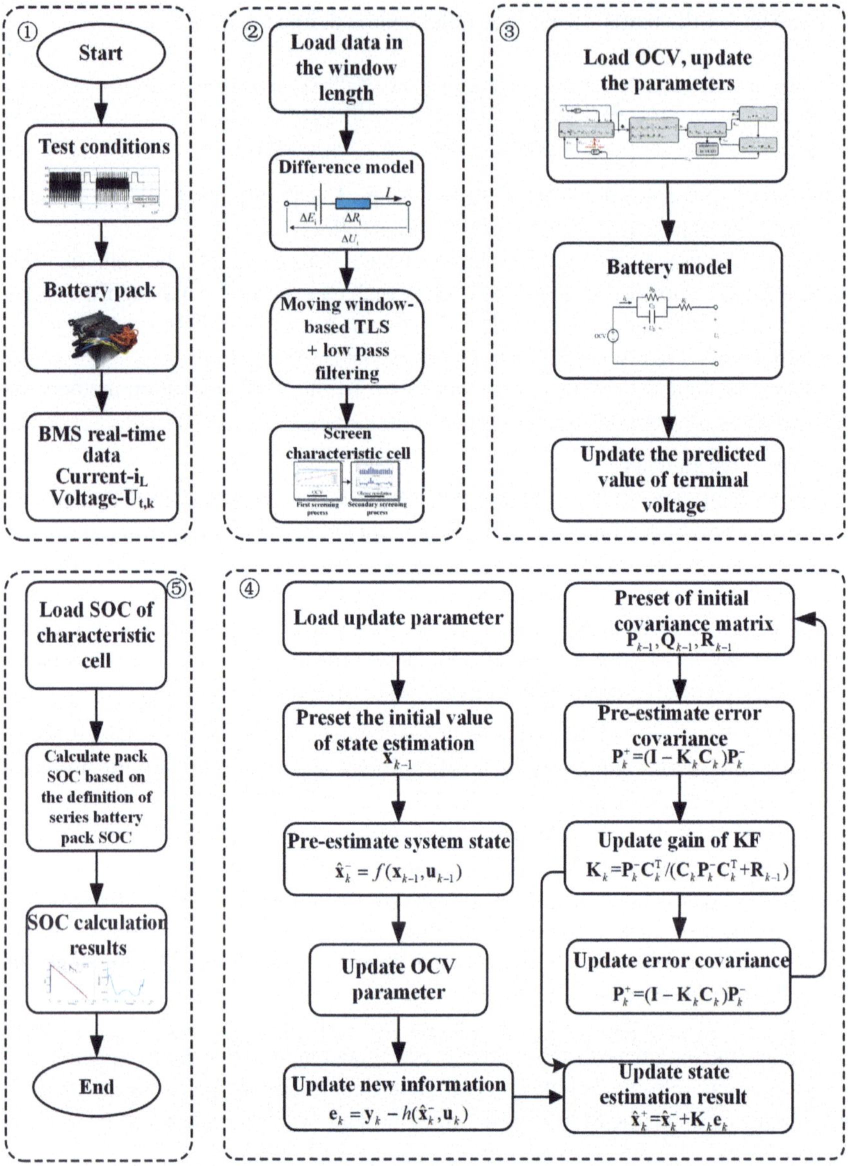

Fig. 5.15 SOC estimation flowchart of the series-connected battery pack

100% and 0, respectively. The reference SOC values of battery pack were obtained by ampere hour integral method whose results are shown in Fig. 5.16.

Figure 5.16 illustrates that the SOC estimation result converges to the reference value with a maximum estimation error less than 1% under the hybrid operation conditions. The detailed statistical results are shown in Table 5.9. The algorithm meets the requirement of application precision and its convergence rate depends on the selected cell with the slowest convergence.

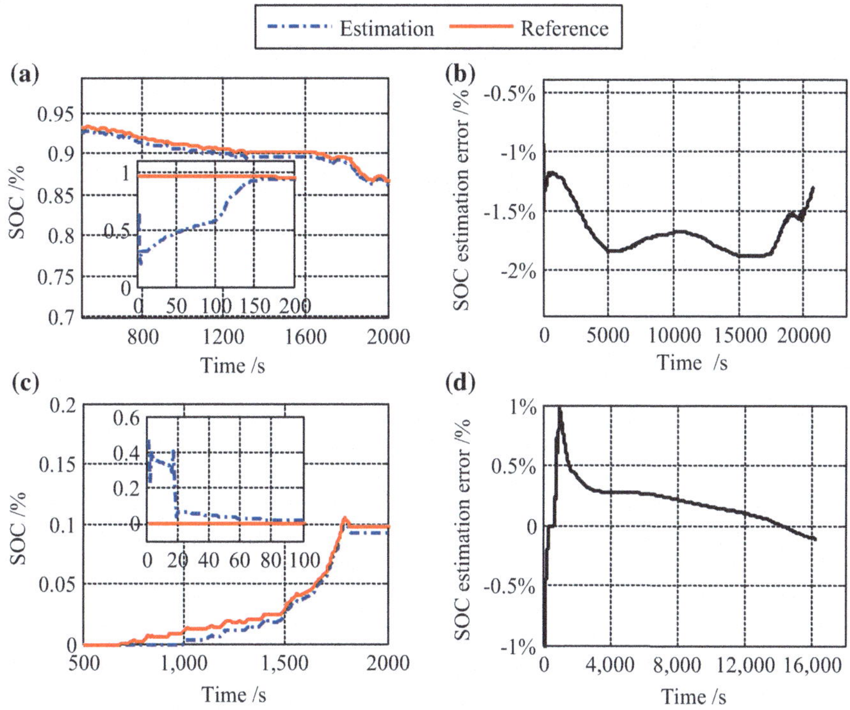

Fig. 5.16 SOC estimation results and estimation errors: **a** result under UDDS; **b** SOC error under UDDS; **c** result under NEDC; **d** SOC error under NEDC

Table 5.9 Estimation error statistics of 7-string pack

	Maximum error (%)	Mean error (%)	Root mean square error (%)
UDDS	0.76	0.47	0.38
NEDC	0.98	0.21	0.18

5.3 SOP Prediction of Batteries

SOP is the maximum power that can be released or absorbed by batteries within a predetermined time interval. SOP estimation is to evaluate the maximum power capability of batteries under different SOC and SOH, and optimize the relationship between battery system and vehicle power performance to meet the acceleration and climbing performances of EVs, and maximize the regenerative braking performance of motors. Influenced by the internal electrochemical dynamics and thermodynamics of the battery, the real-time SOP is subject to its voltage, current, temperature, available capacity, and SOC. This section will elaborate on instantaneous and continuous SOP prediction methods.

5.3.1 Typical Instantaneous SOP Prediction Methods

Instantaneous SOP refers to the maximum power capability that can be charged or discharged of the battery at the following moment. Instantaneous SOP estimation methods mainly include the following four categories:

(1) The HPPC method is to calculate the instantaneous peak current and power based on the upper and lower cutoff voltages of the battery. This method is simple and widely used in the field of vehicle energy management.
(2) The SOC-based method [11] is to calculate the instantaneous peak current and power based on the maximum and minimum SOC constraints. It is often used in combination with other methods.
(3) The voltage-based method is to overcome the shortcomings of the HPPC method and take into account the change of OCV in peak power prediction cycle, so the prediction results are more reliable.
(4) The multi-constrained dynamic (MCD) method [11] is to predict SOP of the battery system in real time, integrating the terminal voltage, current, and SOC. Meanwhile, the influence of electrochemical dynamics, thermodynamics, hysteresis effect, and other dynamic response characteristics on SOP prediction are comprehensively considered.

The implementation details of the above methods are as follows:

HPPC method

HPPC method is a calculation method proposed by Idaho National Engineering and Environmental Laboratory under the cooperation framework of "Partnership for a New Generation of Vehicles (PNGV)" in the United States. This method uses Rint model to estimate SOP of the battery cell. The terminal voltage of each cell in the battery system can be expressed as

$$U_{\mathrm{t,m}}(t) = U_{\mathrm{oc,m}}(z_{\mathrm{m}}(t)) - R_{\mathrm{i}} i_{\mathrm{L,m}}(t) \tag{5.17}$$

where m refers to the mth cell. $U_{oc,m}(z_m(t))$ is the OCV of the mth cell at the SOC of $z_m(t)$, $i_{L,m}(t)$ is the load current of the mth cell, and R_i is the internal resistance of each cell during charging or discharging whose value depends on the direction of current. Considering that the terminal voltage of the battery is limited by the upper and lower cutoff voltages, the peak current of charge and discharge of the mth cell can be calculated as

$$\begin{cases} i_{max,m}^{chg,HPPC} = \frac{U_{oc,m}-U_{t,max}}{R_{chg}} \\ i_{min,m}^{dis,HPPC} = \frac{U_{oc,m}-U_{t,min}}{R_{dis}} \end{cases} \tag{5.18}$$

where $U_{t,max}$ and $U_{t,min}$ are the charging and discharging cutoff voltages of the mth cell, respectively, $i_{min,m}^{chg,HPPC}$ and $i_{max,m}^{dis,HPPC}$ are the peak charging and discharging currents of the mth cell, respectively. It is assumed that the battery pack consists of n cells in which n_s modules are connected in series, and each module is composed of n_p cells in parallel. Then the peak power of the battery pack can be expressed as

$$\begin{cases} P_{min}^{chg} = n_s n_p \max_m (U_{t,max} i_{min,m}^{chg,HPPC}) \\ P_{max}^{dis} = n_s n_p \min_m (U_{t,min} i_{max,m}^{dis,HPPC}) \end{cases} \tag{5.19}$$

HPPC method can predict the instantaneous SOP of the battery system, but it cannot predict the continuous SOP. In addition, the Rint model cannot accurately describe the dynamic voltage characteristics of battery, which leads to overcharge or over-discharge issues caused by the overestimated SOP, threatening the safety in applications. Therefore, this method has two main shortcomings: (1) it is not suitable for the prediction of continuous SOP; (2) it ignores the constraints of current, power, and SOC, which affects the safety of EVs.

SOC-based method

This method obtains the maximum charging and discharging currents based on the limitation of maximum or minimum SOC, and then calculates the SOP. Starting from the current time t, the battery discharges (or charges) at a constant current within a given time interval Δt, and the SOC of the mth cell at $t + \Delta t$ can be expressed as

$$z_m(t + \Delta t) = z_m(t) - i_{L,m}(t)\left(\frac{\eta_i \Delta t}{C_{max}}\right) \tag{5.20}$$

where η_i is the efficiencies of charging and discharging. Considering the upper and lower limits of SOC (z_{max} and z_{min}), the maximum charge/discharge current of the mth cell is

$$\begin{cases} i_{min,m}^{chg,SOC} = \frac{z_m(t)-z_{max}}{\eta_i \Delta t / C_{max}} \\ i_{max,m}^{dis,SOC} = \frac{z_m(t)-z_{min}}{\eta_i \Delta t / C_{max}} \end{cases} \tag{5.21}$$

The peak current of the battery system based on SOC is

$$\begin{cases} i_{\min}^{\text{chg,SOC}} = \max\limits_{m}(i_{\text{min,m}}^{\text{chg,SOC}}) \\ i_{\max}^{\text{dis,SOC}} = \max\limits_{m}(i_{\text{max,m}}^{\text{dis,SOC}}) \end{cases} \tag{5.22}$$

This method assumes that the SOC of each cell in the battery system is accurate. If this assumption is not satisfied, the peak currents of charge and discharge can be estimated approximately by the following formula:

$$\begin{cases} i_{\min}^{\text{chg,SOC}} \approx \frac{z(t)-z_{\max}}{\eta_i \Delta t / C_{\max}} \\ i_{\max}^{\text{dis,SOC}} \approx \frac{z(t)-z_{\min}}{\eta_i \Delta t / C_{\max}} \end{cases} \tag{5.23}$$

When the battery is used in a wide SOC range, using only SOC as a constraint leads to a larger SOP result. This method is generally used in SOP estimation when SOC is close to the operating boundary, combined with other methods.

Voltage-based method

Voltage-based method is suitable for continuous SOP prediction. Taking the Rint model as an example, the Eq. (5.17) is rewritten as follows:

$$U_{\text{t,m}}(t+\Delta t) = U_{\text{oc,m}}(z_m(t+\Delta t)) - R_i i_{\text{L,m}}(t) \tag{5.24}$$

Since OCV is the nonlinear function of current, the peak current is hard to be solved analytically. To address this problem, it is necessary to expand the OCV by Taylor series:

$$\begin{aligned} U_{\text{oc,m}}(z_m(t+\Delta t)) &= U_{\text{oc,m}}\left(z_m(t) - i_{\text{L,m}}(t)\frac{\eta_{\text{i}}\Delta t}{C_{\max}}\right) \\ &= U_{\text{oc,m}}(z_{\text{m}}(t)) - i_{\text{L,m}}(t)\frac{\eta_{\text{i}}\Delta t}{C_{\max}} \frac{\partial U_{\text{oc}}(z)}{\partial z}\bigg|_{z=z_m(t)} \\ &+ R_n\left(z_m(t), i_{\text{L,m}}(t)\frac{\eta_{\text{i}}\Delta t}{C_{\max}}\right) \end{aligned} \tag{5.25}$$

$R_n(\cdot)$ is the remainder of the first-order expansion, because the SOC changes very little in each sampling interval, and $R_n(\cdot)$ can be ignored. The derivative of OCV $\partial U_{oc}(z)/\partial z$ can be determined by the OCV-SOC relationship. Take the LFP battery as an example, and $\partial U_{oc}(z)/\partial z$ is shown in Fig. 5.17.

Given a voltage constraint, the peak current can be expressed as

$$\begin{cases} i_{\text{max,m}}^{\text{dis,volt}} = \frac{U_{\text{oc}}(z_m(t))-U_{\text{t,min}}}{\frac{\eta_i \Delta t}{C_{\max}} \frac{\partial U_{\text{oc}}(z)}{\partial z}\Big|_{z_m(t)} + R_{\text{dis}}} \\ i_{\text{min,m}}^{\text{chg,volt}} = \frac{U_{\text{oc}}(z_{\text{k}}(t))-U_{\text{t,max}}}{\frac{\eta_i \Delta t}{C_{\max}} \frac{\partial U_{\text{oc}}(z)}{\partial z}\Big|_{z_m(t)} + R_{\text{chg}}} \end{cases} \tag{5.26}$$

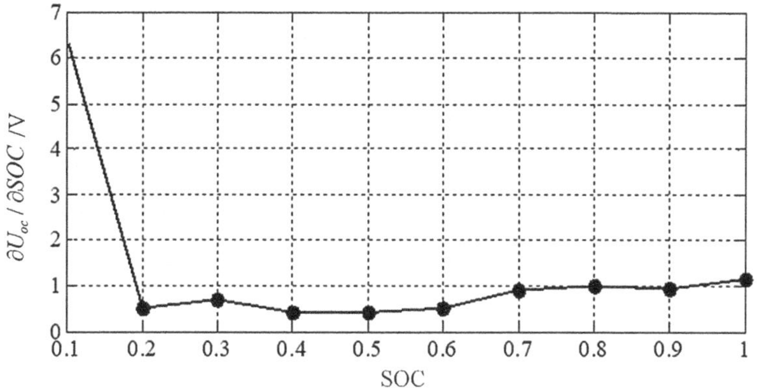

Fig. 5.17 The relationship between SOC and $\partial U_{oc}/\partial z$

where $i_{min,m}^{chg,volt}$ and $i_{max,m}^{dis,volt}$ are the peak charging and discharging currents calculated by the voltage-based method. For most batteries, $\partial U_{oc}(z)/\partial z > 0$ validates in the entire SOC range. Therefore, the peak current calculated by Eq. (5.26) is less than that calculated by Eq. (5.18) under the same SOC. Compared with the HPPC method, the voltage-based method takes into account the changes of SOC and allows longer sampling interval Δt. However, due to the static characteristics of Rint model, the method ignores the influence of dynamic polarization and hysteresis on peak current.

MCD method

Based on the conclusion of Chap. 3, compared with the Rint model, Thevenin model uses RC networks to describe the dynamic voltage characteristics of batteries, which can accurately simulate the polarization and hysteresis of batteries [12–15]. The application in SOP prediction of the battery will further improve the rationality of prediction results. It can be seen from Sect. 5.2.2 that there are some differences in ohmic internal resistance during the charging and discharging process of batteries, which should be considered separately in the SOP prediction. In this section, the Thevenin model in Chap. 3 will be improved, in which the ohmic internal resistance R_i is replaced by the charging and discharging internal resistances R_{chg} and R_{dis}, as shown in Fig. 5.18.

Similarly, this model consists of three parts: open-circuit voltage U_{oc}, internal resistance, and equivalent capacitance. Internal resistance includes ohmic internal resistances R_{chg} and R_{dis}, and polarized internal resistances R_D. Ideal polarized capacitance is used to describe transient response. U_D and U_R are voltage drops flowing through polarization resistance and ohmic resistance, respectively. i_D is the current flowing through the branch of C_D. The state-space equation of the model is

$$\begin{cases} U_t = U_{oc} - U_D - R_i i_L \\ \dot{U}_D = \frac{i_L}{C_D} - \frac{U_D}{C_D R_D} \end{cases} \tag{5.27}$$

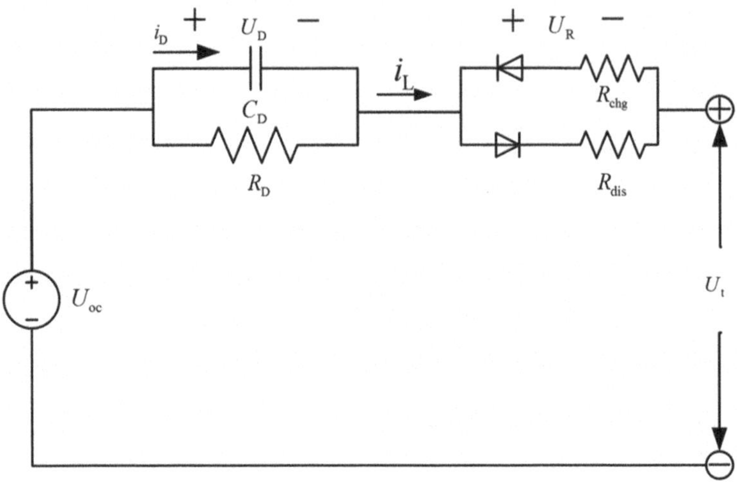

Fig. 5.18 The improved Thevenin model

Equation (5.27) can be further discretized into

$$U_{D,m}(t+\Delta t) = \exp(-\Delta t/\tau)U_{D,m}(t) + R_D(1-\exp(-\Delta t/\tau))i_{L,m}(t) \tag{5.28}$$

where the time constant of polarization can be defined as $\tau = R_D C_D$. Similar to the voltage-based method using Rint model, the peak current based on the improved model is calculated as follows:

$$\begin{cases} i_{max,m}^{dis,MCD} = \dfrac{U_{oc}(z_m(t)) - U_{D,m}(t)\exp(-\Delta t/\tau) - U_{t,min}}{\frac{\eta_i \Delta t}{C_{max}} \frac{\partial U_{oc}(z)}{\partial z}\Big|_{z_m(t)} + R_D[1-\exp(-\Delta t/\tau)] + R_{dis}} \\ i_{min,m}^{chg,MCD} = \dfrac{U_{oc}(z_m(t)) - U_{D,m}(t)\exp(-\Delta t/\tau) - U_{t,max}}{\frac{\eta_i \Delta t}{C_{max}} \frac{\partial U_{oc}(z)}{\partial z}\Big|_{z_m(t)} + R_D[1-\exp(-\Delta t/\tau)] + R_{chg}} \end{cases} \tag{5.29}$$

where $i_{min,m}^{chg,MCD}$ and $i_{max,m}^{dis,MCD}$ are the peak charge current and peak discharge current of the mth cell calculated by MCD method, respectively. The polarization voltage U_D is larger than zero during discharging, and the polarization voltage U_D is less than zero during charging. So when the ohmic internal resistance is the same, the current of Eq. (5.29) is smaller than that of the Eq. (5.26). Because $\partial U_{oc}(z)/\partial z$ is not constant in the whole SOC region, especially in the high and low SOC regions, the peak current calculated by the voltage-constrained method is more accurate and reliable.

Considering the limitation of the cell current, the peak current of the battery system is constrained by the following formula:

$$\begin{cases} i_{max}^{dis} = \min(i_{max}, \min\limits_m i_{max,m}^{dis,SOC}, \min\limits_m i_{max,m}^{dis,MCD}) \\ i_{min}^{chg} = \max(i_{min}, \max\limits_m i_{min,m}^{chg,SOC}, \max\limits_m i_{min,m}^{chg,MCD}) \end{cases} \tag{5.30}$$

where $i_{\max}^{\mathrm{dis}}$ and $i_{\min}^{\mathrm{chg}}$ are the peak discharge and peak charging currents, respectively. The peak current of the battery system is the sum of the peak currents of all parallel cells which can be expressed as follows:

$$\begin{cases} P_{\min}^{\mathrm{chg}} = n_{\mathrm{p}} \sum_{m=1}^{n_s} i_{\min}^{\mathrm{chg}} U_{\mathrm{t,m}}(t+\Delta t) \approx \\ n_{\mathrm{p}} \sum_{m=1}^{n_s} i_{\min}^{\mathrm{chg}} (U_{\mathrm{oc}}(z_m(t) - i_{\min}^{\mathrm{chg}} \frac{\eta_{\mathrm{i}}\Delta t}{C_{\max}}) - \exp(-\Delta t/\tau) \times U_{\mathrm{D,m}}(t) - (R_i + R_D(1-\exp(-\Delta t/\tau)))i_{\min}^{\mathrm{chg}} \\ P_{\max}^{\mathrm{dis}} = n_{\mathrm{p}} \sum_{k=1}^{n_s} i_{\max}^{\mathrm{dis}} U_{\mathrm{t,m}}(t+\Delta t) \approx \\ n_{\mathrm{p}} \sum_{m=1}^{n_s} i_{\max}^{\mathrm{dis}} (U_{\mathrm{oc}}(z_m(t) - i_{\max}^{\mathrm{dis}} \frac{\eta_{\mathrm{i}}\Delta t}{C_{\max}}) - \exp(-\Delta t/\tau) \times U_{\mathrm{D,m}}(t) - (R_i + R_D(1-\exp(-\Delta t/\tau)))i_{\max}^{\mathrm{dis}} \end{cases} \tag{5.31}$$

The power of each parallel cell is equal to the multiplier of the predicted voltage and the maximum allowable current. In addition, the power limitation of cells can also be considered as an extra constraint in the MCD method. It should be noted that the parameters U_{oc}, $C_{\max}$, $U_{\mathrm{t,max}}$, $U_{\mathrm{t,min}}$, $z_{\max}$, $z_{\min}$, $i_{\mathrm{L,max}}$, $i_{\mathrm{L,min}}$, R_{chg}, and R_{dis} in the MCD method may be affected by temperature, SOC and other factors.

Case study

The instantaneous SOP prediction method was evaluated by using a battery pack ($n_{\mathrm{s}} = 1$, $n_{\mathrm{p}} = 9$). The battery parameters are shown in Table 2.6, the capacity of cell is 3 A h and nominal voltage is 3.2 V. The constraints of SOC, terminal voltage, and current are shown in Table 5.10.

In order to comprehensively compare the performance of the HPPC method, SOC-based method, voltage-based method, and MCD method, the accuracy of Thevenin model should be improved. The terminal voltage error is shown in Fig. 5.19 and the statistical results are shown in Table 5.11. The mean absolute error (MAE) is defined as follows:

$$U_{\mathrm{MAE,k}} = \frac{1}{k+1} \sum_{j=0}^{k} \left| e_{\mathrm{j}} - \hat{e}_{\mathrm{j}} \right| \tag{5.32}$$

where e_{j} is the terminal voltage error in step j and $\hat{e}_j$ is the average terminal voltage error up to step k. Figure 5.19b is the result of MAE. It can be seen that the improved Thevenin model can accurately estimate SOP.

The results of peak current estimation are shown in Fig. 5.20. It shows the peak discharging currents predicted by the HPPC method and the voltage-based method. Due to the consideration of $\partial U_{\mathrm{oc}}/\partial z > 0$ shown in Fig. 5.17, the peak current predicted

Table 5.10 Lithium-ion battery constraints

Parameters	Lower limit	Upper limit
SOC	35%	85%
U_t/V	2.65	3.65
i_L/A	−135	270

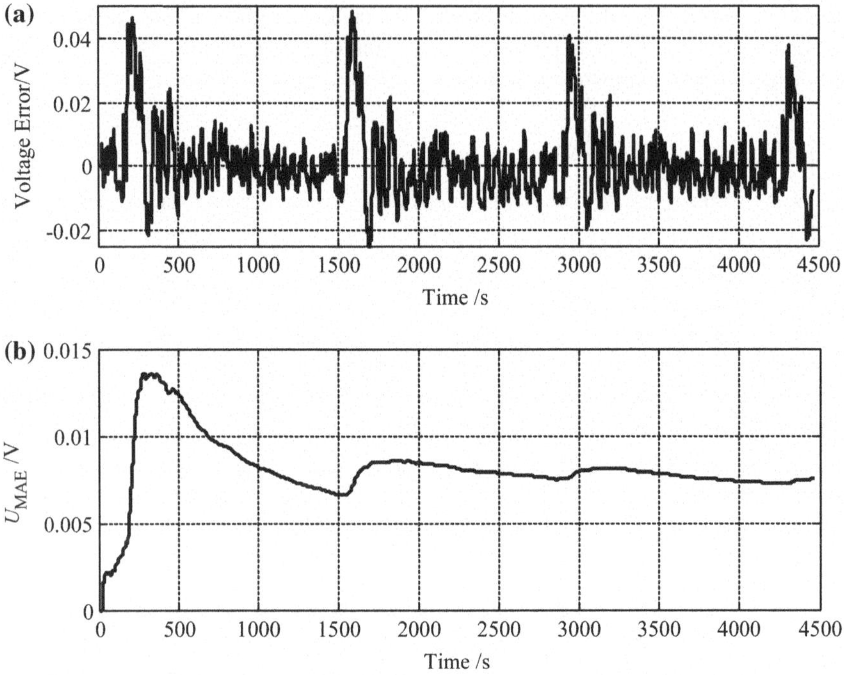

Fig. 5.19 The output curve of the improved Thevenin model under FUDS conditions: **a** voltage error; **b** voltage MAE

Table 5.11 Statistical results of voltage errors

Maximum error (V)	Minimum error (V)	Mean error (V)	Error variance (V^2)
0.0489	−0.0263	8.3797×10^{-4}	1.1987×10^{-4}

by the voltage-based method is also lower than that predicted by HPPC method. As the MCD method takes the polarization characteristics of battery into account and $U_D > 0$ during discharge, the peak current obtained by MCD method is less than that obtained by HPPC method and voltage-based method. Figure 5.20b shows the peak current predicted by SOC-based method is too large. When SOC is close to the upper and lower cutoff constraints, the peak current drops rapidly to a reasonable constraint range, while HPPC method and voltage-based method overshoot the constraints of SOC. Therefore, SOC-based method is generally combined with other methods to improve the reliability of SOP prediction. Figure 5.20c shows that the peak charging current predicted by HPPC method is less than that obtained by voltage-based method, which is also due to the influence of $\partial U_{oc}/\partial z$. Figure 5.20d shows the peak current calculated by MCD method based on the improved Thevenin model. When the SOC reaches the upper limit, the charging peak current decreases and the discharge peak current increases. Therefore, this method can optimize the use range

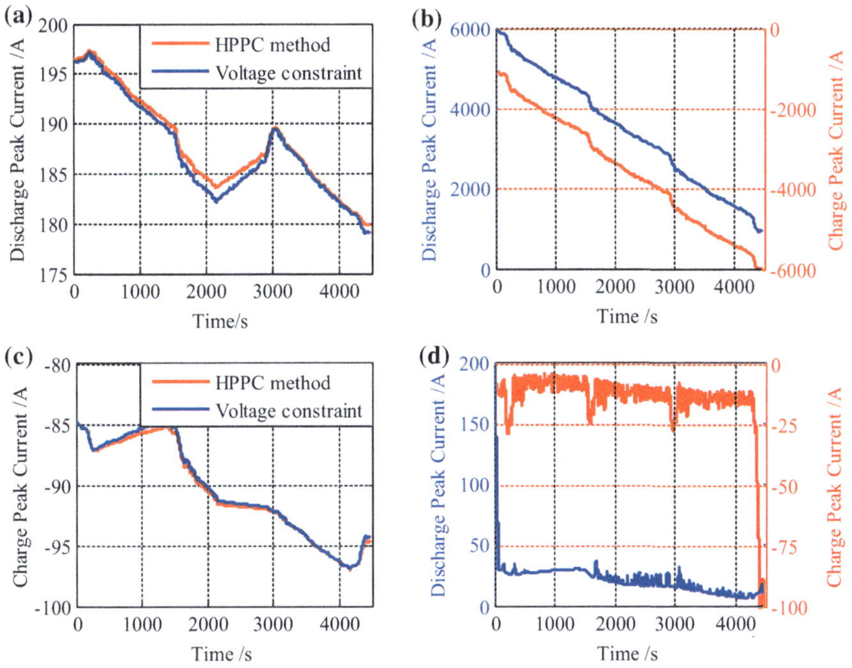

Fig. 5.20 Peak current estimation results: **a** peak current of discharge based on HPPC method and voltage-based method; **b** peak current of charge and discharge based on SOC-based method; **c** peak current of charge and discharge based on HPPC method and voltage-based method; **d** peak current of charge and discharge based on MCD method

of batteries and prolong the life of batteries. At the same time, when the battery discharges at high current rate, the predicted discharge SOP dramatically reduces, while the charge SOP increases, which is in good agreement with the actual use characteristics of batteries.

Figure 5.21 shows the SOP predicted by the HPPC and MCD methods. Figure 5.21a is the case of discharge. It can be seen that the discharge SOP obtained by HPPC method is significantly larger than that obtained by MCD method. This is because the HPPC method only considers the voltage constraints of the battery. In addition, HPPC ignores the limitation of SOC, leading to serious problems in practical application, which also affects the performance of continuous SOP prediction. At the same time, this method does not take the dynamic polarization effect into account. Therefore, this prediction method may lead to over-discharge of battery pack. Unlike the HPPC method, the MCD method takes the full dynamic characteristics of battery system into account in real-time estimation, so the terminal voltage drops sharply undergoing a large current. As shown in Fig. 5.21b, the charge SOP of HPPC method is relatively higher. When a large current occurs, the charge SOP predicted by MCD method becomes larger, but the HPPC estimation result does not change significantly. Because of the dynamic polarization effect, the real-time

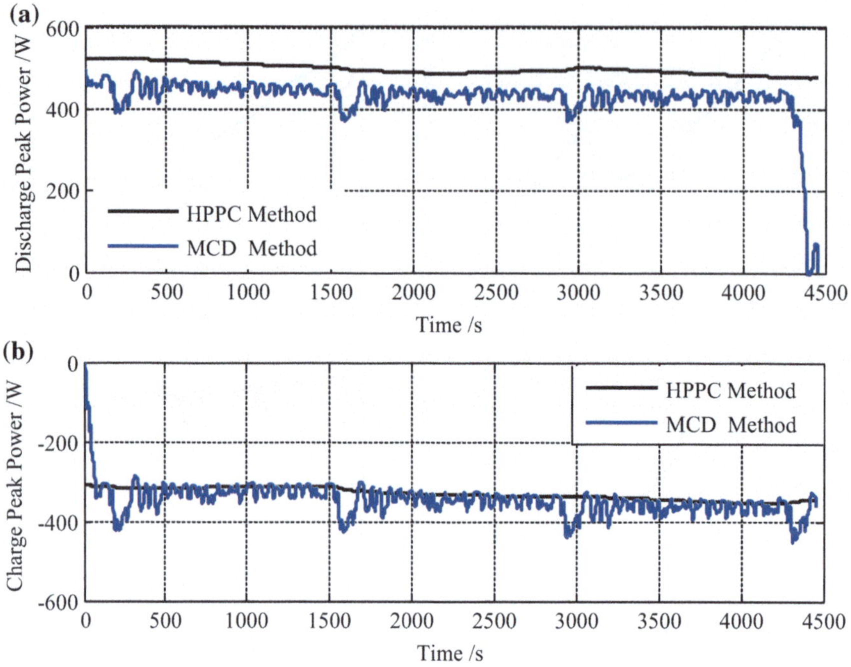

Fig. 5.21 Instantaneous SOP estimation results: **a** discharge SOP; **b** charge SOP

prediction result of MCD method has a certain fluctuation, which accords with the actual working characteristics of batteries. Therefore, the MCD method can achieve a relatively satisfactory SOP prediction.

5.3.2 Continuous SOP Prediction Method

The peak current calculated by Eq. (5.13) is affected by the sampling interval of BMS. If SOP prediction with continuous sampling interval is needed, the above methods fail to work. To solve this problem, this section describes the continuous SOP prediction method with multiple sampling intervals, which is an extension of the MCD method with fixed sampling interval [16].

(1) Estimation of continuous peak current

The OCV in the Thevenin model can be expressed as a function of the SOC, the expression is

$$U_{oc} = K_0 + K_1 \ln SOC + K_2 \ln(1 - SOC) \tag{5.33}$$

The state-space expression equation of the model is

$$\begin{cases} \boldsymbol{X}_{k+1} = \boldsymbol{A}\boldsymbol{X}_k + \boldsymbol{B}\boldsymbol{u}_k + \boldsymbol{\omega}_k \\ \boldsymbol{Y}_{k+1} = \boldsymbol{C}\boldsymbol{X}_{k+1} + \boldsymbol{D}\boldsymbol{u}_k + \boldsymbol{v}_k \end{cases} \tag{5.34}$$

where the Jacobian matrices are defined as

$$\begin{aligned} \boldsymbol{X}_k &= \begin{pmatrix} U_{D,k} \\ z_k \end{pmatrix}, \boldsymbol{Y}_k = U_{t,k}, \boldsymbol{u}_k = I_{L,k}, \boldsymbol{A} = \begin{pmatrix} \exp(\frac{-\Delta t}{\tau}) & 0 \\ 0 & 1 \end{pmatrix}, \boldsymbol{D} = [-R_i] \\ \boldsymbol{B} &= \begin{pmatrix} R_D(1-\exp(\frac{-\Delta t}{\tau})) \\ \frac{\eta_i \Delta t}{C_a} \end{pmatrix}, \boldsymbol{C} = \frac{\partial U_t}{\partial X}\bigg|_{X=\hat{X}_{k+1}} = \left[-1 \quad \frac{dU_{oc}(z)}{dz}\bigg|_{z=z_{k+1}^-} \right] \end{aligned} \tag{5.35}$$

Assuming that the current of the system is constant within the sampling interval $[k, k+L]$ and L is the length of continuous sampling interval, that is $\boldsymbol{u}_{k+L} = \boldsymbol{u}_k$. Thevenin model can be used to predict the voltage at the sampling time of $k+L$th, and the expression is

$$\begin{cases} \boldsymbol{X}_{k+L} = \boldsymbol{A}\boldsymbol{X}_{k+L-1} + \boldsymbol{B}\boldsymbol{u}_{k+L-1} \\ \boldsymbol{Y}_{k+L} = \boldsymbol{C}\boldsymbol{X}_{k+L} + \boldsymbol{D}\boldsymbol{u}_{k+L-1} \end{cases} \tag{5.36}$$

For the convenience of calculation, the battery prediction model in the duration $L \times \Delta t$ is regarded as a time-invariant system, that is, the model parameters remain unchanged, which can be further deduced.

$$\boldsymbol{X}_{k+L} = \boldsymbol{A}^L\boldsymbol{X}_k + \left(\sum_{j=0}^{L-1} \boldsymbol{A}^{L-1-j}\boldsymbol{B}\right)\boldsymbol{u}_k \tag{5.37}$$

In order to solve the maximum discharge current $I_{max,L}^{dis}$ and minimum charge current $I_{min,L}^{chg}$, the output of the battery model should satisfy the following equations:

$$U_{oc}(z_{k+L}) - U_{D,k}\left(\exp\left(\frac{-\Delta t}{\tau}\right)\right)^L - I_{min,L}^{chg}\left(R_i + R_D\left(1-\exp\left(\frac{-\Delta t}{\tau}\right)\right)\sum_{j=0}^{L-1}\left(\exp\left(\frac{-\Delta t}{\tau}\right)\right)^{L-1-j}\right) - U_{t,min} = 0 \tag{5.38}$$

$$U_{oc}(z_{k+L}) - U_{D,k}\left(\exp\left(\frac{-\Delta t}{\tau}\right)\right)^L - I_{max,L}^{dis,EP}\left(R_i + R_D\left(1-\exp\left(\frac{-\Delta t}{\tau}\right)\right)\sum_{j=0}^{L-1}\left(\exp\left(\frac{-\Delta t}{\tau}\right)\right)^{L-1-j}\right) - U_{t,max} = 0 \tag{5.39}$$

The equation for estimating the continuous peak current of battery in duration $L \times \Delta t$ is as follows:

$$\begin{cases} I_{\mathrm{max,L}}^{\mathrm{dis}} = \frac{U_{\mathrm{oc}}(z_{\mathrm{k}}) - U_{\mathrm{D,k}}\left(\exp\left(\frac{-\Delta t}{\tau}\right)\right)^{\mathrm{L}} - U_{\mathrm{t,min}}}{\frac{\eta_{\mathrm{i}} L \Delta t}{C_{\max}} \frac{\mathrm{d}U_{\mathrm{oc}}(z)}{\mathrm{d}z}\Big|_{z_{\mathrm{k}}} + R_D\left(1 - \exp\left(\frac{-\Delta t}{\tau}\right)\right) \sum_{j=0}^{\mathrm{L}-1} \left(\exp\left(\frac{-\Delta t}{\tau}\right)\right)^{\mathrm{L}-1-j} + R_i} \\ I_{\mathrm{min,L}}^{\mathrm{chg}} = \frac{U_{\mathrm{oc}}(z_{\mathrm{k}}) - U_{\mathrm{D,k}}\left(\exp\left(\frac{-\Delta t}{\tau}\right)\right)^{\mathrm{L}} - U_{\mathrm{t,max}}}{\frac{\eta_{\mathrm{i}} L \Delta t}{C_{\max}} \frac{\mathrm{d}U_{\mathrm{oc}}(z)}{\mathrm{d}z}\Big|_{z_{\mathrm{k}}} + R_D\left(1 - \exp\left(\frac{-\Delta t}{\tau}\right)\right) \sum_{j=0}^{\mathrm{L}-1} \left(\exp\left(\frac{-\Delta t}{\tau}\right)\right)^{\mathrm{L}-1-j} + R_i} \end{cases} \tag{5.40}$$

It should be noted that the above equation can be degraded to Eq. (5.29) if L is 1.

(2) Estimation of continuous SOP

In order to ensure the safety of battery system, when SOC approaches the lower limit $z_{\min}$, the maximum discharge current should be constrained and the charging current should be maximized, otherwise the battery system may be at risk of over-discharge. On the contrary, when SOC approaches the upper limit $z_{\max}$, the maximum charging current should be constrained and the discharge current should be maximized, otherwise it will be at risk of overcharging. Using the constraints of SOC is a key principle for peak current estimation, the peak current is expressed as follows:

$$\begin{cases} I_{\mathrm{min,L}}^{\mathrm{chg,SOC}} = \frac{z_{\mathrm{k}} - z_{\max}}{\eta_{\mathrm{i}} L \Delta t / C_{\max}} \\ I_{\mathrm{max,L}}^{\mathrm{dis,SOC}} = \frac{z_{\mathrm{k}} - z_{\min}}{\eta_{\mathrm{i}} L \Delta t / C_{\max}} \end{cases} \tag{5.41}$$

where $I_{\mathrm{min,L}}^{\mathrm{chg,SOC}}$ and $I_{\mathrm{max,L}}^{\mathrm{dis,SOC}}$ are the minimum charging current and the maximum discharge current in the duration of $L \times \Delta t$, respectively. Once the constrained current is obtained, the peak current expression is

$$\begin{cases} I_{\max}^{\mathrm{dis}} = \min\left(I_{\max}, I_{\mathrm{max,L}}^{\mathrm{dis,SOC}}, I_{\mathrm{max,L}}^{\mathrm{dis}}\right) \\ I_{\min}^{\mathrm{chg}} = \max\left(I_{\min}, I_{\mathrm{min,L}}^{\mathrm{chg,SOC}}, I_{\mathrm{min,L}}^{\mathrm{chg}}\right) \end{cases} \tag{5.42}$$

where $I_{\max}$ and $I_{\min}$ are the maximum constraint value of discharge current and the minimum constraint value of allowable charging current of the battery, respectively. $I_{\min}^{\mathrm{chg}}$ and $I_{\max}^{\mathrm{dis}}$ are the minimum charging current and the maximum discharge current under all constraints, respectively. The continuous SOP based on the Thevenin model can be estimated from:

$$\begin{cases} P_{\min}^{\mathrm{chg}} = \max\left(P_{\min}, U_{\mathrm{t,k+L}} I_{\min}^{\mathrm{chg}}\right) \\ P_{\max}^{\mathrm{dis}} = \min\left(P_{\max}, U_{\mathrm{t,k+L}} I_{\max}^{\mathrm{dis}}\right) \end{cases} \tag{5.43}$$

Then, the SOP of charge and discharge are derived:

$$\begin{cases} P_{\min}^{\mathrm{chg}} \approx \max\left(P_{\min}, \left(U_{\mathrm{oc}}(z_{\mathrm{k+L}}) - U_{\mathrm{D,k}}\left(\exp\left(\frac{-\Delta t}{\tau}\right)\right)^{\mathrm{L}} - I_{\min}^{\mathrm{chg}}\left(R_i + R_D\left(1 - \exp\left(\frac{-\Delta t}{\tau}\right)\right) \sum_{j=0}^{\mathrm{L}-1} \left(\exp\left(\frac{-\Delta t}{\tau}\right)\right)^{\mathrm{L}-1-j}\right)\right) I_{\min}^{\mathrm{chg}}\right) \\ P_{\max}^{\mathrm{dis}} \approx \min\left(P_{\max}, \left(U_{\mathrm{oc}}(z_{\mathrm{k+L}}) - U_{\mathrm{D,k}}\left(\exp\left(\frac{-\Delta t}{\tau}\right)\right)^{\mathrm{L}} - I_{\max}^{\mathrm{dis}}\left(R_i + R_D\left(1 - \exp\left(\frac{-\Delta t}{\tau}\right)\right) \sum_{j=0}^{\mathrm{L}-1} \left(\exp\left(\frac{-\Delta t}{\tau}\right)\right)^{\mathrm{L}-1-j}\right)\right) I_{\max}^{\mathrm{dis}}\right) \end{cases} \tag{5.44}$$

where P_{max} and P_{min} are the power constraints of discharge and charge, respectively.

5.3.3 Joint Estimation of SOC and SOP for Batteries

Due to the coupling relationship between the multiple parameters of batteries, it is difficult to accurately estimate the parameters with inaccurate SOC. Accurate SOP prediction is limited by the accuracy of SOC and model parameters. To solve this problem, this section focuses on the joint estimation method of SOC and SOP.

(1) Battery SOC and SOP joint estimation algorithm
According to the model-based SOC estimation method for batteries and the continuous SOP prediction method shown in Eq. (5.44), a model-based SOC and SOP joint estimation method for batteries is proposed as shown in Fig. 5.22 [17, 18]. The specific steps are as follows:

① Real-time battery data acquisition;
② Online parameters identification based RLS algorithm;
③ SOC estimation based on AEKF algorithm;
④ Continuous SOP prediction.

(2) Case study
The DST test of battery 2-cell 02 was conducted to evaluate the performance of the algorithm. In view of the design limitations of new energy vehicles, this section only used the data of SOC in the range of 20–100%. The constraints of the continuous power output within 30 s are shown in Table 5.12.

(1) Accuracy evaluation of SOC estimation
Before the SOP prediction it is necessary to evaluate the estimation accuracy of voltage and SOC in the joint estimation algorithm. Figure 5.23 shows the estimation results of voltage and SOC, where the initial value of SOC is accurate (100% SOC).

It can be seen that the joint estimation method can guarantee the estimation accuracy of voltage and SOC under the accurate initial SOC, in which the prediction error of voltage is less than 15 mV and the SOC estimation error is less than 1%. However, it is difficult to obtain accurate initial SOC in practical applications. In order to evaluate the convergence of the joint estimation method with erroneous initial SOC error, Fig. 5.24 shows the estimated SOC results with the initial 60% SOC (the initial SOC error is 40%).

Figure 5.24 shows that the joint estimation method can still accurately estimate the battery voltage and SOC under the inaccurate initial SOC values. The voltage error and SOC error are consistent with the results obtained under the precise initial SOC values. Therefore, the joint estimation method can ensure the accuracy and reliability of the SOC estimation method based on the fused data model. Therefore, the joint estimation method also has strong robustness against uncertain initial SOC values [19].

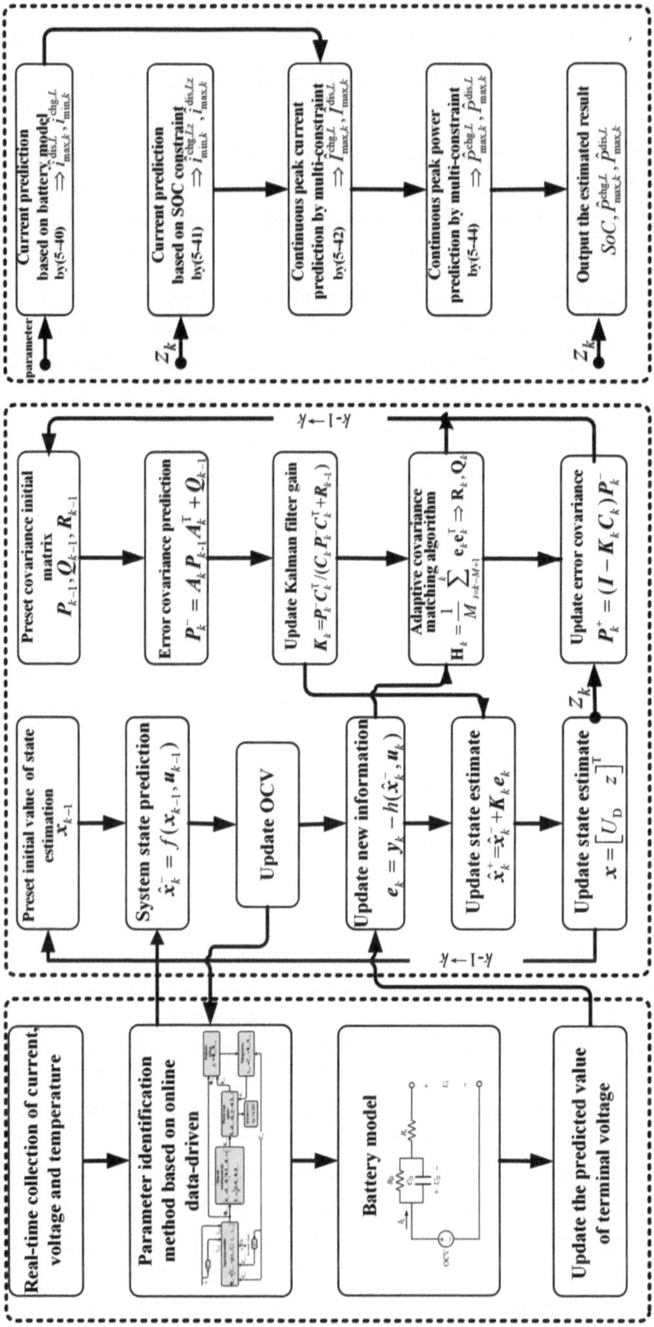

Fig. 5.22 Joint estimation method of adaptive SOC and peak power for battery based on the fused data model

Table 5.12 Design limitations of cell in continuous SOP prediction (30 s)

Parameters	Maximum	Minimum
SOC (z_{max}, z_{min}) (%)	100	20
U_t ($U_{t,max}$, $U_{t,min}$) (V)	4.05	3.0
I_L (I_{max}, I_{min}) (A)	320	−160
P (P_{max}, P_{min}) (W)	1000	−450

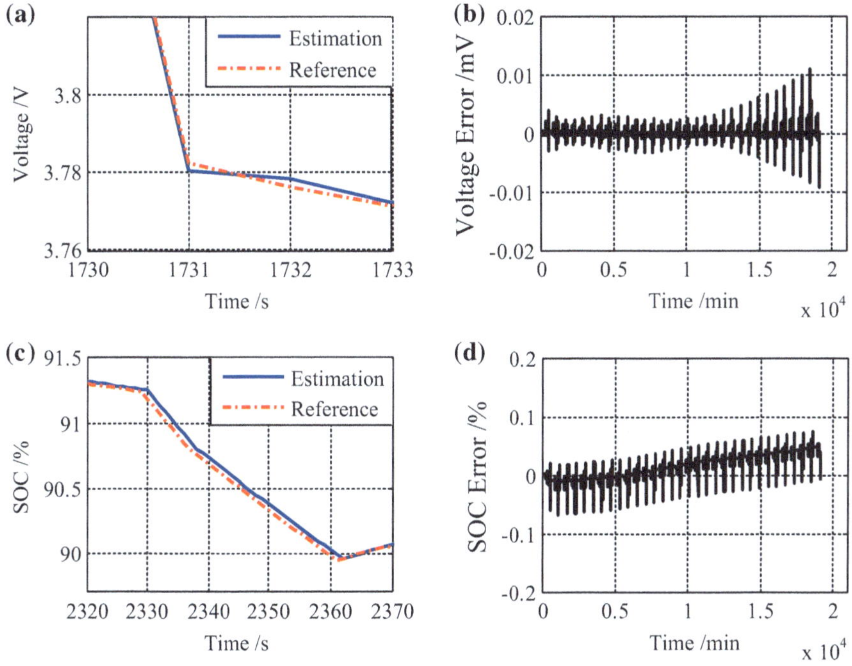

Fig. 5.23 Voltage and SOC estimation results of cell 02 (accurate initial SOC value): **a** comparison between measurement and predicted voltages; **b** voltage error; **c** comparison between reference and estimated SOCs; **d** SOC error

(2) Evaluation of SOP prediction

Figure 5.25 shows the estimation results of the continuous peak charging and discharging current within 30 s of the battery 2-cell 02, in which the initial SOC is 100% (the initial SOC error is 0). Figure 5.25a, b shows the peak current estimation results for SOC-based, voltage-based and multi-constraint dynamic method. The magnitude of the peak current results based on the multi-constraint dynamic method differs greatly from the other two methods. Therefore, the charge and discharge current results for multi-constrained dynamic method are illustrated in Fig. 5.25c, d.

It can be seen that the peak current estimation results for the multi-constrained dynamic method are divided into two parts: one is the voltage-based peak current

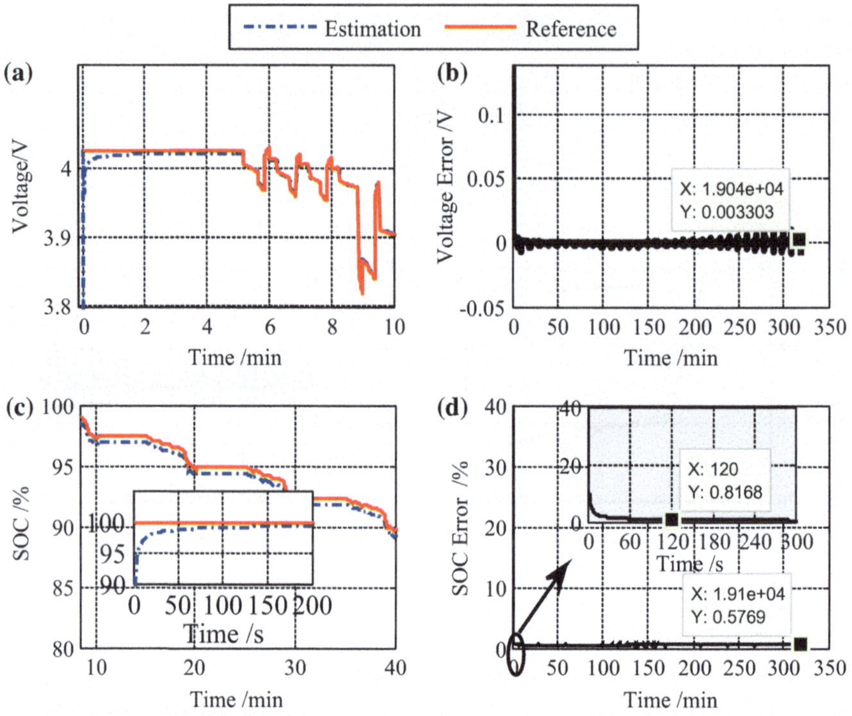

Fig. 5.24 Voltage and SOC estimation results of cell 02 (inaccurate initial SOC value): **a** comparison between the measured value and predicted voltages; **b** voltage error; **c** comparison between reference and estimated SOCs; **d** SOC error

estimation result; the other is the peak current estimation result based on the SOC value. When the actual SOC is close to the maximum limit, the peak current estimation method based on SOC will reduce or even turn off the peak charging current of battery, which can prevent the overcharge of battery. At this time, the value of SOC-based peak current is the main result of multi-constraint peak charging current. Similarly, when the actual SOC value is close to the minimum value, the peak current estimation method based on SOC will reduce or even turn off the peak discharge current to prevent over-discharge of battery. At this time, the value of SOC-based peak current is the main result of multi-constraint peak charging current. Peak current estimation result at other times is mainly obtained from voltage-based estimation, but if the estimation result exceeds the design limit, the estimation result will be dominated by the limits. Thus, the peak current estimation based on SOC can effectively avoid the risks of overcharge and over-discharge, improve the security and reliability of battery application in new energy vehicles. With the decrease of SOC, the peak charging current increases and the peak discharging current decreases [20, 21].

Based on the estimation results of the continuous peak current and the voltage model of the battery, the estimation results of the continuous SOP can be deduced.

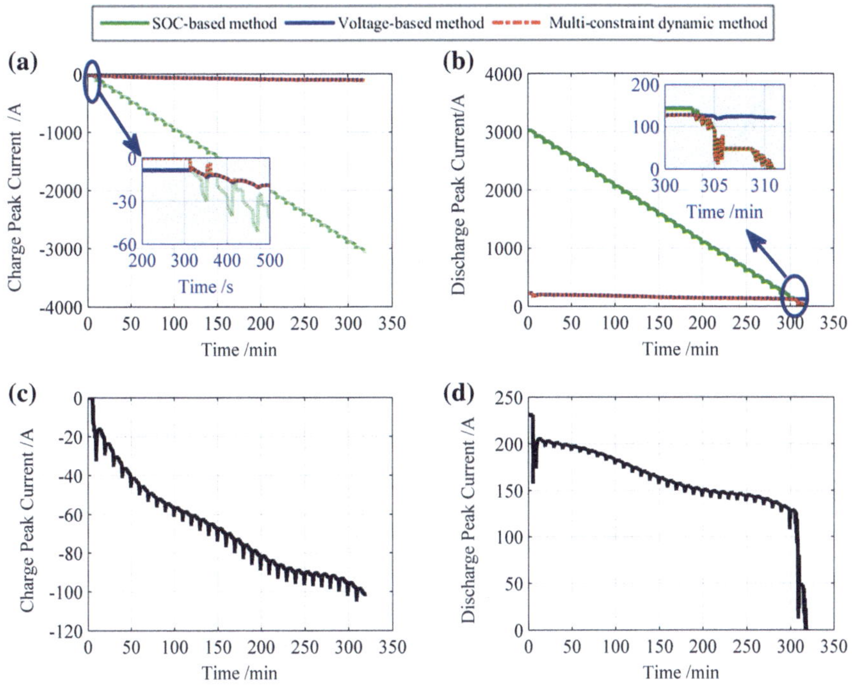

Fig. 5.25 Continuous peak current results (accurate SOC initial value, SOC = 100%, $L = 30$ s): **a** comparison of peak charge current results; **b** comparison of peak discharge current results; **c** multi-constrained peak charge current results; **d** multi-constrained peak discharge current results

The results of continuous SOP prediction within 30 s are shown in Fig. 5.26, which have the same trends as the continuous peak current estimation results obtained in Fig. 5.25. It can be seen that the SOP does not exceed the design limit. In fact, the experimental batteries are secondary batteries that have been used for two years in new energy vehicles whose actual power capacity cannot reach the calibration value.

In order to evaluate the relationship between the charge and discharge SOP and the duration, the peak current and SOP estimation under different durations are carried out. The results are shown in Fig. 5.27.

It can be seen that the charge and discharge current of battery are closely related to the durations. The instantaneous (such as 1 s) peak charge current can reach 190 A, and the peak discharge current can reach 395 A. With the increasing duration, the peak charge/discharge current decreases. Therefore, the peak charge and discharge current are not reliable and even lead to improper matching of the battery system only using the limited conditions given by the manufacturer without considering the durations.

As can be seen from Fig. 5.28, the SOP estimation results based on inaccurate initial SOC values converge rapidly to the estimation results when accurate initial

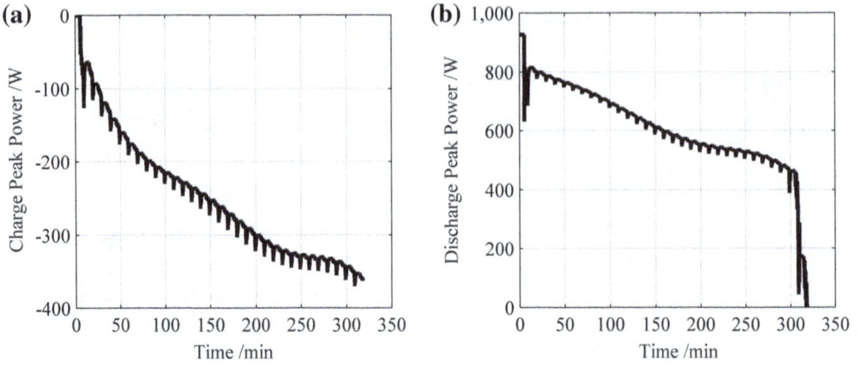

Fig. 5.26 SOP estimation result (accurate SOC initial value, SOC = 100%, L = 30 s): **a** multi-constraint peak charge SOP estimation result; **b** multi-constraint peak discharge SOP estimation result

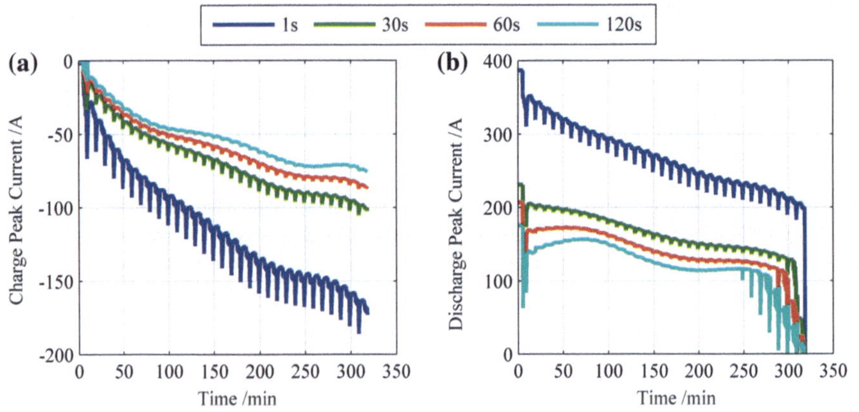

Fig. 5.27 Peak current estimation results (precise SOC initial value, SOC = 100%): **a** multi-constrained peak charging current estimation results; **b** multi-constrained peak discharge current estimation results

SOC values are used. Therefore, this joint estimation method can robust against uncertain initial SOC values.

In summary, this joint estimation method can obtain reliable and accurate SOP and SOC under uncertain initial SOC values.

5.3.4 Introduction of SOP Evaluation Method

The calculation results of SOP vary with different calculation methods. In order to evaluate the validity of the calculation results, China has formulated the Technical

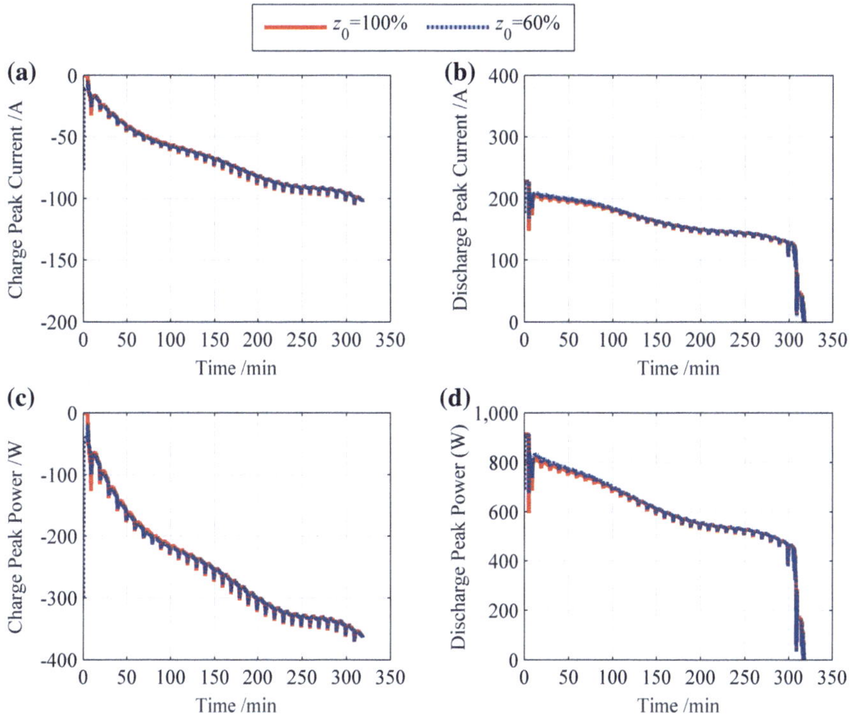

Fig. 5.28 Estimation results of peak current and peak power for uncertain initial SOC ($L = 30$ s): **a** multi-constrained peak charging current estimation; **b** multi-constrained peak discharge current estimation; **c** multi-constrained peak charging power estimation; **d** multi-constrained peak discharge power estimation

Conditions of BMS for Electric Vehicles (Draft). According to the general rules of SOP test in Appendix C, the test environment adopts three temperatures (10 °C ± 2 °C, 25 °C ± 2 °C, 40 °C ± 2 °C), and the SOC is set to 80, 50, and 30%. The preprocessing should be carried out before SOP test: the objects need to rest at the predefined ambient temperature until the difference between the surface temperature of the cell and the ambient temperature less than 2 °C (usually more than 16 h). Taking the 10 s continuous SOP test as an example, and the test flowchart of this standard is shown in Fig. 5.29. The specific steps are as follows:

① Fully charge the battery system with the charging specification described in Chap. 2;
② Rest for no less than 30 min but no more than 60 min;
③ Set initial SOC value of the algorithm to 100%;
④ Discharge the battery to the predefined SOC point by the current $I_{1,1}$;
⑤ Rest for no less than 30 min but no more than 60 min;

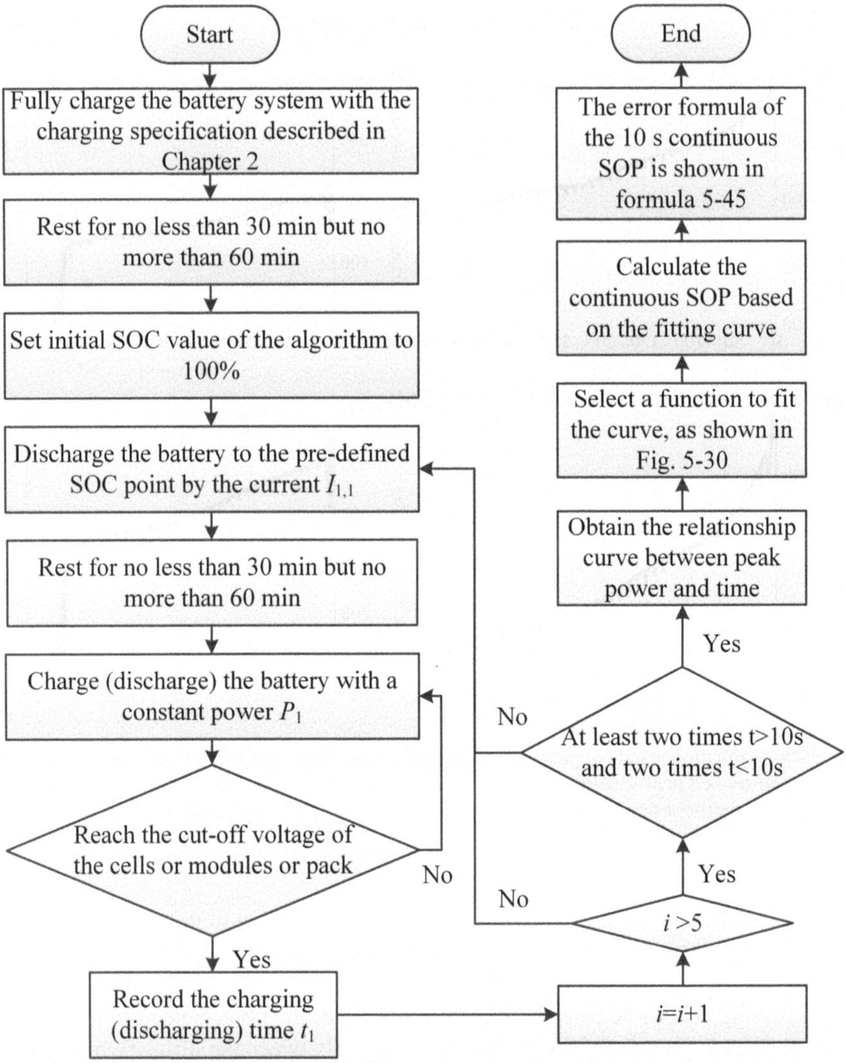

Fig. 5.29 Flowchart of SOP evaluation method of China standard

⑥ Charge (discharge) the battery with a constant power P_1 until it reaches the upper/lower cutoff voltage of the cells or modules, or battery pack in the battery system. Record the charging (discharging) time t_1;

⑦ Charge the battery with current $I_{1,1}$ to the predefined SOC point of step ④;

⑧ Rest for no less than 30 min but no more than 60 min;

⑨ According to the time of t_1, charge or discharge the battery with constant power P_2 until one of the following conditions is reached: upper/lower cutoff voltage of

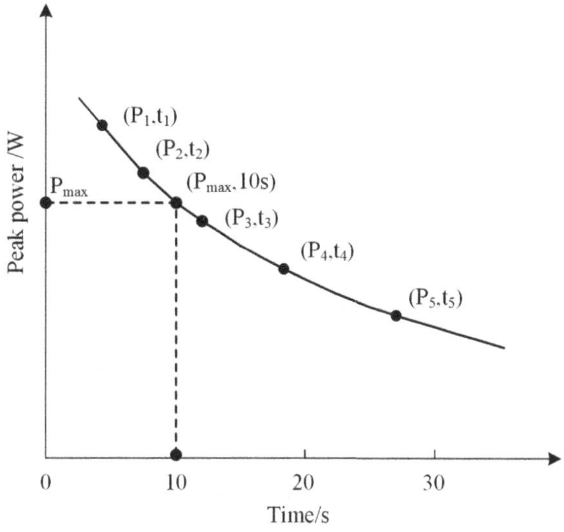

Fig. 5.30 Fitting curve of constant SOP pulse charge (discharge) test

the cells or modules, or battery pack in the battery system. Record the charging (discharging) time t_2;

⑩ Repeat peak power test 5 times or more, at least two recording times are longer than 10 s, and two times are less than 10 s. Thus, the relationship curve between peak power of charge (discharge) and charge (discharge) time of the battery at the SOC point is obtained. According to the trend of the curve, the corresponding function is selected to fit the curve, as shown in Fig. 5.30;

⑪ Calculate the continuous SOP based on the fitting curve. For example, the 10 s continuous SOP is f (10), the error formula of the 10 s continuous SOP is

$$Error_{SOP} = \frac{(f(10) - SOP_{Algorithm})}{f(10)} \times 100\% \tag{5.45}$$

where $SOP_{Algorithm}$ is the SOP predicted by the algorithm.

5.4 Conclusion

This chapter mainly discusses the state estimation and SOP prediction of battery packs. In order to improve the efficiency and reliability of the battery system, the performance analysis of battery packs is carried out. The results show that battery packs in parallel–series have more advantages than battery packs connected in series–parallel. The analysis theory of evolutionary mechanisms and causes of inconsistency in·battery packs are established, and the rapid grouping method of battery system is carried out by the dynamic parameter screening method. The results show that this method significantly improves the consistency of the battery system. In terms of the modeling of battery pack, a combined modeling method of battery pack based on the "average model" and "difference model" is established, and the inconsistency is quantified by using the difference model. As for the state estimation of battery pack, the dual estimator algorithms are used to estimate the SOC of battery pack. The results show that the SOC estimation error is less than 1%. In the aspect of SOP prediction for battery packs, four typical SOP prediction methods are elaborated in detail, and the characteristics of instantaneous SOP prediction methods are compared and analyzed through application examples. Because of the importance of continuous peak power in vehicle applications, this chapter also deeply discusses the prediction methods of continuous peak current and continuous SOP, and focuses on the co-estimation theoretical framework of SOC and SOP. The results show that the SOC estimation error is less than 1%, the voltage prediction error is less than 15 mV, and the reliable continuous SOP can be predicted. Finally, the China battery test standard (draft) is introduced, and the SOP testing process is elaborated in detail for the reference of SOP evaluation.

References

1. Ye M, Song X, Xiong R, Sun F (2019) A novel dynamic performance analysis and evaluation model of series-parallel connected battery pack for electric vehicles. IEEE Access 7(1):14256–14265
2. Xiong R (2014) Estimation of battery pack state for electric vehicles using model-data fusion approach. Ph.D. dissertation. Beijing Institute of Technology, Beijing
3. Liang L (2018) Study on the optimization of battery pack based on inconsistency analysis. M.A. thesis. Beijing Institute of Technology, Beijing
4. Xiong R, Sun F, Gong X (2013) Adaptive state of charge estimator for lithium-ion cells series battery pack in electric vehicles. J Power Sources 242(22):699–713
5. Sun F, Xiong R, He H (2016) A systematic state-of-charge estimation framework for multi-cell battery pack in electric vehicles using bias correction technique. Appl Energy 162:1399–1409
6. Gong X, Xiong R, Mi CC (2016) A data-driven bias-correction-method-based lithium-ion battery modeling approach for electric vehicle applications. IEEE Transa Ind Appl 52(2):1759–1765
7. Chen X, Lei H, Xiong R (2018) A bias correction based state-of-charge estimation method for multi-cell battery pack under different working conditions. IEEE Access 6:78184–78192
8. Sun F, Xiong R (2015) A novel dual-scale cell state-of-charge estimation approach for series-connected battery pack used in electric vehicles. J Power Sources 274:582–594
9. Wang J, Xiong R, Li L, Fang Y (2018) A comparative analysis and validation for double-filters-based state of charge estimators using battery-in-the-loop approach. Appl Energy 229:648–659
10. Xiong R, Gong X, Mi CC, Sun F (2013) A robust state-of-charge estimator for multiple types of lithium-ion batteries using adaptive extended Kalman filter. J Power Sources 243:805–816
11. Sun F, Xiong R, He H (2012) Model-based dynamic multi-parameter method for peak power estimation of lithium–ion batteries. Appl Energy 96(3):378–386
12. He H, Xiong R, Fan J (2011) Evaluation of lithium-ion battery equivalent circuit models for state of charge estimation by an experimental approach. Energies 4(4):582–598
13. Roscher MA, Sauer DU (2011) Dynamic electric behavior and open-circuit-voltage modeling of $LiFePO_4$-based lithium ion secondary batteries. J Power Sources 196(1):331–336

14. Kim J, Lee S, Cho BH (2011) Discrimination of li-ion batteries based on hamming network using discharging–charging voltage pattern recognition for improved state-of-charge estimation. J Power Sources 196(4):2227–2240
15. He H, Xiong R, Chang YH (2010) Dynamic modeling and simulation on a hybrid power system for electric vehicle applications. Energies 3(11):1821–1830
16. Xiong R, He H, Sun F (2013) Model-based state of charge and peak power capability joint estimation of lithium-ion battery in plug-in hybrid electric vehicles. J Power Sources 229(9):159–169
17. Xiong R, Sun F, He H (2013) A data-driven adaptive state of charge and power capability joint estimator of lithium-ion polymer battery used in electric vehicles. Energy 63(1):95–308
18. Xiong R, He H, Sun F, Zhao K (2012) Online estimation of peak power capability of Li-Ion batteries in electric vehicles by a hardware-in-loop approach. Energies 5(5):1455–1469
19. Sun F, Xiong R, He H (2014) Estimation of state-of-charge and state-of-power capability of lithium-ion battery considering varying health conditions. J Power Sources 259:166–176
20. Lu J, Chen Z, Yang Y, Lv M (2018) Online estimation of state of power for lithium-ion batteries in electric vehicles using genetic algorithm. IEEE Access 6:20868–20880
21. Lu, J (2018) Research on state of power estimation of Lithium battery and its applications in energy management. M.A. thesis. Northeastern University, Shenyang

Chapter 6
Remaining Useful Life Prediction of Lithium-Ion Batteries

The internal mechanism of lithium-ion batteries is very complicated. There are many reasons for performance degradation, and various factors are coupled with each other, which eventually leads to an extremely challenging engineering problem [1]. The battery performance degrades throughout the whole process of use and maintenance. As the charging and discharging cycles increase, some irreversible chemical reactions occur inside the battery, resulting in an increase of internal resistance and declines of maximum available capacity, energy, and SOP, which reduces the driving mileage of EVs, even leading to some safety hazards [2–4]. A reliable remaining useful life (RUL) prediction method can ease the user's anxiety about the remaining driving mileage and safety issues, and ensure the safe and efficient operation of the batteries. It can also guarantee the safety and reliability of the EV during operation, reduce failure rate and operating costs, improve user experience and avoid accidents. Therefore, the battery RUL prediction is one of the core tasks of BMS. After introducing the related concepts of battery RUL prediction, this chapter summarizes and classifies the mainstream RUL prediction methods, and finally introduces two representative battery RUL prediction methods.

6.1 Overview of RUL Prediction

6.1.1 Problem Description

The battery RUL refers to the cycles required for the maximum available capacity degrade to a specified threshold under a given charge and discharge rules. RUL prediction is to mathematical calculate the battery RUL based on the historical data. With the wide applications of battery in various fields, the RUL prediction technologies have attracted widely attention [5].

Currently, data-driven method is the main approach of battery RUL prediction. The key of this method lies in mining and generalization of the battery capacity

R. Xiong, *Battery Management Algorithm for Electric Vehicles*,
https://doi.org/10.1007/978-981-15-0248-4_6

degradation trajectories and historical data. The first priority of this method is to mine the inherent law of battery degradation from the capacity data of the battery aging experiments. Then, generalize and extend the degradation tendency, and finally realize the prediction of the future battery degradation trajectory. In general, with the advantages of simple process, low calculation amount and without considering complicated mechanism of battery, RUL prediction based on data-driven method effectively reduces the operating load of the BMS and is suitable for the operating conditions of vehicles.

6.1.2 Method Classification

RUL prediction based on data-driven method can be divided into three categories: fitting-based method, filter observation-based method and sequence prediction-based method as shown in Table 6.1. Each category is described in detail as shown below.

Fitting-based method
This method assumes that the battery capacity degradation trajectory follows some inherent mathematical relationship. The battery RUL prediction problem can be solved with a mathematical model that can reasonably describe the battery aging characteristics. From the perspective of data fitting, the mathematical expressions with the number of cycles as input and the maximum available capacity as output help to describe the battery aging characteristics. To obtain an appropriate mathematical model, it is necessary to fit the battery capacity degradation trajectory repeatedly with different mathematical expressions and select the most suitable mathematical expression as the battery life empirical model. The mathematical expressions commonly used as fitting-based models for RUL prediction are listed in Table 6.2 [6–10], where n represents the cycles and C_0 represents the nominal capacity of battery.

Table 6.1 Classification of battery RUL prediction methods

Methods	Principle	Commonly used models or algorithms
Fitting-based method	Data fitting	Single-exponential model, dual-exponential model, linear model, polynomials model, verhulst model
Filter observation-based method	State estimation	KF, EKF, UKF, PF, unscented particle filter (UPF), spherical cubature particle filter (SCPF)
Sequence prediction-based method	Time series prediction	Gray prediction, autoregressive integrated moving average model, neural network, support vector machine, relevance vector machine

Table 6.2 Mathematical expressions of battery empirical models

Methods	Mathematical expressions	Model parameters
Single-exponential model	$C_{\max} = a_1 \exp(a_2 n) + a_3$	a_1, a_2, a_3
Dual-exponential model	$C_{\max} = b_1 \exp(b_2 n) + b_3 \exp(b_4 n)$	b_1, b_2, b_3, b_4
Linear model	$C_{\max} = c_1 n + c_2$	c_1, c_2
Polynomials model	$C_{\max} = d_1 n^2 + d_2 n + d_3$	d_1, d_2, d_3
Verhulst model	$C_{\max} = \frac{e_1/e_2}{1+(e_1/(e_2 C_0)-1)\exp(-e_1 n)}$	e_1, e_2

The flowchart of fitting-based method for battery RUL prediction is shown in Fig. 6.1.

When the RUL prediction procedure starts, the BMS extracts historical capacity data of battery from the storage. In order to select a suitable empirical model, it is usually required to perform offline aging experiments on the same type of batteries and store the appropriate empirical model into the BMS based on the experimental results. Considering that the suitable mathematical models may vary under different charge/discharge profiles or different aging stages, it is better to store several empirical models. To capture the battery capacity degradation trajectory, the BMS employs the corresponding empirical model according to the actual situation and fits the historical capacity data to obtain the parameters of the empirical model. The data fitting samples can use all or partial historical capacity data. After the data fitting completed, a given capacity failure threshold is substituted into the empirical model to solve the battery RUL. It is worth noting that the empirical model not only achieve the battery RUL prediction, but also predict the battery life trajectory. Generally, the fitting-based method is simple and less computing. However, since the data fitting is sensitive to the fluctuation of sample data, the prediction is prone to diverge.

Filter observation-based method

This method is the most commonly used method for the battery RUL prediction [11–14]. From the perspective of state estimation, the empirical model can be updated and corrected in real time through the observation data, which can effectively settle the divergence problem of RUL prediction caused by simple data fitting in the fitting-based method. Therefore, an empirical model is still demanded in this method, which can be regard as an improvement of the fitting-based method.

The flowchart of the filter observation-based method for battery RUL prediction is shown in Fig. 6.2.

When the RUL prediction procedure starts, the BMS extracts the historical capacity data of battery in the storage. Different from the fitting-based method, this method

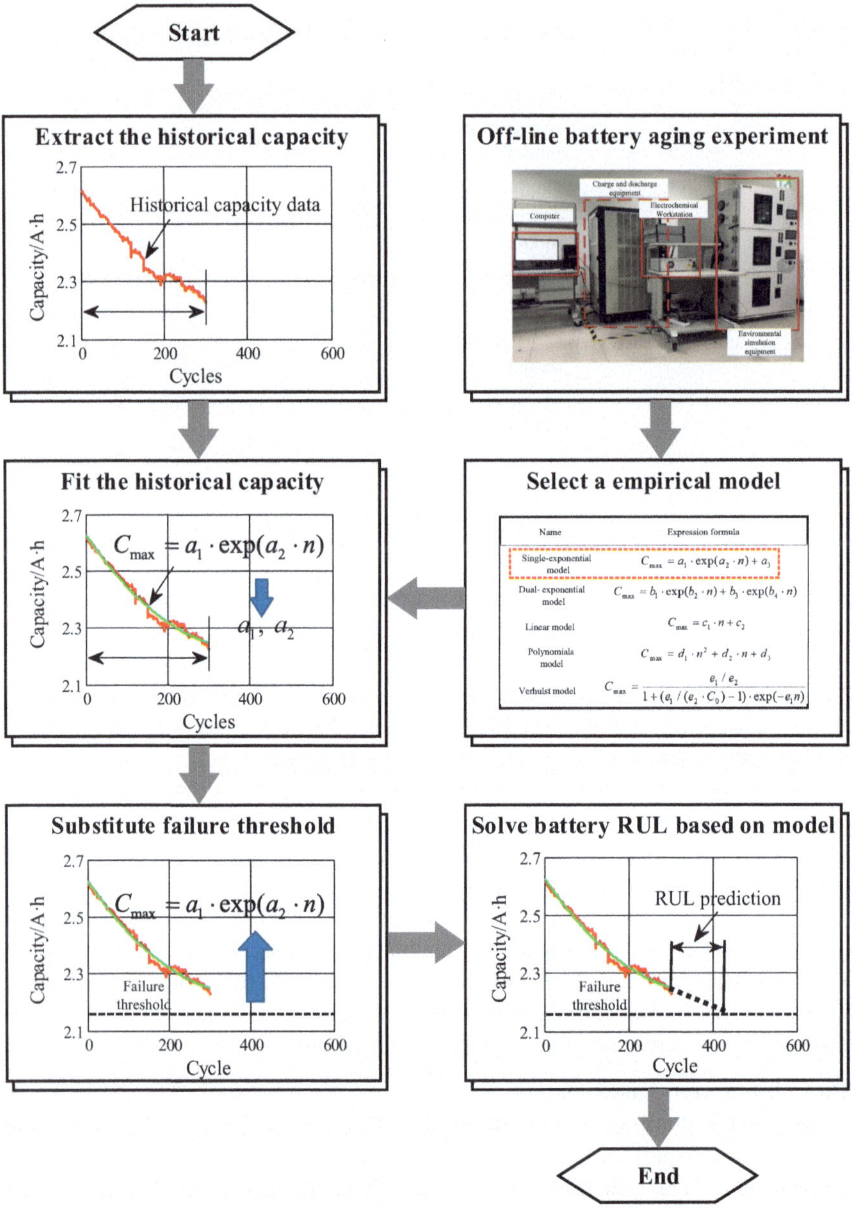

Fig. 6.1 Flowchart of fitting-based method for battery RUL prediction

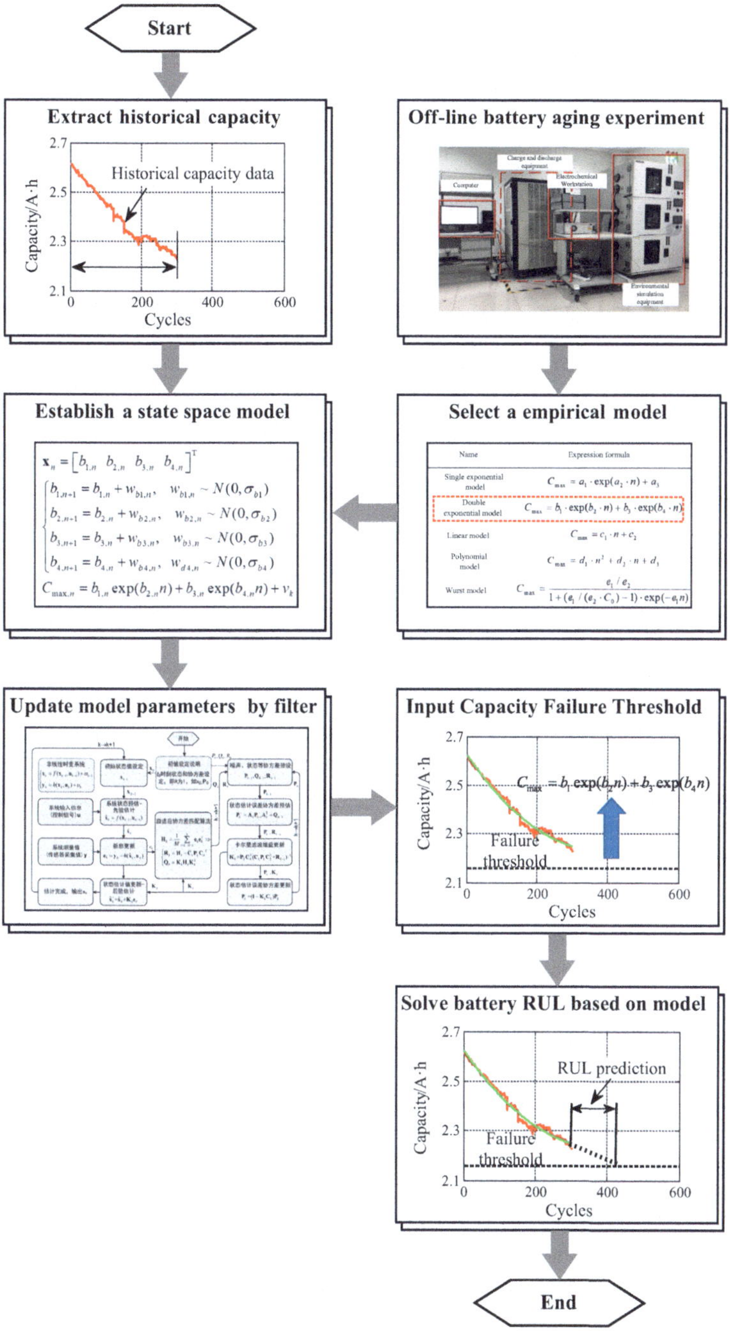

Fig. 6.2 Flowchart of filter observation-based method for battery RUL prediction

is to establish a corresponding state space equation based on the model rather than directly fit data after selecting a suitable empirical model. Taking the dual-exponential model in Table 6.2 as an example [5], the corresponding state equation is

$$\mathbf{x}_n = \left[b_{1,n}\, b_{2,n}\, b_{3,n}\, b_{4,n}\right]^{\mathrm{T}}$$
$$\begin{cases} b_{1,n+1} = b_{1,n} + w_{b1,n}, & w_{b1,n} \sim N(0, \sigma_{b1}) \\ b_{2,n+1} = b_{2,n} + w_{b2,n}, & w_{b2,n} \sim N(0, \sigma_{b2}) \\ b_{3,n+1} = b_{3,n} + w_{b3,n}, & w_{b3,n} \sim N(0, \sigma_{b3}) \\ b_{4,n+1} = b_{4,n} + w_{b4,n}, & w_{d4,n} \sim N(0, \sigma_{b4}) \end{cases} \quad (6.1)$$

where $\mathbf{x}_n$ is the state vector, $w_{\mathrm{b}1} \sim w_{\mathrm{b}4}$ are the noises with a mean of 0, and $\sigma_{\mathrm{b}1} \sim \sigma_{\mathrm{b}4}$ are the variances of the parameter noises. The corresponding observation equation is

$$C_{\max,n} = b_{1,n} \exp(b_{2,n} n) + b_{3,n} \exp(b_{4,n} n) + v_k, \quad v_n \sim N(0, \sigma_v) \quad (6.2)$$

where v_n is the observed noise with a mean of 0, and σ_v is the variance of the observed noise. After the state space equation established, the BMS uses the filtering algorithm to update and correct the state vector based on the historical capacity data, thereby obtaining the parameters of the empirical model. The detailed filter calculation process is described in Chap. 4.

The battery RUL can be solved by substituting a given capacity failure threshold into the empirical model after the parameters have been updated and corrected. The convergence of the empirical model and the RUL prediction accuracy are greatly improved, but the algorithm is highly dependent on the model accuracy. The commonly used algorithms for RUL prediction include KF, EKF, UKF, PF, UPF, and SCPF, etc.

Sequence prediction-based method

Different from the above two methods, this method does not rely on the empirical model. From the perspective of time series prediction, battery RUL prediction can be treated as a time-series prediction problem, which can be formulated as:

$$C_{\max,n+1} = f(C_{\max,n}, C_{\max,n-1}, \ldots, C_{\max,n-m}), m < n \quad (6.3)$$

where the function $f(\cdot)$ is a sequence model.

The time-series model is usually used to predict the future values. The core idea of this method is to quantitatively predict the future developing trend based on the development of past time series. As for batteries, the capacity degradation trajectory is essentially a long-term time series [15]. The prediction of RUL is the expansion and extension of time series. Therefore, the time-series model can be applied to deal with the RUL prediction problem.

It is can be observed from Eq. (6.3) that the battery capacity after the $n + 1$th cycle can be predicted by the historical capacity sequence from n–m th to nth cycles based on the time-series model $f(\cdot)$, which illustrates that the capacity after any cycles can be recursively predicted using Eq. (6.3). Therefore, selecting a suitable sequence model is the key to this method. The time-series models commonly used for RUL prediction include gray prediction [16], autoregressive integrated moving average model [17], [18], neural network, support vector machine [19], and relevance vector machine [20].

The flowchart of sequence prediction-based method for RUL prediction is shown in Fig. 6.3.

When the RUL prediction procedure starts, the BMS extracts the historical capacity data of battery in the storage. Subsequently, the BMS builds the time-series model, which consists of two steps: the first step is to construct the input and output samples based on the historical capacity data and initialize model structure and its parameters, the second step is to train the time-series model based on the input and output samples. After time-series model is built, the battery capacity after the next cycle can be predicted which is also called one-step ahead prediction. After completing the one-step ahead prediction, the BMS terminates the prediction if the result is less than the pre-defined capacity failure threshold, and the output RUL value is the amount of one-step ahead prediction before termination. Otherwise the BMS continues to perform the next one-step ahead prediction.

In summary, this prediction method can predict the battery RUL simply using the historical capacity data in the BMS storage without considering the rationality of the selected empirical model.

6.1.3 Probability Distribution

Any prediction method has a certain error. The model error in RUL prediction comes from the deviation between mathematical modeling and actual problems, which is hard to avoid and estimate. In addition, the influence of battery data on the model can also bring about certain errors (measurement error or truncation error) to the system. The uncertainty caused by these errors would spread with the running of the prediction algorithm and eventually influence the RUL prediction results. Thus, it is of great significance to describe the uncertainty of the battery RUL prediction results.

The probability distribution of the battery RUL prediction refers to the uncertainty distribution discipline of the prediction for a given prediction method, which is usually described by a probability density function of RUL. The probability density function not only calculates the confidence of the RUL prediction, but also obtains its distribution law and confidence interval, which contributes to the maintenance and recycling of the battery for EVs. Therefore, the probability density function of the battery RUL is more valuable than RUL itself in practical applications.

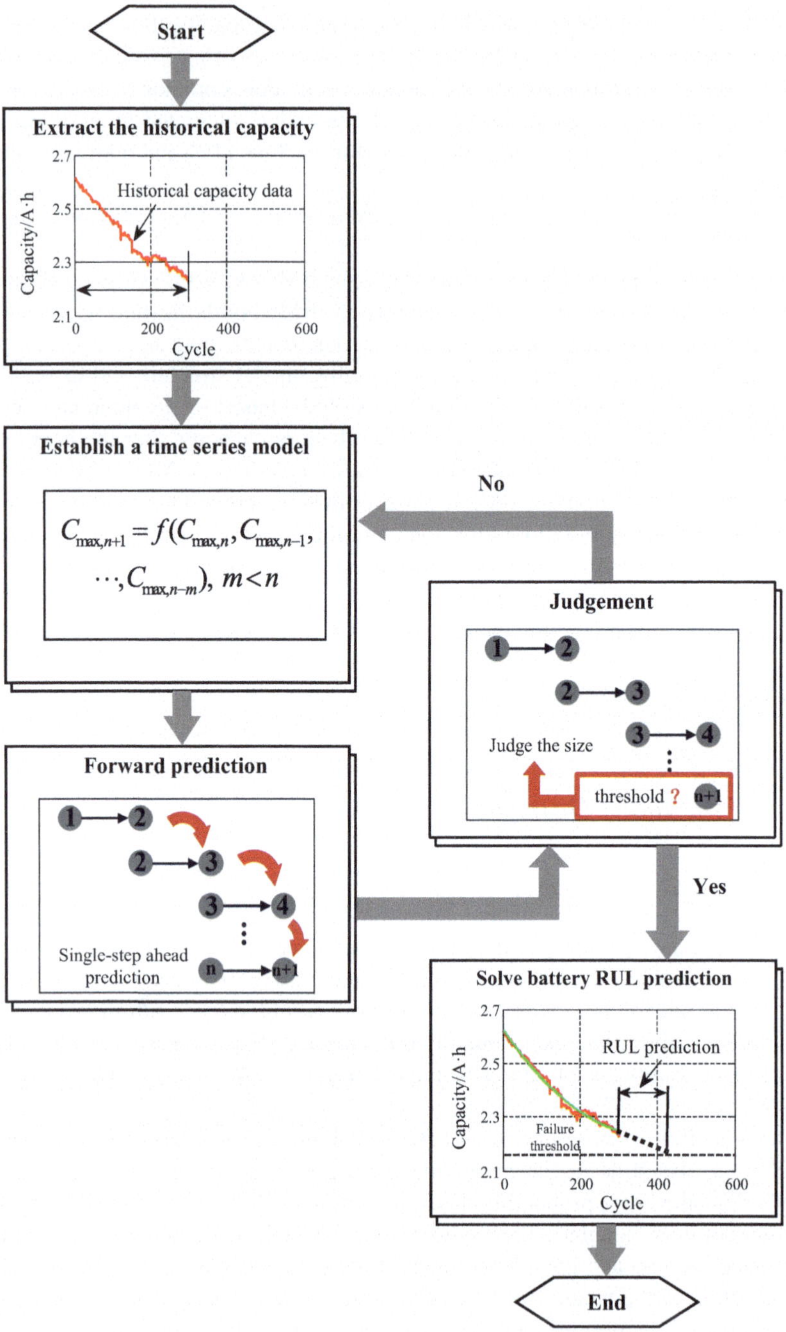

Fig. 6.3 Flowchart of sequence prediction-based method for RUL prediction

Table 6.3 Main uncertainties generation processes

Methods	Main uncertainties generation process
Fitting-based method	Fitting process of historical capacity data based on empirical model
Filter observation-based prediction	State vector initialization process of the state space equation
Sequence prediction-based method	Time series model input process of battery RUL prediction starting point

The Monte Carlo (MC) method is often adopted to calculate the probability density function of the battery RUL in combination with different prediction methods. The main idea of MC method is to numerically calculate the results of complex processes by repeated randomized trials. The basic steps of calculating the battery RUL probability density function using the MC method are described in detail below:

① Determine the main source of uncertainty for the battery RUL prediction method. It is demanded to focus on the main generation processes of uncertainty during the calculation. Table 6.3 lists the main uncertainties generation processes in the fitting-based method, the filter observation-based method, and the sequence prediction-based method.

② Determine the distribution of main uncertainties. For the fitting-based method, the distribution rule of uncertainty can be determined by the means and variances of the parameters in the fitting process of empirical model. The initial state vector of the filter observation-based method is typically determined by fitting historical capacity data with empirical model. Therefore, the determination method of uncertainty distribution rule of the filter observation-based method is similar to that of the fitting-based method. For the prediction method based on time series, the uncertainty distribution rule can be determined by the distribution rule of historical capacity around the prediction starting point.

③ MC simulation. First, several samples are randomly generated based on the distribution rule of the main uncertainties. The sample types generated by the above three methods are the parameter samples of empirical model, the initial state vector samples, and the initial input vector samples of the sequence model, respectively. Then, the simulation prediction is performed based on each individual of samples according to the selected prediction method to obtain several simulated RUL.

④ Calculate the probability density function of the RUL prediction. According to all the simulation results from step ③, the probability density function of the RUL prediction can be calculated by:

$$\hat{f}_h(c_{\max}) = \sum_{i=1}^{N}\left[K_{\mathrm{p}}\left(\frac{c_{\max} - c_{\max,i}^{-}}{h_{\mathrm{p}}}\right) + K_{\mathrm{p}}\left(\frac{c_{\max} - c_{\max,i}}{h_{\mathrm{p}}}\right) + K_{\mathrm{p}}\left(\frac{c_{\max} - c_{\max,i}^{+}}{h_{\mathrm{p}}}\right)\right] \tag{6.4}$$

where $\hat{f}_h(c_{\max})$ represents the probability density function of the RUL prediction, $K_p(\cdot)$ is the Gaussian kernel function, h_p is the bandwidth, $c^{-}_{\max,i}$ and $c^{+}_{\max,i}$ can be calculated by:

$$\begin{cases} c^{-}_{\max,i} = 2L_c - c_{\max,i} \\ c^{+}_{\max,i} = 2U_c - c_{\max,i} \end{cases} \tag{6.5}$$

where U_c and L_c are the upper and lower bounds of the MC simulation results, respectively, and $c_{\max,i}$ is the ith simulated RUL.

6.2 RUL Prediction Based on Box-Cox Transformation

The fitting-based method performs well in online computation, but the poor prediction performance fails to meet the actual needs of EVs. Filter observation-based method facilitates the accuracy and convergence of the fitting-based method, but enhances the dependence on the model and increases the complexity of the calculation. Furthermore, both fitting-based method and observation-based prediction method are demanded to analyze massive battery off-line aging experimental data to ensure the adaptability of empirical model, which consumes expensive resources and time. Therefore, a new fitting-based method based on Box-Cox transformation is introduced to settle these problems in this section. This new method not only improves the accuracy and convergence of RUL prediction, but also eliminates numerous battery off-line aging experiments, which improves the engineering applicability of traditional fitting-based methods.

6.2.1 *Box-Cox Transformation Technology*

As can be seen from Table 6.1, apart from linear model, the others have the characteristics of strong nonlinear and complex expressions. Consequently, the traditional fitting-based method is highly susceptible to data fluctuations (especially the influence of terminal data fluctuations) when fitting the historical capacity data of battery, leading to divergent RUL prediction. Although the linear model has the most stable fitting performance and the least parameters, it is not suitable for predicting nonlinear capacity degradation trajectory. From this perspective, an effective solution comes up based on Box-Cox transformation.

Proposed by George Box and Sir David Cox in 1964, the Box-Cox transformation is a parametric generalized power transformation method based on the basic assumptions of linearity, normality and homoscedasticity, which is often applied to reduce irregularities such as non-additive, non-normal and heteroscedasticity in statistical modeling [21]. Box-Cox transformation technology has been widely concerned and applied since it was proposed, especially the empirical function determination process in various fields. The core idea of Box-Cox transformation is to create a family of functions, which transform the data monotonically by introducing some parameters to reduce deviation of samples from normal characteristics. As for battery RUL prediction, Box-Cox transform can linearize the nonlinear capacity degradation trajectory of battery with a transform coefficient, which effectively reduces the difficulty of RUL prediction and makes it possible for accurate RUL prediction with linear model [15]. The battery RUL prediction process based on Box-Cox transformation is described in detail below.

6.2.2 RUL Prediction Based on Box-Cox Transformation

The flowchart of battery RUL prediction based on Box-Cox transformation is shown in Fig. 6.4.

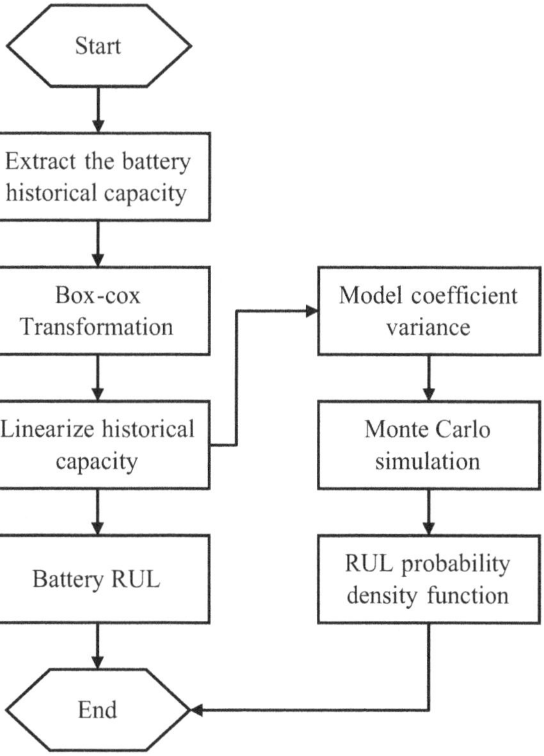

Fig. 6.4 Flowchart of battery RUL prediction based on Box-Cox transformation

Firstly, Box-Cox transform is to linearize the historical capacity degradation trajectory to obtain the transform coefficient and the transformed historical capacity. The expression is:

$$\tilde{C}_{\max} = \begin{cases} \dfrac{(C_{\max})^{\lambda_{\mathrm{bc}}} - 1}{\lambda_{\mathrm{bc}}}, \ \lambda_{\mathrm{bc}} \neq 0 \\ \log C_{\max}, \ \lambda_{\mathrm{bc}} = 0 \end{cases}, \tag{6.6}$$

where $\tilde{C}_{\max}$ represents the transformed historical capacity, λ_{bc} represents the Box-Cox transform coefficient which is calculated by the maximum likelihood estimation method. The calculation is equivalent to finding a suitable value of λ_{bc} which can maximize the following expression:

$$L^{*}(\lambda_{\mathrm{bc}}) = \frac{n_{\mathrm{s}}}{2} \log(\hat{\sigma}^{2}(\lambda_{\mathrm{bc}})) + (\lambda_{\mathrm{bc}} - 1) \sum_{i=1}^{n_{\mathrm{s}}} \log(C_{\max,i}) \tag{6.7}$$

where $L^{*}(\lambda_{\mathrm{bc}})$ is a logarithmic natural function, n_{s} is the sample size of the historical capacity observation, and $\hat{\sigma}^{2}(\lambda_{\mathrm{bc}})$ can be calculated as follows:

$$\hat{\sigma}^{2}(\lambda_{\mathrm{bc}}) = \frac{[\mathbf{C} - \mathbf{K}\hat{\boldsymbol{\beta}}]^{\mathrm{T}}[\mathbf{C} - \mathbf{K}\hat{\boldsymbol{\beta}}]}{n_{\mathrm{s}}} \tag{6.8}$$

where $\mathbf{C}$ donates the vector of historical capacity, $\mathbf{K}$ donates the cycle vector corresponding to the historical capacity, and the expression of $\hat{\boldsymbol{\beta}}$ is

$$\hat{\boldsymbol{\beta}} = (\mathbf{K}^{\mathrm{T}}\mathbf{K})\mathbf{K}^{\mathrm{T}}\mathbf{C} \tag{6.9}$$

The historical battery capacity data has been substantially linearized after the Box-Cox transformation. Considering the nonlinearity of the capacity degradation trajectory is weakened and the RUL prediction difficulty is reduced, adopting the linear model as the empirical model can satisfy the accuracy requirement. The fitting expression of the linear model is

$$\tilde{C}_{\max} = c_1 + n \cdot c_2 \tag{6.10}$$

where the coefficients (c_1 and c_2) of the linear model can be solved by the least squares method. The solution expression is

$$\begin{cases} c_2 = \frac{\sum_{i=1}^{n_{\mathrm{s}}} (n_i - \bar{n})(\tilde{C}_{\max,i}(\lambda_{\mathrm{bc}}) - \bar{\tilde{C}}(\lambda_{\mathrm{bc}}))}{\sum_{i=1}^{n} (n_i - \bar{n})} \\ c_1 = \bar{\tilde{C}}(\lambda_{\mathrm{bc}}) - c_2 \bar{n} \end{cases} \tag{6.11}$$

where $\bar{\tilde{C}}$ is the average of the transformed historical capacity, $\bar{n}$ is the average of historical cycles.

After obtaining the coefficient of the linear model, Box-Cox transformation is performed on the given capacity failure threshold, and the transformation result is substituted into the Eq. (6.10) to calculate the cycles corresponding to the capacity failure threshold. The battery RUL can be obtained by subtracting the current cycles.

The detailed process using the MC method to calculate the RUL probability density function is described below.

The prediction uncertainty of the fitting-based method is mainly generated from the fitting process of the empirical model. Therefore, the BMS first needs to randomly generate several parameter samples according to the uncertainty of the fitting process. In order to solve the variances of c_1 and c_2 during the fitting process, the following equation need to be solved:

$$\begin{cases} Var(c_1) = \frac{s^2}{n_s} + \frac{s^2 \bar{n}^2}{\sum_{i=1}^{n} (n_i - \bar{n})^2} \\ Var(c_2) = \frac{s^2}{\sum_{i=1}^{n_s} (n_i - \bar{n})^2} \end{cases} \tag{6.12}$$

where $Var\ (c_1)$ and $Var\ (c_2)$ represent the variance of c_1 and c_2 respectively, and s^2 is the estimated value of the error term variance whose expression is

$$s^2 = \frac{SSR}{n_s - 2} \tag{6.13}$$

where SSR is the sum of the squares of the residuals which can be calculated by

$$SSR = \sum_{i=1}^{n_s} \left[C_{\max,i}(\lambda) - (\beta_0 + \beta_1 n_i) \right]^2 \tag{6.14}$$

According to the MC method, N parameter samples of linear model are randomly generated, where c_1 and c_2 are treated as means, and $Var\ (c_1)$ and $Var\ (c_2)$ are treated as variances. Then, according to Eq. (6.10), RUL prediction simulation is performed based on each individual of samples to obtain N RUL prediction results. At this time, the probability density function of the battery RUL can be estimated according to Eqs. (6.4) and (6.5). It is worth noticing that increasing the parameter samples of the linear model helps to improve the reliability of the probability density function, but it also significantly increases the computational burden.

6.2.3 Case Study

Battery 6-cell 25 and battery 6-cell 26 were taken to evaluate the performances of the above method. For convenience, the traditional fitting-based method and the Box-Cox transformation-based method are hereinafter referred to as FIT and BCT, respectively.

Comparison with FIT

The RUL prediction results of BCT and FIT are shown in Fig. 6.5 [15], among which the results of FIT is obtained based on the dual-exponential model and the polynomials model in Table 6.2.

Figure 6.5a shows the RUL prediction results of cell 25 based on the first 300 cycles, BCT can effectively linearize the nonlinear capacity degradation trajectory. It is calculated that the Pearson correlation coefficient between the transformed capacity and the cycles reaches –0.9888, indicating a strong linear relationship. The results predicted by BCT shows that the cell will undergo 505 cycles before reaching the given failure threshold, which is only 5 cycles more than the true value with only 1% prediction error. In addition, the 95% confidence interval is [495, 515] covering only 20 cycles, which justifies that BCT enjoys a high reliability.

Figure 6.5b illustrates the RUL prediction results of cell 26 based on the first 310 cycles. The Pearson correlation coefficient reaches –0.9769. Compared with

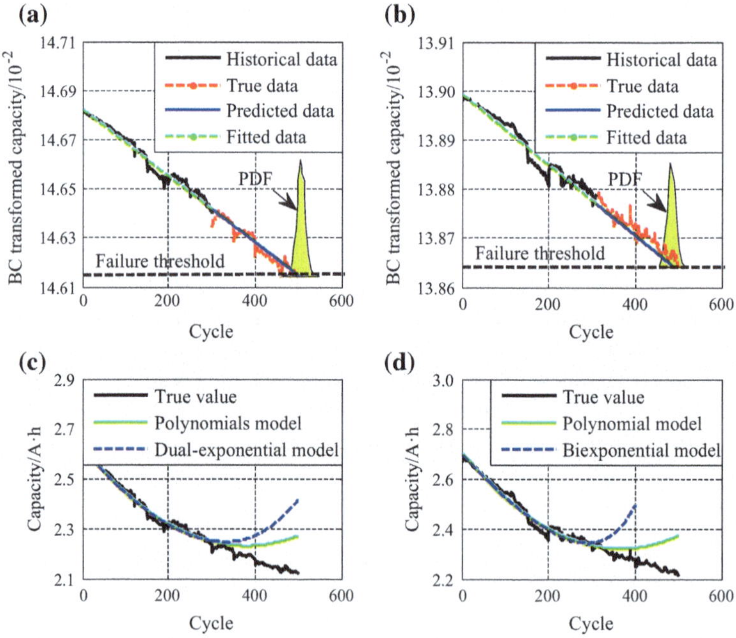

Fig. 6.5 RUL prediction results: **a** BCT for cell 25; **b** BCT for cell 26; **c** FIT for cell 25; **d** FIT for cell 26

the true value, the prediction error of BCT is only 12 cycles, namely the prediction error is 2.4%. The 95% confidence interval is [474, 502] covering only 28 cycles. It can be seen that this method can perform excellently on two different battery cells. The above RUL predictions are performed in the MATLAB R2016b environment of the Intel Core i7-6700HQ 3.5 GHz processor. The calculation time does not exceed 2.25 s indicating that BCT has good online operation capability.

Figure 6.5c, d show the RUL prediction results of FIT using the polynomials model and dual-exponential model, respectively. Noted that with the increase of cycles, the predictive effects of polynomials model and dual-exponential model deteriorate gradually and the capacity degradation trajectory even rises abnormally. Affected by data fluctuations, FIT produces divergent prediction results and is brought to the failure in correctly predicting the battery capacity degradation trajectory.

In summary, BCT can effectively reduce the prediction difficulty of the battery RUL, and greatly improve both the prediction accuracy of the linear model and the prediction performance of FIT.

Performance evaluation of online prediction

The BCT coefficient directly affects the linearization of the capacity degradation trajectory. In general, negative transform coefficient errors will result in over-prediction, while positive transform coefficient errors will be responsible for under-prediction. Therefore, it is of great significant to evaluate the accuracy of the BCT coefficients.

Figure 6.6a illustrates the results of RUL prediction adopting BCT, in which the battery starts from the 50th cycle and runs a BCT every 10 cycles.

Figure 6.6a, b show the prediction results of BCT coefficients of two cells under different cycles, in which the reference values are obtained based on offline aging experimental data. The upper boundary is set 2 higher than the reference value while the lower boundary is set 1 lower than the reference value. The results indicate that the estimated BCT coefficients fluctuate at the initial stage of the battery aging and start to converge rapidly to the reference value after the 180th cycle. The estimated BCT coefficients for cells 25 and 26 approach the upper and lower boundaries after 250th and 310th cycle, respectively.

The RUL prediction results in Fig. 6.6c, d present the same trends as the BCT coefficients. The predicted RUL for cell 25 converges to the true value after approximately 60 cycles. It is worth noting that once converging to the true value, the error of the battery RUL is mostly controlled within 10 cycles. The RUL prediction result of the cell 26 shows a stable trend at 330th cycle. Figure 6.6e, f indicate that BCT can effectively control the standard deviation of the RUL prediction within 7 cycles, which decreases with the increasing of the historical aging data.

In summary, BCT realizes the stability and accuracy of RUL prediction.

Comparison with filter observation-based prediction

Both the BCT method and the filter observation-based method can be regarded as an improvement scheme for the fitting-based method under different prediction ideas. Therefore, a comparative evaluation between them is performed below. The fitting-based method adopts the PF for state estimation. For the convenience of discussion,

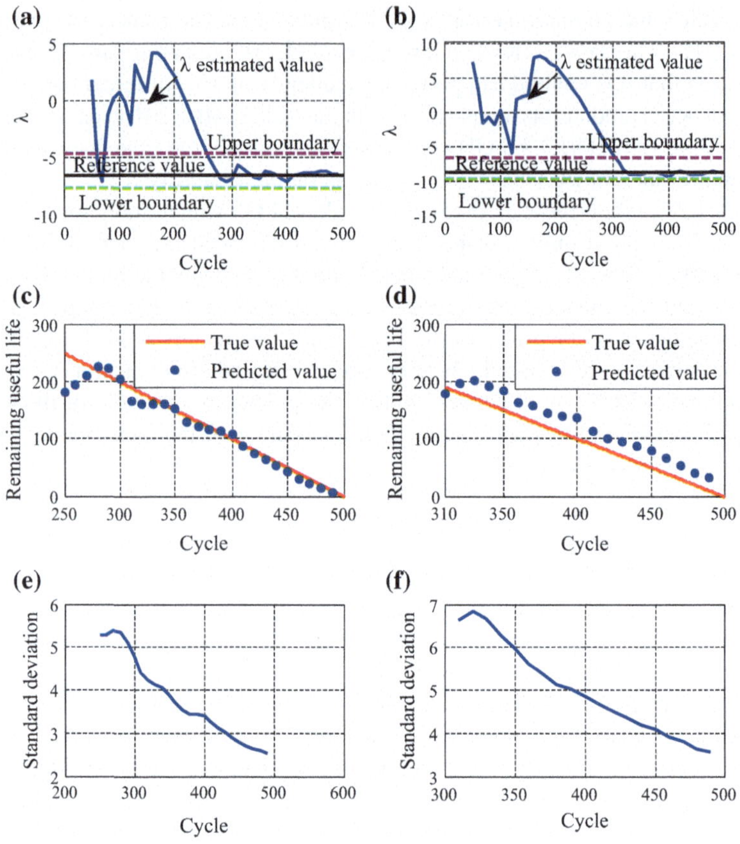

Fig. 6.6 Battery RUL prediction based on BCT: **a** estimated λ for cell 25; **b** estimated λ for cell 26; **c** predicted result for cell 25; **d** prediction result for cell 26; **e** prediction standard deviation for cell 25; **f** prediction standard deviation for cell 26

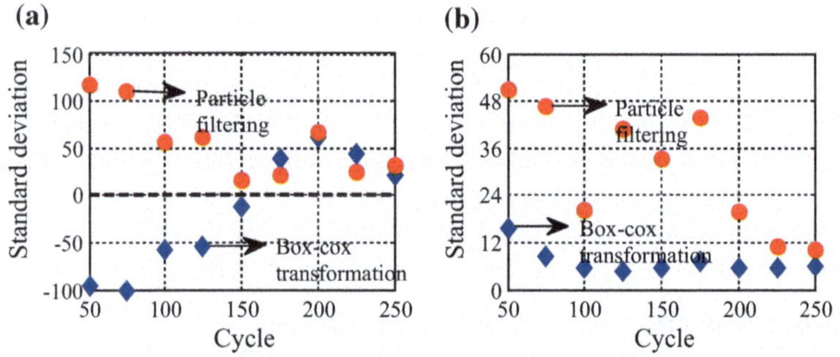

Fig. 6.7 RUL prediction of BCT and PF (cell 25): **a** prediction error; **b** prediction standard deviation

the PF-based method is simply referred to PFM. Figure 6.7 shows the RUL prediction results for the cell 25 by BCT and PFM, respectively.

Figure 6.7a shows that both BCT and PFM need to go through about 130 cycles before converging to the true value of RUL. In most cases, the prediction accuracy of BCT is slightly higher than that of PFM. The average prediction errors of BCT and PFM are 53.7 cycles and 56.5 cycles, respectively (the test value of the RUL is 500 cycles). It can be seen that the prediction accuracy of BCT is basically higher than that of PFM.

Figure 6.7b illustrates the RUL prediction standard deviations of these two methods. It can be seen that the standard deviations decrease with the increase of the aging data, but the standard deviation of BCT is significantly lower than of PFM, and can maintain a lower level through the whole process. The average prediction standard deviations of BCT and PFM are 7.2 cycles and 30.8 cycles, respectively.

In summary, BCT has superior performance in accuracy and precision than PFM.

6.3 RUL Prediction Based on Long Short-Term Memory Recurrent Neural Network

From the perspective of time series, the aging of batteries typically takes a long time, covering thousands of cycles, and battery capacity between each cycle is highly correlated. Even though it is possible to fit and reconstruct the battery capacity degradation trajectory based on historical data, the weights of historical data are often difficult to determine in the reconstruction process. Noise or even abnormal data are hard to avoid in the process of collection of historical data despite the continuous advancement of onboard measurement technology. Therefore, quickly and efficiently extracting the key information from battery historical capacity data is essential to settle the RUL prediction problem.

In view of the above problems, this section is to elaborate a battery RUL prediction method based on time series prediction, namely the Long Short-term Memory Recurrent Neural Network (LSTM RNN) prediction method.

6.3.1 LSTM RNN

LSTM RNN is a recurrent neural network with deep learning capabilities raised by Sepp Hochreiter and Jürgen Schmidhuber in 1997 [22], designed for long-term dependency information. In recent years, LSTM RNN performs well in many fields such as language modeling, machine translation, image recognition and language recognition. Unlike the traditional Simple Recurrent Neural Network (SimRNN), LSTM RNN introduces a logic gate to control and protect information, effectively solving the insensitivity to long-term input of SimRNN cycle layer.

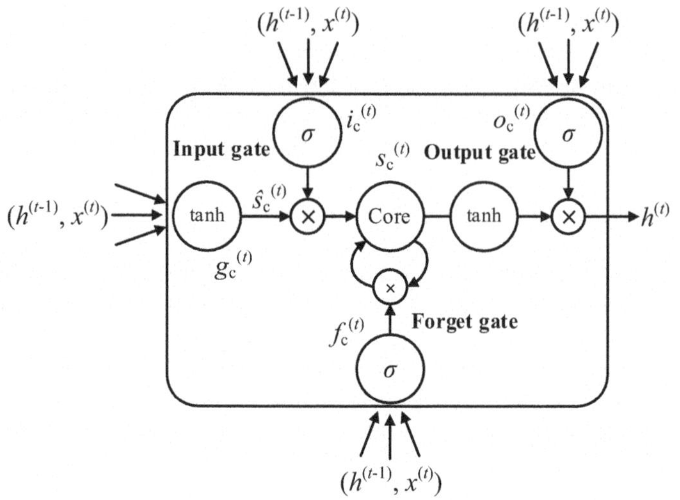

Fig. 6.8 Network architecture of LSTM RNN

Figure 6.8 shows the neuron structure of LSTM RNN, where $x^{(t)}$ donates neuron input at time t, and $h^{(t-1)}$ donates the hidden layer information at time t-1. The key of LSTM RNN lies in the internal state quantity s_{c}, which is located at the center of each neuron and activated by a linear function. The internal state of LSTM RNN takes charge of the storing or forgetting historical information. The three logical gates in LSTM RNN are described in detail below, namely the forget gate, the input gate, and the output gate.

The primary problem to be solved by LSTM RNN is deciding which information needs to be forgotten, controlled by the forget gate. The forget gate performs based on the σ function transformation. By inputting $x^{(t)}$ and $h^{(t-1)}$, a value between 0 and 1 is returned, which directly determines the internal state value s_{c}. The values 0 and 1 represent complete forgetting and memory, respectively. The expression is

$$f^{(t)} = \sigma\left(W^{fX}x^{(t)} + W^{fh}h^{(t-1)} + b_f\right) \tag{6.15}$$

where f_{c} represents the output of the forget gate, W^{fX} and W^{fh} represent the forget gate weights of inputs $x^{(t)}$ and $h^{(t-1)}$, respectively, and b_f represents the forget gate threshold. It should be noted that for the RUL prediction, the forget gate helps to discard the noise of the capacity data, the abnormal data, and the redundant information among the adjacent cycle data, thereby improving the prediction accuracy.

After determining the forgotten information, LSTM RNN needs to determine which information to be stored in the internal state, controlled by the input gate. The realization of function by the input gate consists of two parts: one is to determine which inputs to be updated by the σ function transformation, and the other is to generate alternative update information by the tanh function transformation. The expression is

$$i_c^{(t)} = \sigma\left(W^{iX}x^{(t)} + W^{ih}h^{(t-1)} + b_i\right) \tag{6.16}$$

$$g^{(t)} = \tanh\left(W^{gX}x^{(t)} + W^{gh}h^{(t-1)} + b_g\right) \tag{6.17}$$

where $i_c^{(t)}$ and $g^{(t)}$ are the input gate information under the σ function transformation and the tanh function transformation, respectively. W^{iX} and W^{gX} respectively are the weights of input gate for $x^{(t)}$ in the above two-part transformations. W^{ih} and W^{gh} are the weights of input gate for $h^{(t-1)}$ in the above two-part transformations, respectively. b_i and b_g are the input gate thresholds under the σ function transformation and the tanh function transformation, respectively.

By multiplying $i_c^{(t)}$ and $g^{(t)}$ point by point, the internal state of LSTM RNN can be updated eventually. The expression is

$$s_c^{(t)} = i_c^{(t)} \circ g^{(t)} + f^{(t)} \circ s_c^{(t-1)} \tag{6.18}$$

where the symbol $\circ$ represents a multiplication by element.

After the selective memory and update of the information is completed, LSTM RNN needs to determine the current output information based on the output gate. In this process, the internal state s_c transformed by the tanh function is multiplied by the output gate $o_c^{(t)}$ to determine the output information. The expression is

$$o_c^{(t)} = \sigma\left(W^{oX}x^{(t)} + W^{oh}h^{(t-1)} + b_o\right) \tag{6.19}$$

$$h^{(t)} = \tanh\left(s^{(t)}\right) \circ o_c^{(t)} \tag{6.20}$$

where W^{oX} and W^{oh} donate the weights of output gate of the inputs $x^{(t)}$ and $h^{(t-1)}$ respectively, and b_o donates the output gate threshold.

Based on the above three unique logic gate structures, LSTM RNN can effectively learn long-term battery capacity degradation data and capture the hidden information about historical capacity sequences to achieve accurate prediction of future capacity sequences.

6.3.2 RUL Prediction Process of LSTM RNN

The flowchart of LSTM RNN for RUL prediction is shown in Fig. 6.9 [23].

When the RUL prediction starts, the BMS requires to perform a structured initialization on LSTM RNN, including the settings of parameters such as the input, output, number of neurons, number of hidden layers, and type of activation function for the deep network. Subsequently, the BMS extracts battery historical capacity data and constructs training samples of LSTM RNN according to Eq. (6.3). For example, if the capacity of the $n + 1$th cycle is predicted by the capacity data of the nth cycle,

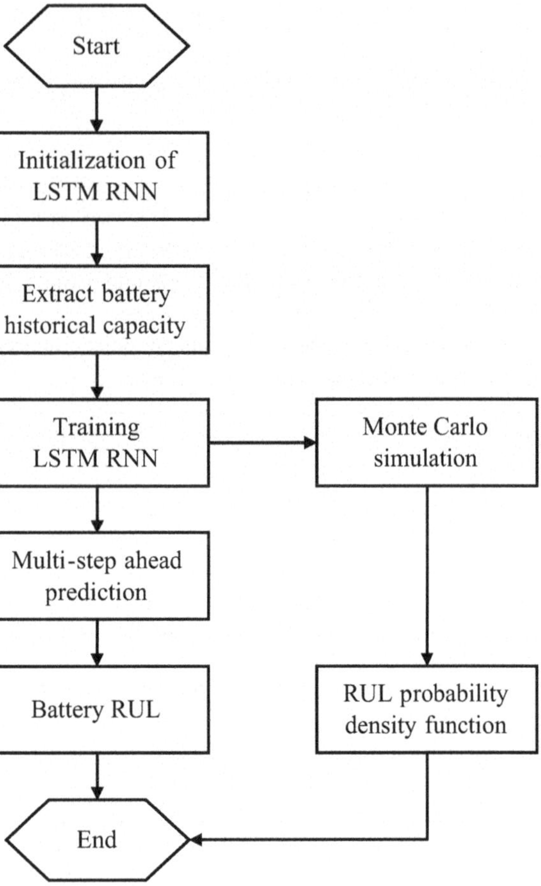

Fig. 6.9 Flowchart of LSTM RNN for battery RUL prediction

each training sample should take the capacity data corresponding to the $n-1$th cycle as an input and the capacity data corresponding to the nth cycle as an output.

The time series model $f(\cdot)$ in Eq. (6.3) is obtained by training the LSTM RNN. Traditional training methods based on batch gradient descent or random gradient descent are often difficult for the requirement of deep learning, which may easily cause low convergence rate of the network weight coefficient and affect the real-time computing efficiency of RUL prediction. In addition, overfitting tends to be particularly severe in LSTM RNN training. Therefore, a neural network parameter optimization method based on Root Mean Square prop (RMSprop) which can effectively avoid the above problems is introduced here. It effectively guarantees the convergence rate of the network weight coefficient. The neural network parameter θ is updated by

$$g_t = \nabla_{\theta} J\left(\theta_t; x^{(i:i+N)}; y^{(i:i+N)}\right) \tag{6.21}$$

$$E\left[g^2\right]_t = \gamma_{\text{lstm}} E\left[g^2\right]_{t-1} + (1 - \gamma_{\text{lstm}}) g_t^2 \tag{6.22}$$

$$\theta_{t+1} = \theta_t - \frac{\eta_{\text{lstm}}}{\sqrt{E\left[g^2\right]_t + \varepsilon_{\text{lstm}}}} g_t \tag{6.23}$$

where x and y donate the input and output of the network respectively. $(x^{(i:i+N)}, y^{(i:i+N)})$ donates N samples in each small batch, $J(\theta, x^{(i:i+N)}, y^{(i:i+N)})$ donates the objective function of each small batch, g_t donates the derivative of the objective function at time t, γ_{lstm} donates the coefficient that determines the average of the squared gradient, η_{lstm} donates the learning rate of the training algorithm, and $\varepsilon_{\text{lstm}}$ donates the smoothing term that avoids the divisor to zero.

In terms of preventing overfitting, the L1 and L2 regularization methods are two commonly used methods. The main idea of these two methods is to standardize weight learning by adding extra items to the objective function of the batch sample. However, the L1 and L2 regularization methods add the computational burden of summation and quadratic sum of weights, respectively. To settle this problem, the Dropout technology proposed by Srivastava randomly discard some neurons (along with their connections) during deep neural network training. And each neuron has a probability of being discarded independently of other neurons. It is illustrated in Fig. 6.10 that the neurons in dashed lines and their corresponding connections between input and output are temporarily removed from the network. The Dropout technology is equivalent to train a sparse network which extracts from the original deep network. Briefly, the neural network with the Dropout technology can be regarded as a large-scale sparse network with a large number of sharing weight neurons, in which the probability that each element in the neural network is selected is relatively low. Therefore, the Dropout technology effectively reduces the sensitivity of neurons to a specific weight in the neural network, thereby preventing the overfitting.

After training the LSTM RNN, the BMS needs to input the historical capacity data into the network for multi-step ahead recursive prediction. The prediction is terminated until the predicted capacity falls below the failure threshold. Subsequently, the number of recursive steps experienced is counted, regarding as the battery RUL.

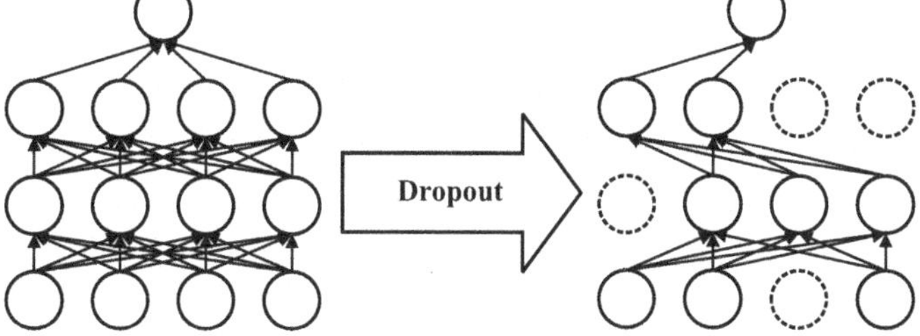

Fig. 6.10 Schematic diagram of the dropout technology

For instance, inputting the capacity $C_{\max,n}$ of the nth cycle to the LSTM RNN network to predict the capacity $C_{\max,n+1}$ of the $n + 1$th cycle. $C_{\max,n+1}$ is then used as the new input to predict the capacity $C_{\max,n+2}$ of the $n + 2$th cycle. The recursion does not terminate until the prediction result $C_{\max,n+x}$ falls below the failure threshold, and the value of x is recorded as the output of RUL.

The calculation of the prediction probability density function in this method is similar to that of the sequence prediction-based method. N capacity sequence samples are randomly generated around the predicted starting point according to the statistical characteristics of the historical capacity data. Then, each individual of samples is input into LSTM RNN based on the MC method to perform the multi-step ahead prediction simulation, thereby obtaining N values of RUL. Finally, the probability density function of the RUL prediction can be calculated according to Eqs. (6.4) and (6.5).

6.3.3 Case Study

The battery 6-cell 27 and battery 6-cell 28 were taken to evaluate the performance of the above method in detail [23]. In particular, the LSTM RNN in this case adopted a dual LSTM layers consisting of 50 and 100 neurons respectively with a Dropout discard probability of 20%.

Prediction performance evaluation at different aging stages

Figure 6.11 shows the RUL results of LSTM RNN for cells 27 and 28. Figure 6.11a shows the RUL results of LSTM RNN for cell 27 based on the first 253 cycles, which presents that the predicted capacity degradation trajectory manages to be basically close to the test result with only 253 historical capacity data as training samples. Figure 6.11b shows the result for cell 27 based on the first 354 cycles. The predicted starting point is located in the middle and later stages of the battery aging process with more historical training data compared with Fig. 6.11a. It can be seen that the future capacity degradation trajectory can be basically predicted by this method. It is worth noting that the shape of the probability density function is significantly higher than that of Fig. 6.11a with a narrower distribution range, indicating that the increase of historical capacity data facilitates the reliability of LSTM RNN.

Figure 6.11c, d show the RUL results for cell 28 with both 285 and 399 cycles as a starting point, respectively. Similar to the results in Fig. 6.11a, b, the predicted capacity degradation trajectories are consistent with the test results and the distribution range of the probability density function tends to be narrower with the increase of historical data. Therefore, LSTM RNN is justified for an impressive adaptability to RUL predictions in different aging stages.

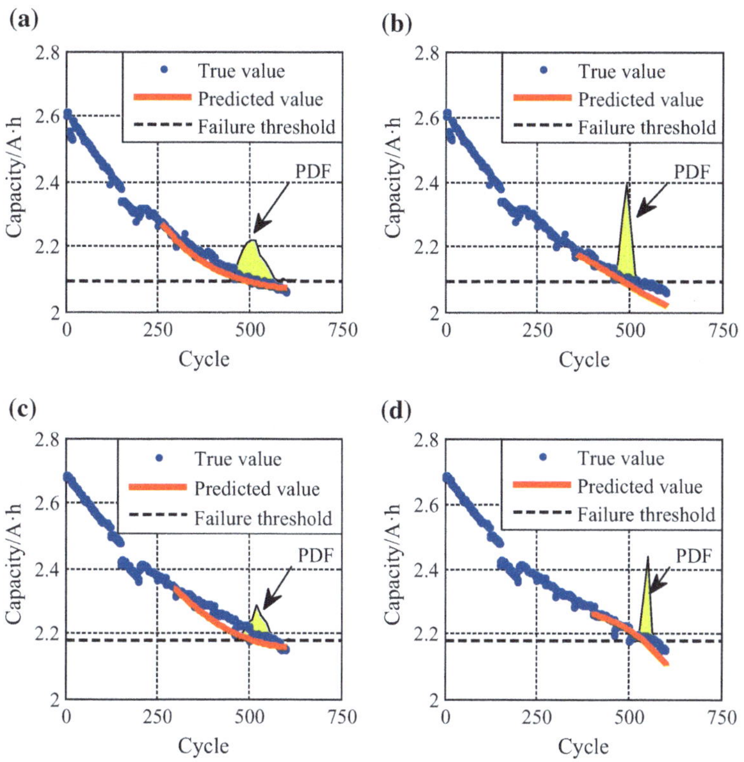

Fig. 6.11 RUL results of LSTM RNN: **a** prediction result at 253 cycles for cell 25; **b** prediction result at 354 cycles for cell 25; **c** prediction result at 285 cycles for cell 26; **d** prediction result at 399 cycles for cell 26

Comparison with simple recurrent neural network

Tables 6.4 and 6.5 describe the prediction results corresponding to Fig. 6.11. The SimRNN network is a common sequence prediction-based method, which is introduced as an evaluation benchmark for LSTM RNN. Table 6.4 shows that the RUL prediction error of LSTM RNN method is only 3 cycles with an error of only 0.6%, and the 95% confidence interval covers 94 cycles in the initial aging stage with a

Table 6.4 RUL prediction results of battery 6-cell 25 (the actual RUL is 506)

Methods	Starting cycle	Error	Relative error (%)	95% confidence bound	Training time (s)
LSTM RNN	253	−3	0.6	[470, 564]	20.74
	354	15	3.0	[473, 511]	23.04
SimRNN	253	135	26.7	[358, 388]	44.15
	354	78	15.4	[395, 470]	55.25

Table 6.5 RUL prediction results of battery 6-cell 26 (the actual RUL is 571)

Methods	Starting cycle	Error	Relative error (%)	95% confidence bound	Training time (s)
LSTM RNN	285	48	8.4	[487, 585]	20.49
	399	26	4.6	[532, 561]	28.50
SimRNN	285	195	34.2	[368, 392]	41.73
	399	95	16.6	[442, 514]	56.86

prediction starting point of 253. In contrast, the SimRNN network has a prediction error of 135 cycles with an error of 24.7% under the same conditions. Although the probability density function of SimRNN is narrower (covering 30 cycles), the probability density function has been far from the RUL test results due to the low prediction accuracy. Starting from 354 cycles, the prediction errors of LSTM RNN and SimRNN are 15 and 78 cycles, respectively, with the prediction errors of 3% and 14.3%, respectively. Compared with the results starting from 253 cycles, the prediction error of SimRNN is reduced by 11.3%, but it is still much larger than that of the LSTM RNN. In addition, with the change of the prediction starting point from 253 to 354 cycles, the amount of the historical capacity data for training increases. The span of the probability density function of LSTM RNN decreases from 94 to 38 cycles, while that of SimRNN increases from 30 to 75 cycles. The main reason is that it is difficult for SimRNN to store important information of capacity degradation for a long term without the help of logic gates, which ultimately induces a decrease in the reliability of prediction. It can be seen that the long and short-term memory structure of LSTM can effectively learn the long-term dependent capacity degradation data, which improves not only the long-term prediction performance of the recurrent neural network, but also the stability and accuracy of the prediction. The RUL prediction performance of the two methods presented in Table 6.5 is basically the same as Table 6.4.

The case is simulated in the environment of Python 3.5. For deep learning, the combination of CPU and GPU provides the best system performance. Therefore, this case adopts the Intel Core i7-6700 HQ (up to 3.50 GHz) processor and the NVIDIA Quadro M1000 M (4 Gb) graphics card for simulation. Tables 6.4 and 6.5 show that the training time using LSTM RNN is about 50% of that using SimRNN, meeting the requirements of EV applications.

6.4 Conclusion

The battery RUL prediction is a vital function of the BMS for EVs. This chapter first introduces the basic concepts of the battery RUL prediction, and then summarizes the typical methods of RUL prediction, which can be divided into three categories including fitting-based method, filter observation-based method and sequence prediction-based method. Subsequently, the concepts of the RUL probability distribution of battery is introduced, and the flowchart of probability density function by MC method is expounded for the above three types of prediction methods.

Aiming at the problems of low accuracy and poor convergence of traditional FIT, a method based on BCT is described in this chapter. The experimental results indicate that this method has admirable online performance and convergence characteristics. For the predicted starting point from the middle and later aging stage, the RUL prediction error of the method is less than 2.4%.

From the perspective of time series, a battery RUL prediction method based on LSTM RNN that effectively learn long-term capacity information is explained in this chapter. The experimental results show that the proposed method performs well for different aging stages, facilitating the long-term prediction performance of the recurrent neural network and improving the stability and accuracy of RUL prediction as well.

References

1. Xiong R, Cao J, Yu Q, He H, Sun F (2017) Critical review on the battery state of charge estimation methods for electric vehicles. IEEE Access 6:1832–1843
2. Xiong R, Li L, Tian J (2018) Towards a smarter battery management system: A critical review on battery state of health monitoring methods. J Power Sour 405:18–29
3. Xiong R, Li L, Li Z, Yu Q, Mu H (2018) An electrochemical model based degradation state identification method of lithium-ion battery for all-climate electric vehicles application. Appl Energy 219:264–275
4. Lu J, Chen Z, Yang Y, Lv M (2018) Online estimation of state of power for lithium-ion batteries in electric vehicles using genetic algorithm. IEEE Access 6:20868–20880
5. He W, Williard N, Osterman M, Pecht M (2011) Prognostics of lithium-ion batteries based on Dempster-Shafer theory and the Bayesian Monte Carlo method. J Power Sour 196(23):10314–10321
6. Saha B, Goebel K, Poll S, Christophersen J (2009) Prognostics methods for battery health monitoring using a bayesian framework. IEEE Trans Instrum Meas 58(2):291–296
7. Miao Q, Xie L, Cui H, Liang W, Pecht M (2013) Remaining useful life prediction of lithium-ion battery with unscented particle filter technique. Microelectron Reliab 53(6):805–810
8. Xiong R, Zhang Y (2018) Lithium-ion battery health prognosis based on a real battery management system used in Electric Vehicles. IEEE Trans Veh Technol 68(5):4110–4121
9. Xing Y, Ma M, Tsui KL, Pecht M (2013) An ensemble model for predicting the remaining useful performance of lithium-ion batteries. Microelectron Reliab 53(6):811–820
10. Xian W, Long B, Li M, Wang H (2013) Prognostics of lithium-ion batteries based on the verhulst model, particle swarm optimization and particle filter. IEEE Trans Instrum Meas 63(1):2–17
11. Jing Z, Wang G, Zhang S, Qiu C (2017) Building tianjin driving cycle based on linear discriminant analysis. Transp Res Part D Transp Environ 53:78–87
12. Song Y, Liu D, Hou Y, Yu J, Peng Y (2018) Satellite lithium-ion battery remaining useful life estimation with an iterative updated RVM fused with the KF algorithm. Chinese J Aeronaut 1(1):31–40
13. Zheng X, Fang H (2015) An integrated unscented kalman filter and relevance vector regression approach for lithium-ion battery remaining useful life and short-term capacity prediction. Reliab Eng Syst Saf 144:74–82
14. Zhang Y, Xiong R, He H, Pecht M (2019) Validation and verification of a hybrid method for remaining useful life prediction of lithium-ion batteries. J Clean Prod 212(1):240–249

15. Zhang Y, Xiong R, He H, Pecht M (2018) Lithium-ion battery remaining useful life prediction with Box-Cox transformation and Monte Carlo simulation. IEEE Trans Ind Electron 66(2):1585–1597
16. Zhou Y, Huang M (2016) Lithium-ion batteries remaining useful life prediction based on a mixture of empirical mode decomposition and ARIMA model. Microelectron Reliab 65:265–273
17. Liu D, Wang H, Peng Y, Xie W, Liao H (2013) Satellite lithium-ion battery remaining cycle life prediction with novel indirect health indicator extraction. Energies 6(8):3654–3668
18. Liu D, Xie W, Liao H, Peng Y (2015) An integrated probabilistic approach to lithium-ion battery remaining useful life estimation. IEEE Trans Instrum Meas 64(3):660–670
19. Patil MA (2015) A novel multistage support vector machine based approach for li ion battery remaining useful life estimation. Appl Energy 159:285–297
20. Liu D, Zhou J, Liao H (2015) A health indicator extraction and optimization framework for lithium-ion battery degradation modeling and prognostics. IEEE Trans Syst Man Cybern Syst 45(6):915–928
21. Box GEP, Cox DR (1964) An analysis of transformations. J R Stat Soc B (Methodological) 26(2):211–252
22. Hochreiter S, Schmidhuber J (1997) Long short-term memory. Neural Comput 9(8):1735–1780
23. Zhang Y, Xiong R, He H, Pecht M (2018) Long short-term memory recurrent neural network for remaining useful life prediction of lithium-ion batteries. IEEE Trans Veh Technol 67(7):5695–5705

Chapter 7
Low-Temperature Heating and Optimal Charging Methods for Lithium-Ion Batteries

With the promotion and popularization of new energy vehicles, the problems of short driving range accompanied by the difficulty in starting and charging during winter in high altitude regions are becoming severer nowadays. The working characteristics of lithium-ion batteries are sensitive to both working conditions and aging paths [1, 2]. In particular, the driving range will be significantly reduced in the low-temperature environment, which will also be obviously worse with the use of air conditioning. In addition, the battery is difficult to reach the fully charged state in low temperature, and the low-temperature charging may cause lithium deposition on anode, leading to permanent damage to the battery. Therefore, the low-temperature heating and optimized battery charging methods are key techniques to guarantee the normal operation of new energy vehicles in all climates and working conditions.

7.1 Classification of Low-Temperature Heating Methods for Batteries

As shown in Fig. 7.1, the studies on the low-temperature heating of batteries could be mainly divided into two categories: external heating and internal heating [3, 4]. External heating methods comprise a convective heating method, wide-wire metal film heating method and other external heating methods. Internal heating methods mainly include alternating current (AC) heating method and internal self-heating method.

7.1.1 Convective Heating Method

Heating the battery pack in hot air is adopted in this method [5–7]. The energy source to heat the air comes from the onboard air conditioning system. A common

R. Xiong, *Battery Management Algorithm for Electric Vehicles*,
https://doi.org/10.1007/978-981-15-0248-4_7

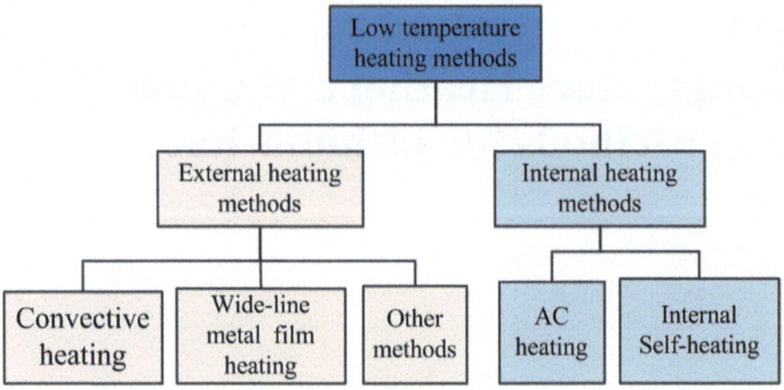

Fig. 7.1 Classification of low-temperature heating methods

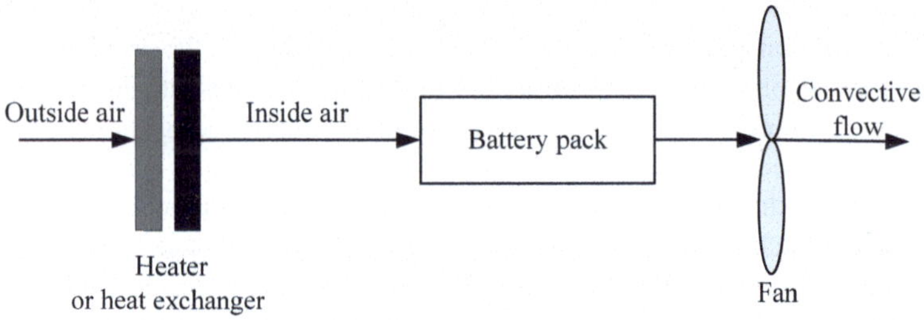

Fig. 7.2 Schematic diagram of the battery thermal management system with air as the medium

convective heating strategy is shown in Fig. 7.2. At low temperature, the heater or heat exchanger warms outside air, and the fan enhances heat transfer from the heater to air by creating a convective flow.

7.1.2 *Wide-Line Metal Film Heating Method*

This method heats the battery with the heat generated by the current when it flows through the wide-line metal film [8–10]. As shown in Fig. 7.3, the wide-line metal

Fig. 7.3 Schematic diagram of wide-line metal film heating method

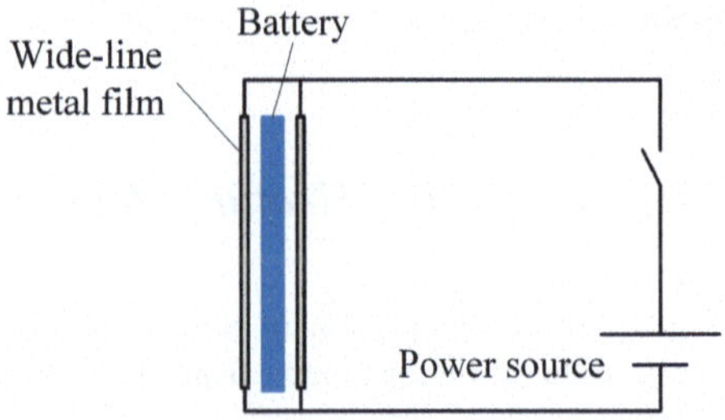

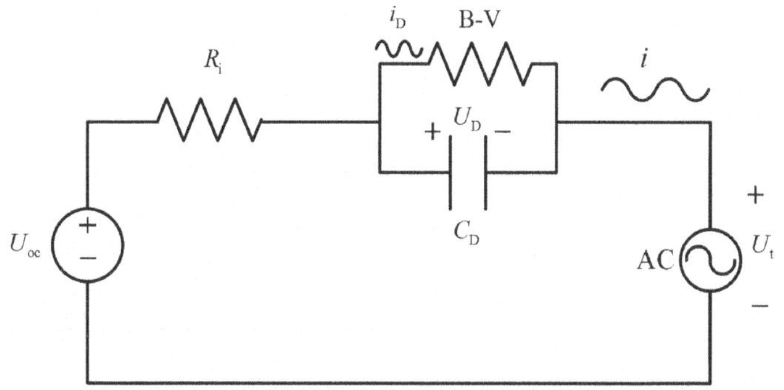

Fig. 7.4 Schematic diagram of the battery internal alternating current heating method

film is made of a 1 mm thick FR4 plate, both sides of which are covered by a copper film with a thickness of 0.035 mm. One side is a complete rectangle and the other side is continuous copper wire with a certain width. The power source is connected to both ends of the copper wire, and the surfaces of the two copper films are covered with an insulating layer. When the current flows through the wire, the heat generated by the copper wire is uniformly transmitted to the battery.

7.1.3 Battery Internal Alternating Current Heating Method

This heating method directly heats the inside of the battery by alternating current [11–15]. As shown in Fig. 7.4, an AC power supplies AC current at low temperature, which flows continuously through the internal impedance of the battery to generate heat, thereby heating the inside of the battery. The AC heating is a kind of low-temperature fast heating method with great research significance, whose temperature rise rate can reach 3 °C/min [16]. In Sect. 7.2, the heating method is taken as an example, and the design of the heating strategy is described in detail.

7.1.4 Battery Internal Self-heating Method

This method heats the battery itself by the current flowing through a nickel piece inside the battery to generate ohmic heat. A piece of nickel is added inside the battery and the structure is shown in Fig. 7.5. When the temperature is lower than a certain temperature, the switch is turned off, and the current flows through the nickel foil to generate heat. Conversely, when the temperature overtakes certain temperature, the switch is turned on and the heating is stopped. In this way, the controllability of low-temperature heating for lithium-ion batteries is achieved. It was proposed by Prof. Chaoyang Wang [17, 18] at Pennsylvania State University. This method has

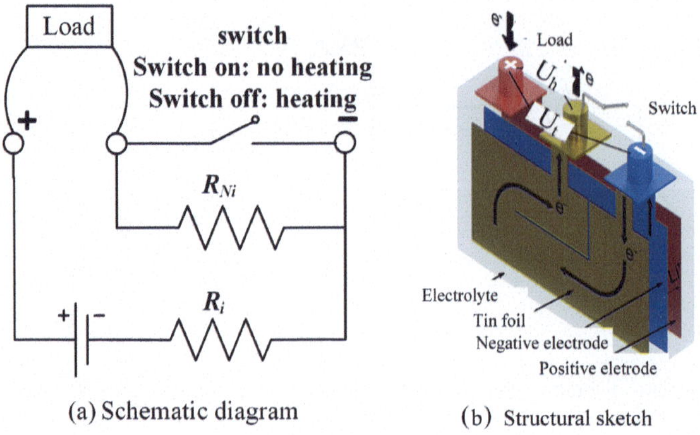

Fig. 7.5 Schematic diagram and structure of the battery internal self-heating method

an excellent temperature rising rate which can heat the battery from −30 °C to over 0 °C within 1 min, as verified by the experimental results. The method is a promising heating method. However, this method needs to change the internal structure of the battery and has not yet been fully commercialized.

7.1.5 Other Heating Methods

In addition to the heating strategies described above, there are also other low-temperature heating strategies such as Peltier effect heating and heat pump heating [19–22]. The Peltier effect heating shown in Fig. 7.6 is conducted on account of the Peltier effect. When the current flows through a loop composed of different semiconductors, the heat absorption and heat release occur, respectively, at the two joints with different current directions. The heat pump heating shown in Fig. 7.7 uses the heat pump to heat the battery by consuming electrical energy to absorb heat from

Fig. 7.6 Schematic diagram of the Peltier effect heating method

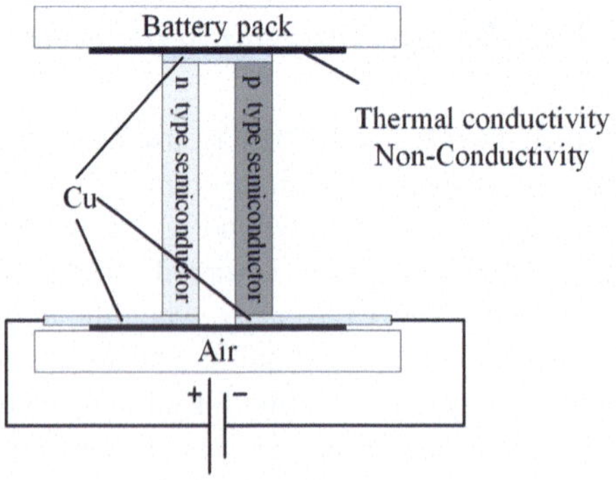

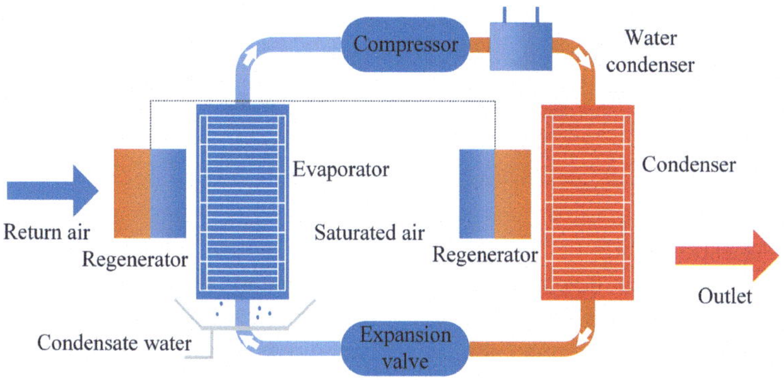

Fig. 7.7 Schematic diagram of heat pump heating method

the outside air [23]. The heat pump is a mechanical device that forces heat from a low-temperature object to a high-temperature object in a reversible cycle manner. It can obtain a large amount of heat supply with only a little work consumption, which can effectively make use of the thermal energy that is difficult to apply, achieving energy saving.

7.2 AC Heating Principle

7.2.1 *The Heat Generation Mechanism of Lithium-Ion Batteries*

The heat generation of lithium-ion batteries is composed of reaction heat, Joule heat, polarization heat, and side reaction heat.

(1) Reaction heat
The reaction heat [24] is generated by the electrochemical reaction in the process of charge and discharge. In the process, lithium ions are inserted into and removed from the electrode. The expression is

$$Q_{\mathrm{r}} = \int i_{\mathrm{L}} T \frac{dE_{\mathrm{ocv}}}{dT} dt \tag{7.1}$$

where $\frac{dE_{\mathrm{ocv}}}{dT}$ is the entropy thermal coefficient of the electrode material, and E_{ocv} is the open=-circuit potential of the electrode material.

(2) Joule heat
The Joule heat is generated by the ohmic internal resistance of the battery when the current flows through the battery [25]. The ohmic internal resistance of the battery includes the contact resistance, the internal resistance of the electrode

and the internal resistance of the electrolyte. Joule heat is the main part of the heat generation during the process of battery charging and discharging. The expression is

$$Q_J = \int i_L^2 R_i dt \tag{7.2}$$

(3) Polarization heat
When the current flows through a lithium-ion battery, the electrode potential of the battery deviates from its equilibrium state due to the load current, and the heat generated in the process is the polarization heat [26]. The expression is

$$Q_D = \int i_D^2 R_D dt \tag{7.3}$$

(4) Side reaction heat
The side reaction heat includes heat generation in the process of electrolyte decomposition, self-discharge, overcharge, over-discharge [27, 28], which is negligible in the normal operating condition of the battery.
The total heat generation of a lithium-ion battery is

$$Q_Z = Q_r + Q_J + Q_D = \int i_L^2 R_i dt + \int i_D^2 R_D dt + \int i_L T \frac{dE_{ocv}}{dT} dt \tag{7.4}$$

7.2.2 AC Heating Mechanism

AC heating applies AC excitation to heat the battery at a certain frequency and amplitude, which generates heat by the AC excitation on the impedance of the battery. This method can warm the battery efficiently and uniformly.

Low temperature can lead to the poor lithium ions diffusion in graphite, sluggish kinetics of charge transfer, slow electrolyte conductivity, and increased resistance of the battery. In this case, the DC charging may cause lithium deposition, which would seriously affect the cycle life and safety of the battery. The AC heating method alternates between the lithium intercalation process and the de-intercalation process in one cycle, so that the lithium plating can be effectively avoided.

When the battery is charged on DC condition in low temperature, the model of electrode reaction mechanisms is shown in Fig. 7.8. The solid phase diffusion coefficient of lithium in the graphite anode active material particles decreases. Thus, the lithium ions generated by the electrochemical reaction cannot diffuse into the interior of the particles in time. They accumulate on the surface of the negative active material particles, contributing to the lithium deposition (only occurs in the negative electrode) [14].

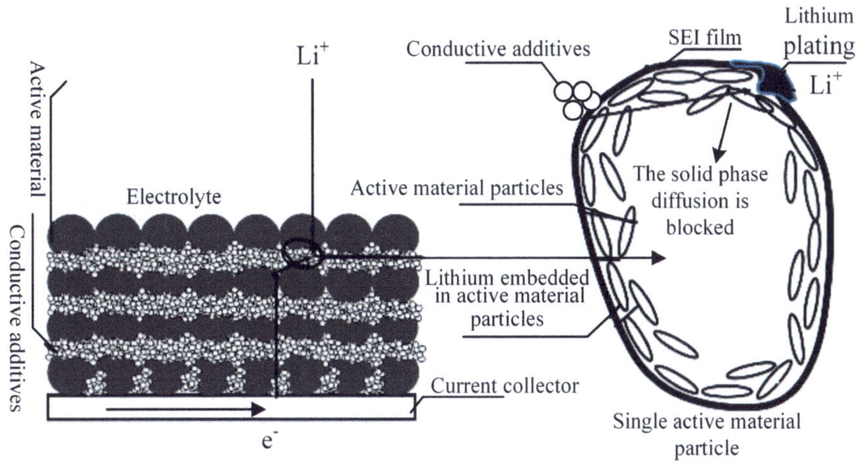

Fig. 7.8 Electrode reaction mechanism of the battery during DC charging

When a battery is excited by AC current, the de-intercalation reaction occurs immediately after the lithium intercalation, and consumes the lithium generated by the intercalation reaction. Since the lithium ions generation and consumption in each cycle are balanced, the battery capacity is not apparently damaged, demonstrating that the excitation method does not affect the battery life significantly.

However, if the excitation current is too large, the terminal voltage of the battery will exceed its maximum allowable threshold, which would influence the battery service life. Moreover, a large scale current forms a strong electric field at some uneven places on the surface of the negative electrode material, resulting in a large amount of lithium precipitation therein. Failed to be oxidized to lithium ions in the second half cycle, this part of lithium metal is regarded as "dead lithium". The process results in a decrease of available lithium ions in the electrode and further causes irreversible attenuation of the battery capacity. Therefore, it is necessary to limit the excitation current and control the terminal voltage of the battery within the allowable range in the process of AC excitation.

Figure 7.9 shows the mechanism of the electrode reaction model under low-frequency AC excitation. Lithium intercalation and de-intercalation are alternately performed at this time. However, a large amount of lithium is precipitated and even forms lithium dendrites in the process due to the low excitation frequency and the long period of the electrochemical reaction. During the de-intercalation process, parts of plated lithium near the anode start to react, result in some lithium escape from the anode surface. It can no longer participate in the electrode reaction, becoming the so-called "dead lithium", which leads to permanent damage to the battery capacity and decrease of the battery service life.

Under the premise of limiting the amplitude of the excitation current, the electrochemical reaction becomes faster as the frequency of the AC excitation increases. Less reaction product of the lithium is accumulated in the first half cycle, which can

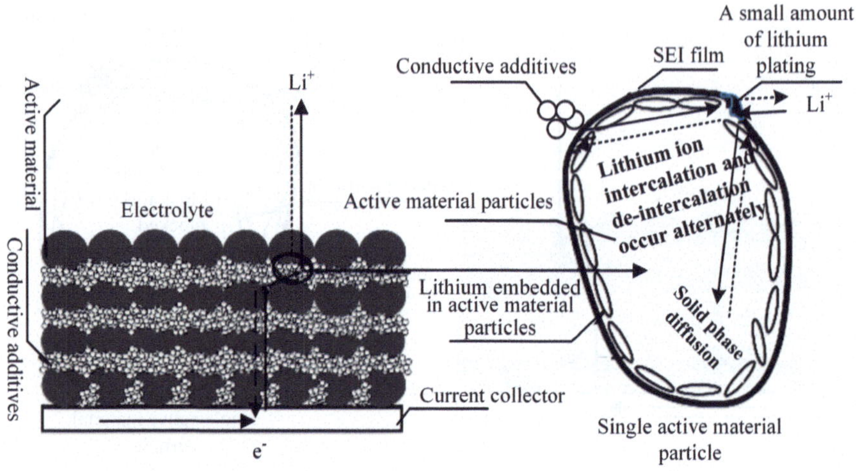

Fig. 7.9 Electrode reaction mechanism of the battery under low-frequency AC excitation

be completely consumed in the latter half cycle to form a reversible lithium precipitation. Therefore, the lithium precipitation will not occur in the battery during the whole AC excitation. The reaction mechanism model is shown in Fig. 7.10.

Under the premise of limiting the amplitude of the excitation current, the period of AC excitation continues to shrink with the increase of the frequency. The lithium produced by the lithium intercalation in the first half of the cycle, which is too little to reach the concentration of lithium precipitation, is consumed during the second half of the de-intercalation. There is no lithium precipitation during the entire AC

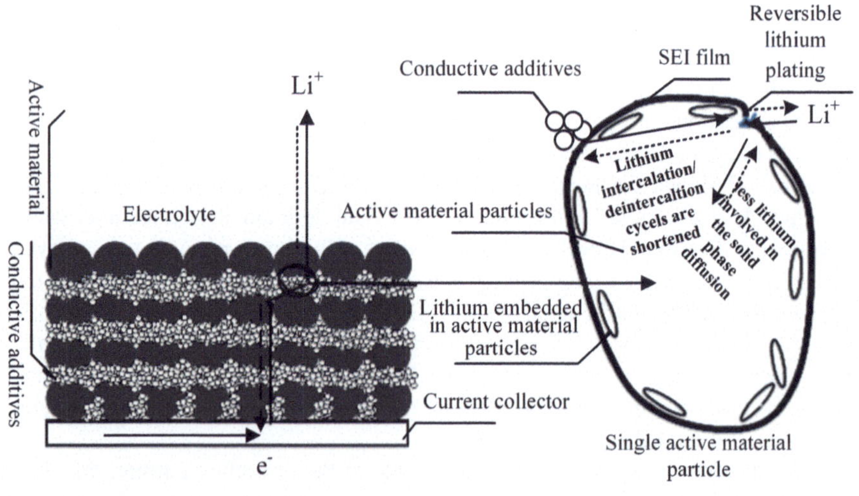

Fig. 7.10 Electrode reaction mechanism of the battery under mid-frequency AC excitation with a limited current

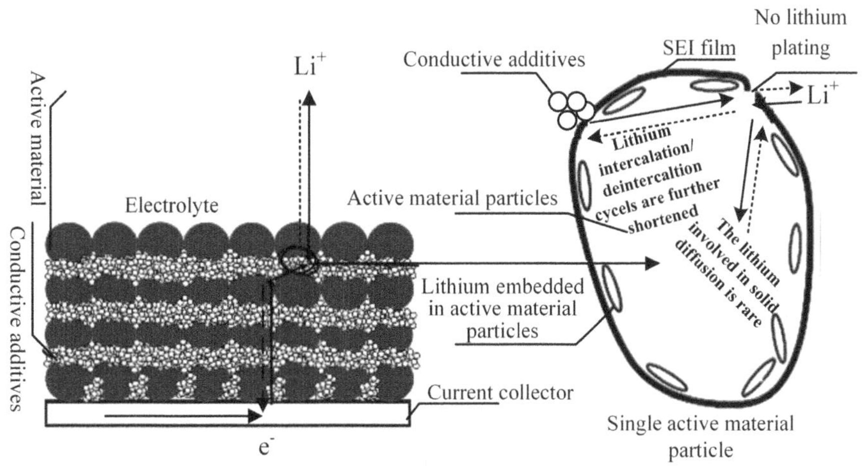

Fig. 7.11 Electrode reaction mechanism of high-frequency AC-excited excitation battery with limited current

excitation process, causing no permanent damage to the battery capacity. The reaction mechanism model is shown in Fig. 7.11.

7.3 Echelon AC Heating Method

7.3.1 Problem Description

Figure 7.12 shows that the battery temperature rise rate grows as the amplitude of the excitation current increases during the low-temperature AC heating process [4]. However, if the AC current amplitude is too large or the battery terminal voltage exceeds the allowable voltage value, it will affect the battery cycle life, and can lead to thermal runaway in severe cases.

On the other hand, as shown in Fig. 7.13, the heat generation rate increases as the frequency of the AC current decreases. However, if the frequency of the AC current is too low, lithium deposition may occur in the low-temperature environment. In severe cases, it may even cause internal short circuit of the battery, threatening the battery safety.

Therefore, when batteries are heated using AC excitation, the amplitude and the frequency value of the AC current should not only consider the influence on temperature rise rate but also focus on the safety and health of batteries. Compared with the frequency of AC current, as shown in Figs. 7.12 and 7.13, the amplitude of the AC current has a greater influence on the temperature rise rate of the battery. To reduce the computational cost and consider both the temperature rise rate and battery safety, the next section will detail the fixed frequency echelon AC heating method.

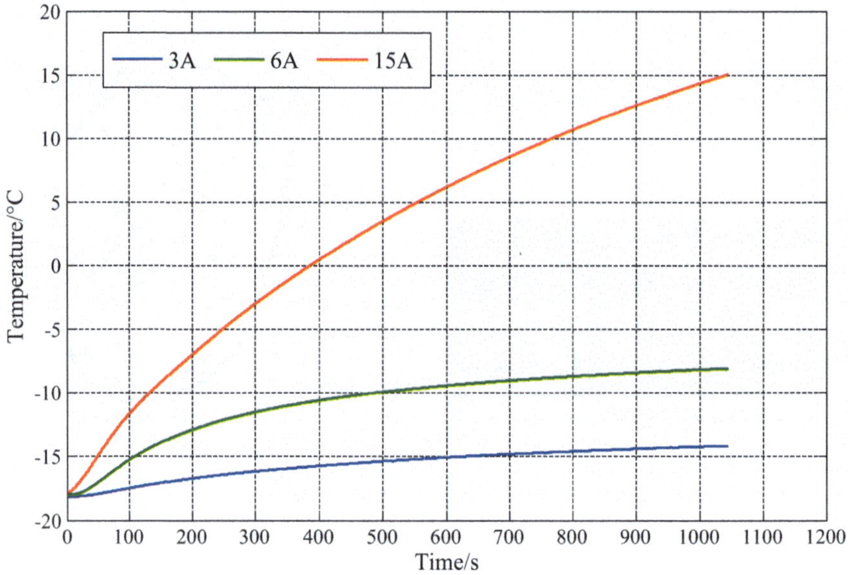

Fig. 7.12 Effect of excitation current amplitude on temperature rise rate

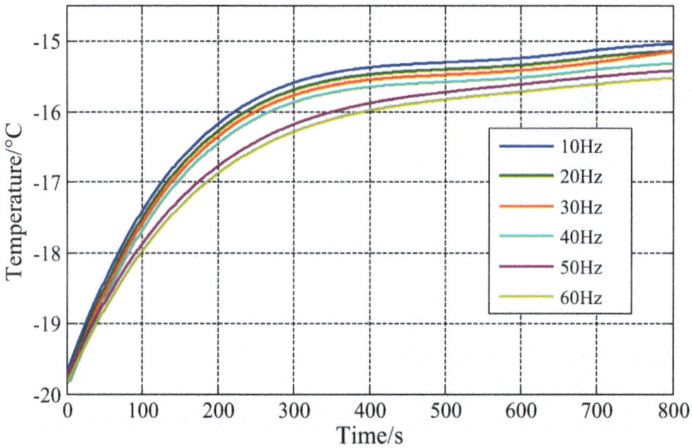

Fig. 7.13 Effect of excitation frequency on temperature rise rate

7.3.2 *Adaptive Echelon AC Heating Method*

(1) Heat balance equation of the battery

The 18650 battery can be regarded as a unit because the temperature rise rates of the internal and surface of the battery are basically the same. According to the Fourier thermodynamic equation, the temperature rise rate is expressed as [11]

$$mc_{\mathrm{p}}\frac{\partial T}{\partial t} = q - q_{\mathrm{n}} \tag{7.5}$$

where m is the mass of the cell, c_{p} is the specific heat capacity, T is the temperature, t is the time, q is the heat generation rate, and q_{n} is the heat dissipation rate.

The calculation equation of q_{n} is

$$q_{\mathrm{n}} = hS_{\mathrm{sur}}(T - T_{\mathrm{amb}}) \tag{7.6}$$

where h is the convection coefficient, S_{sur} is the battery surface area, and T_{amb} is the ambient temperature.

(2) Calculation of the optimal AC current amplitude

The equation for calculating the heat production rate q in AC excitation is:

$$q = \left(\frac{i}{\sqrt{2}}\right)^2 R_Q \tag{7.7}$$

where i is the excitation current and R_{Q} is the real impedance (temperature dependent, acquired by EIS offline).

According to the B-V equation [29], the polarization current i_{D} can be obtained as

$$i_{\mathrm{D}} = Si_0\left\{e^{\frac{\alpha_{\mathrm{a}}F\eta}{RT}} - e^{-\frac{\alpha_{\mathrm{c}}F\eta}{RT}}\right\} \tag{7.8}$$

As $e^{\frac{\alpha_{\mathrm{a}}F}{RT}\eta} \gg e^{-\frac{\alpha_{\mathrm{a}}F}{RT}\eta}$, Eq. (7.8) can be transformed into

$$\ln\left(\frac{i_{\mathrm{D}}}{i_0S}\right) = \frac{\alpha_{\mathrm{a}}F}{RT}\eta \tag{7.9}$$

According to the principle of rational approximation, the above equation can be transformed into

$$\ln\left(\frac{i_{\mathrm{D}}}{Si_0}\right) = \frac{2\frac{i_{\mathrm{D}}}{Si_0} - 2}{1+\frac{i_{\mathrm{D}}}{Si_0}} \tag{7.10}$$

$$\frac{2\frac{i_{\mathrm{D}}}{Si_0}-2}{1+\frac{i_{\mathrm{D}}}{Si_0}}=\frac{\alpha_{\mathrm{a}}F}{RT}\eta \tag{7.11}$$

$$i_{\mathrm{D}}=\frac{i_0S\left(2+\frac{\alpha_{\mathrm{a}}F}{RT}\eta\right)}{2-\frac{\alpha_{\mathrm{a}}F}{RT}\eta} \tag{7.12}$$

The polarization voltage U_{D} can be written as [30]

$$\eta\approx U_{\mathrm{D}}=U_{\mathrm{oc}}-U_{\mathrm{t}}-R_{\mathrm{i}}i \tag{7.13}$$

$$C_{\mathrm{D}}\frac{dU_{\mathrm{D}}}{dt}=i-i_{\mathrm{D}} \tag{7.14}$$

The terminal voltage U_{t} and the polarization voltage U_{D} at time $t+\Delta t$ can be expressed as

$$\frac{\left(U_{\mathrm{D},t+\Delta t}-U_{\mathrm{D},t}\right)}{\Delta t}=\frac{i_t}{C_{\mathrm{D}}}-\frac{i_0S\left(2+\frac{\alpha_{\mathrm{a}}F}{RT}U_{\mathrm{D},t}\right)}{C_{\mathrm{D}}\left(2-\frac{\alpha_{\mathrm{a}}F}{RT}U_{\mathrm{D},t}\right)} \tag{7.15}$$

$$U_{\mathrm{t},t+\Delta t}=U_{\mathrm{oc},t+\Delta t}-U_{\mathrm{D},t+\Delta t}-i_{t+\Delta t}R_{\mathrm{i}} \tag{7.16}$$

From Eqs. (7.15) and (7.16), the AC excitation current $i_{t+\Delta t}$ at time $t+\Delta t$ can be calculated as

$$i_{t+\Delta t}=\frac{(U_{oc,t+\Delta t}-U_{\mathrm{t},t+\Delta t})}{R_{\mathrm{i}}}-\frac{U_{\mathrm{D},t}}{R_{\mathrm{i}}}-\frac{i_t\Delta t}{C_{\mathrm{D}}R_i}+\frac{i_0S(2+\frac{\alpha_{\mathrm{a}}F}{RT}U_{\mathrm{D},t})\Delta t}{C_{\mathrm{D}}R_{\mathrm{i}}\left(2-\frac{\alpha_{\mathrm{a}}F}{RT}U_{\mathrm{D},t}\right)} \tag{7.17}$$

$$U_{\mathrm{t,min}}\le U_{\mathrm{t}}\le U_{\mathrm{t,max}} \tag{7.18}$$

Combining the Eqs. (7.17) and (7.18), the upper and lower permissible excitation currents $i_{\mathrm{upper\_max},t+\Delta t}$, $i_{\mathrm{lower\_max},t+\Delta t}$ are obtained as

$$i_{\mathrm{upper\_max},t+\Delta t}=\frac{(U_{oc,t+\Delta t}-U_{\mathrm{t,max}})}{R_{\mathrm{i}}}-\frac{U_{\mathrm{D},t}}{R_{\mathrm{i}}}-\frac{i_t\Delta t}{C_{\mathrm{D}}R_{\mathrm{i}}}+\frac{i_0S(2+\frac{\alpha_{\mathrm{a}}F}{RT}U_{\mathrm{D},t})\Delta t}{C_{\mathrm{D}}R_{\mathrm{i}}\left(2-\frac{\alpha_{\mathrm{a}}F}{RT}U_{\mathrm{D},t}\right)} \tag{7.19}$$

$$i_{\mathrm{lower\_max},t+\Delta t}=\frac{(U_{oc,t+\Delta t}-U_{\mathrm{t,min}})}{R_{\mathrm{i}}}-\frac{U_{\mathrm{D},t}}{R_{\mathrm{i}}}-\frac{i_t\Delta t}{C_{\mathrm{D}}R_{\mathrm{i}}}+\frac{i_0S(2+\frac{\alpha_{\mathrm{a}}F}{RT}U_{\mathrm{D},t})\Delta t}{C_{\mathrm{D}}R_{\mathrm{i}}\left(2-\frac{\alpha_{\mathrm{a}}F}{RT}U_{\mathrm{D},t}\right)} \tag{7.20}$$

The optimal excitation current $i_{\mathrm{opt},t+\Delta t}$ is

$$i_{\mathrm{opt},t+\Delta t}=\min(i_{\mathrm{upper\_max},t+\Delta t},i_{\mathrm{lower\_max},t+\Delta t}) \tag{7.21}$$

where R_{i}, C_{D}, and i_0 are obtained by the parameter identification method similar to that used in Chap. 3.

Discrete expression of model [31] is

$$\begin{cases} U_{\mathrm{t}}(t) = U_{\mathrm{oc}} - i(t)R_i - U_{\mathrm{D}}(t) \\ U_{\mathrm{D}}(t) = U_{\mathrm{D}}(t-\Delta t) + \Delta t\frac{i(t-\Delta t)}{C_{\mathrm{D}}} - \Delta t\frac{i_0}{C_{\mathrm{D}}}\sinh\left(\frac{\alpha_{\mathrm{a}}F}{RT}U_{\mathrm{D}}(t-\Delta t)\right) \end{cases} \tag{7.22}$$

7.3.3 Adaptive Echelon AC Heating Process

The adaptive echelon AC heating method is to minimize the energy loss of the energy supply system and shorten the heating time under the premise of ensuring the service life and safety. The amplitude of the AC current is calculated within the permissible current range of the battery and updated in time, and then applied into the battery, which can improve the low-temperature performance and ensure the reliable operation of the battery [4]. The heating process is shown in Fig. 7.14.

① Collect the battery data in real time. Before performing AC heating, the current temperature of the battery and ambient temperature are measured.

② Pre-judge and determine whether AC heating is required. If the battery temperature of the battery is higher than the threshold, there is no need for AC heating and the vehicle can start or work normally. While the battery temperature is lower than the threshold (generally 0 °C/5 °C/10 °C, etc.), the AC heating is required.

③ Obtain the system parameters when the algorithm starts. After the current SOC, temperature, terminal voltage, and other information of the battery are obtained, the impedance value can be calculated based on the model and experimental data pre-stored in the BMS. The vehicle usually parks at least 6 h before restarting, so the terminal voltage of the battery approaches the open-circuit voltage. Therefore, the terminal voltage before the warm-up is taken as the initial value of the open-circuit voltage on the present moment.

④ Calculate the optimal excitation current. According to Eqs. (7.19)–(7.21), the optimal AC current is calculated.

⑤ Perform AC excitation. The optimal AC current obtained in step 4 is applied to the battery to perform AC excitation.

⑥ Determine the heating termination conditions. If the battery has reached the target temperature, stop AC heating. If no, go to step 3.

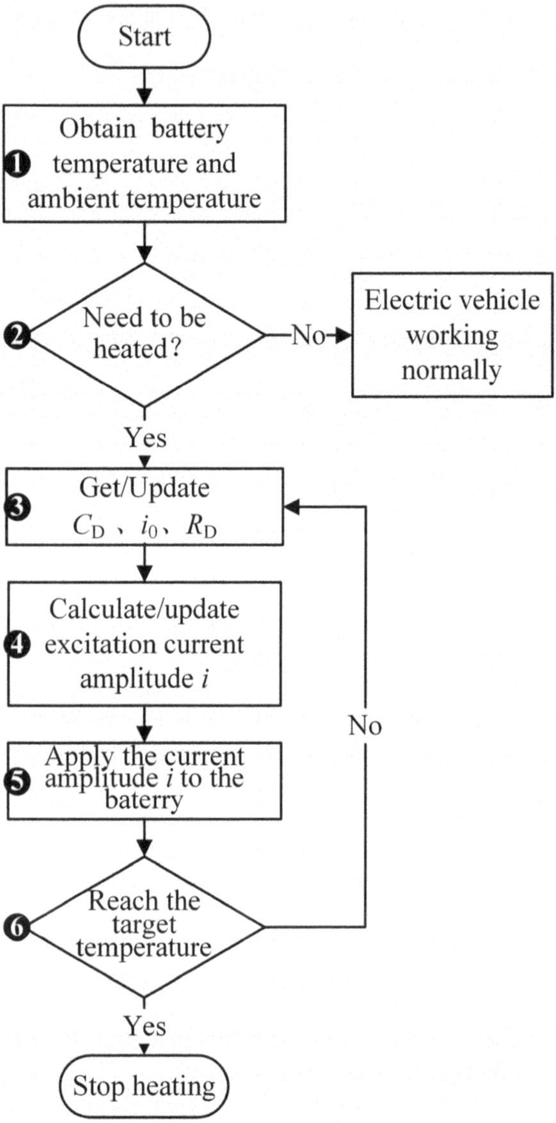

Fig. 7.14 Adaptive echelon heating flowchart

7.3.4 Case Study

The battery 8-cell 01 and a series-connected battery pack with battery 8-cells 02–05 was employed to evaluate the adaptive echelon heating method at a fixed frequency of 10 Hz. The initial SOC of each cell is 50% and the fixed heating frequency is 10 Hz to ensure a faster temperature rise rate and the battery safety at the same time. This case study uses the hardware platform shown in Fig. 7.15. The bipolar power

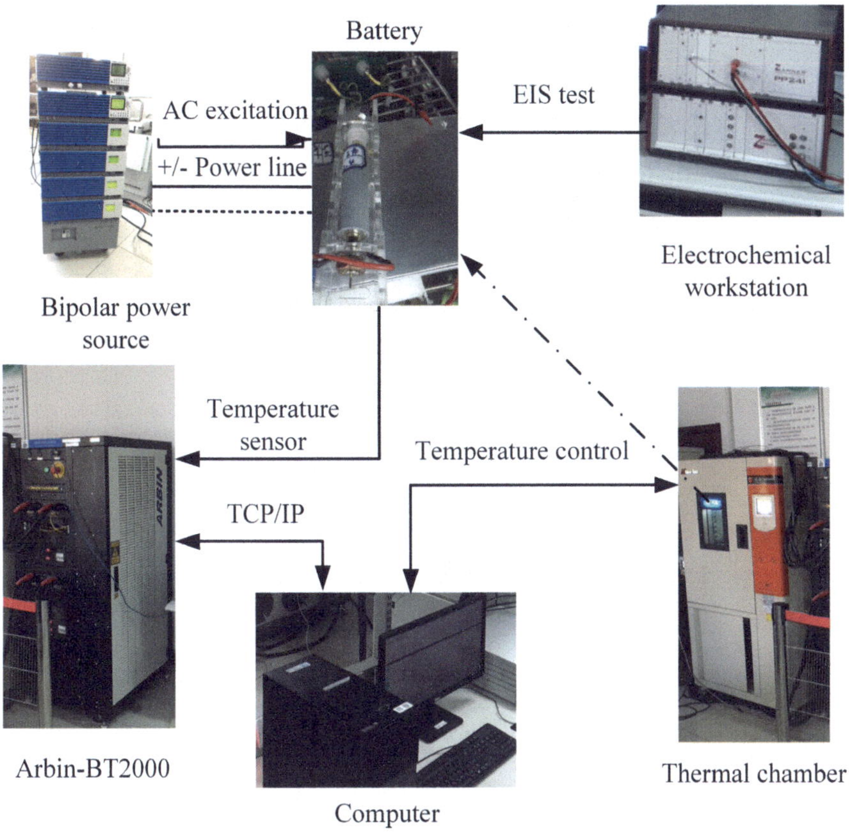

Fig. 7.15 Low-temperature heating test platform

source provides AC excitation and the thermal chamber regulates the temperature. The Arbin test equipment measures the current, voltage and temperature in real time.

The impedance spectra (shown in Fig. 7.16) were obtained at a temperature range of −20 to 10 °C (interval 2 °C). The offline experimental data was pre-stored in the BMS to calculate the real impedance of the battery at 10 Hz at different temperatures.

Before AC heating, the battery was soaked at −20 °C for more than 4 h to ensure the temperature equipartition of the battery. Then an evaluation of the adaptive echelon heating method needed to be developed to record the temperature rise of the battery. The battery capacity was tested once every 10 heating cycles, and the experimental results were observed and analyzed. Figure 7.17 shows the AC current curve of the battery 8-cell 01 at different temperatures, and Fig. 7.18 shows the heating temperature rise curve.

Figure 7.19 shows the heating temperature rise curve of the series-connected battery pack and Fig. 7.20 shows the attenuation of the charge and discharge capacity of the battery after 40 heating cycles.

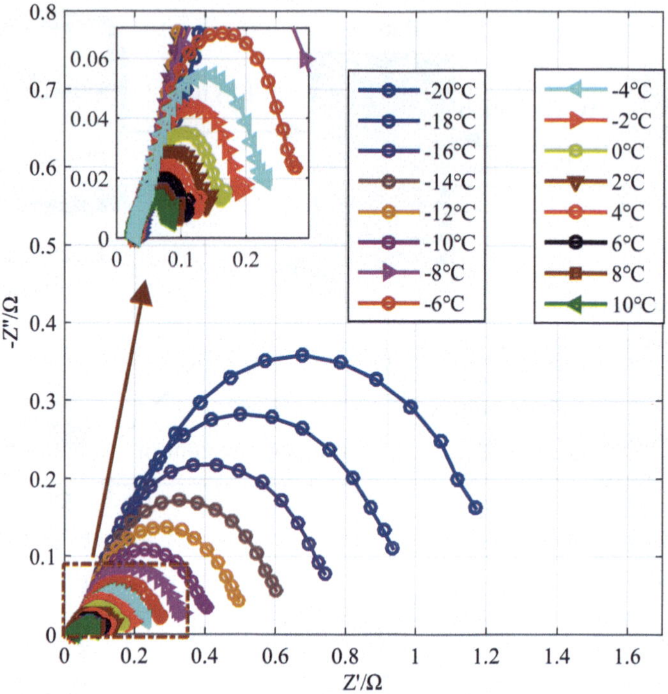

Fig. 7.16 Impedance spectra at different temperatures

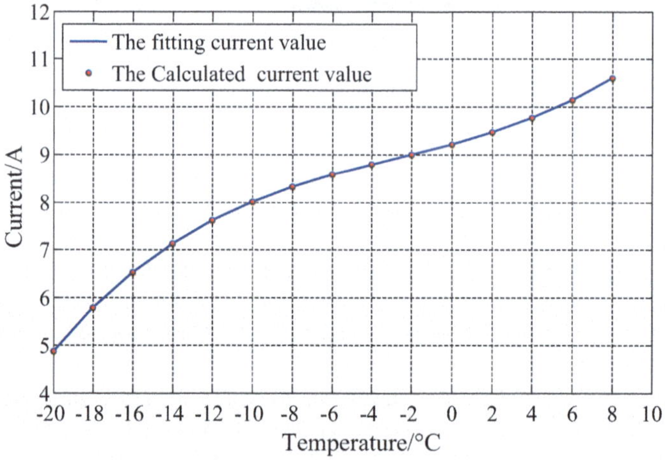

Fig. 7.17 Excitation current of the battery 8-cell 01

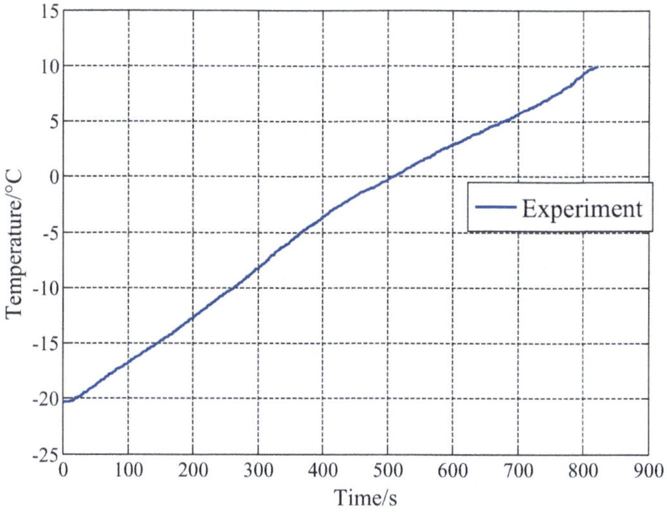

Fig. 7.18 The heating temperature rise curve of the battery 8-cell 01

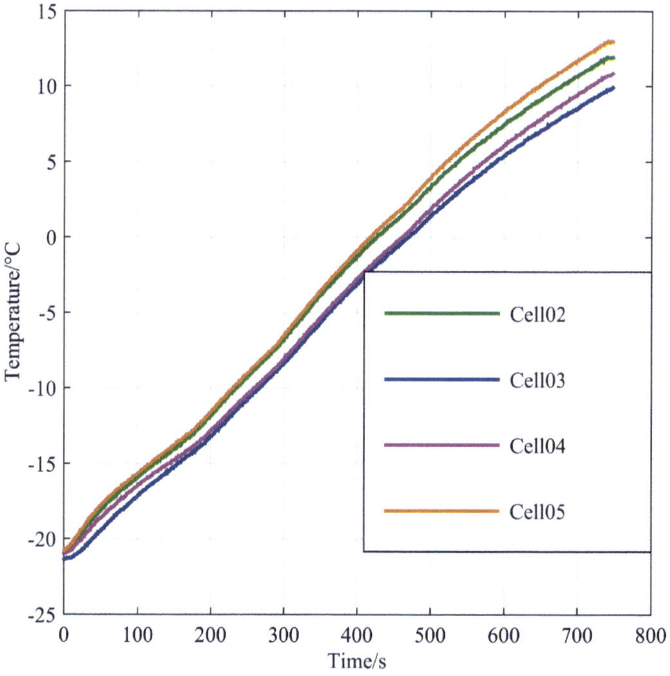

Fig. 7.19 Heating temperature rise curve of the battery pack

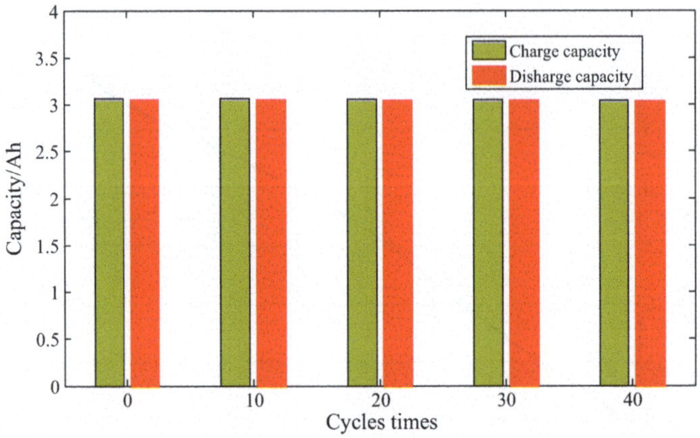

Fig. 7.20 Charge and discharge capacity of the battery after cyclic heating

As shown in Figs. 7.18 and 7.19, using the adaptive echelon heating method, the cell and the battery pack were heated from −20.3 °C to 10.02 °C in 13.7 min and 12.4 min, respectively. The corresponding rate of temperature rise is 2.21 °C/min and 2.47 °C/min, respectively. The latter one has a better temperature rise consistency. As shown in Fig. 7.20, this method has no significant effect on the charge and discharge capacity of the battery. Therefore, the adaptive echelon heating method can achieve an ideal short-time and high-efficiency temperature rise rate and has no obvious influence on the service life of the battery. It can be considered as a promising preheating method for the low-temperature environment application of the battery.

7.4 Battery Optimal Charging

Optimal charging methods are a series of improved charging schemes for lithium-ion batteries with long charging time, short life cycle and temperature sensitivity. The charging method mainly includes [32]: CCCV charging, multi-stage constant-current charging, pulse charging, and smart charging, etc.

7.4.1 CCCV Charging

The battery is charged with the preset constant current. When the battery voltage reaches the preset value, it shifts to CV charging and completes the charging process as the charging current decreases to a predetermined value. The charging curve is shown in Fig. 2.11. Combining the advantages of CC charging and CV charging, the method overcomes the problem of undercharging by CC charging, and avoids damage to the battery caused by large current in the initial stage of CV charging. As the main way of charging, it has been researched for a long time around the world [33], which is mainly about the effects of cutoff charging voltage and charging rate on the battery service life.

The method of increasing the charging speed of CCCV can be divided into the following types:

① Increase the CC charging rate. It can effectively shorten the charging time, while reducing the charging capacity.
② Raise the upper cutoff voltage of the battery. Different upper cutoff voltages have a great impact on the charging speed of the battery. CV charging with a higher upper cutoff voltage may shorten the battery life.

Therefore, increasing the current value of the CC stage is a relatively common charging method for shortening the charging time among the improved CCCV methods.

7.4.2 *Multi-stage Constant Current Charging*

In order to solve the disadvantage of the long CV time during CCCV, a multi-stage CC charging method is proposed. The method divides the whole charging process into several stages (generally 4–5 stages) based on the declining trend of charging current. The current boundary of each stage is preset. The current at each stage is determined by setting the optimization target, and selecting the optimization algorithm. When the cutoff voltage is reached during the charging, it shifts to the next stage, and the charging curve is shown in Fig. 7.21. The calculation process of the current value at each stage is as follows:

- Establish an objective function. It can be a single objective function or a multi-objective function of charging time, temperature rise, and life attenunation.

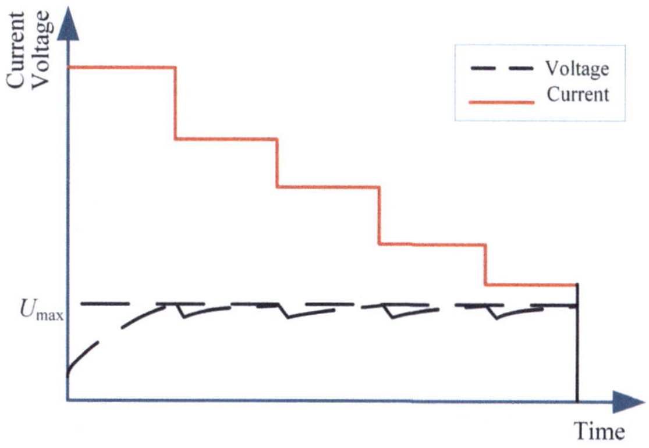

Fig. 7.21 Fifth-stage constant-current optimized charging current curve

- Set the range of current at each stage. The criterion is to reach the upper cutoff voltage, so the current amplitude of each step is gradually reduced. In order to ensure the charged capacity of the battery, the charging current of the final stage cannot be set too large.
- Use the optimization algorithm to minimize the objective function and obtain the optimized charging current curve.

In order to fully charge the battery, the current at the last stage should be reduced which prolongs the charging time. Therefore, the multi-stage CC charging is mostly adopted in the SOC range of 0–80%.

7.4.3 Pulse Charging

This method was initially applied to the lead–acid battery. The polarization voltage of the lead–acid battery is high during the charging and discharging process. A discharging or a rest process is inserted to the CC charging to eliminate oxygen and hydrogen generation. Thereafter, the charging method is also applied to the lithium-ion battery [34]. Some special advantages of pulse charging are:

① Pulse charging can reduce the polarization voltage by short-time discharging or rest processes [35]. The internal ion concentration of the battery tends to be equalized, which ensures a relatively balanced battery state during the charging process. It is beneficial to increase the charging/discharging power and shorten the battery charging time.

② During the pulse charging, the impedance and the heat production of the battery are relatively low, and the charging efficiency is high [36].

Figure 7.22 shows three different types of pulse charging method with constant amplitude and constant frequency [14]. The experimental results indicate that neither the three methods could significantly shorten the charging time. Considering the charging time, researchers proposed a constant-amplitude and variable-frequency charging mode and a variable-amplitude and constant-frequency charging mode shown in Fig. 7.23. Experimental tests show that these improved modes significantly improve the battery service life and shorten the charging time. To further accelerate

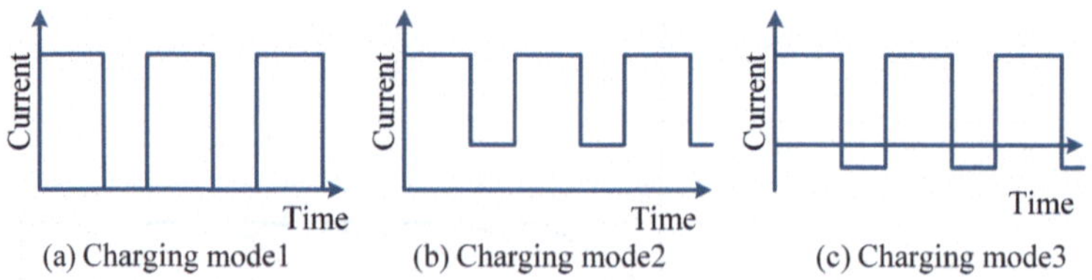

Fig. 7.22 Three different forms of pulse charging method with constant amplitude and constant frequency: **a** rest **b** small current gap **c** discharging gap

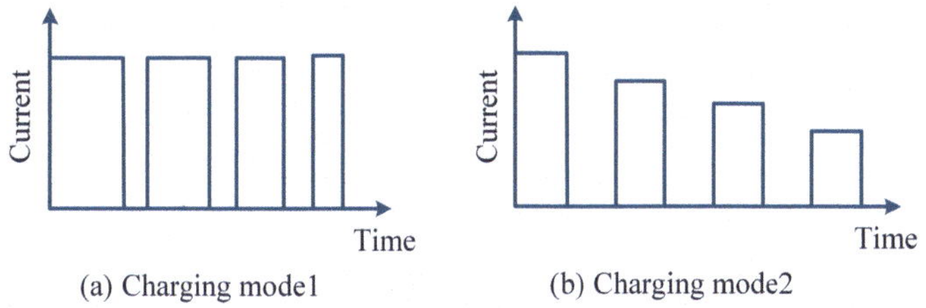

Fig. 7.23 Pulse charging mode with variable-frequency and variable amplitude: **a** variable frequency and constant amplitude **b** constant frequency and variable amplitude

the charging speed, they applied charging method with adaptive frequency conversion and variable duty cycle. This method could calculate the acceptable current, bandwidth, adaptive update frequency and duty cycle based on the battery model. It further shortened the charging duration without generating more heat.

7.4.4 Model-Based Charging Method

An optimal charging current is obtained in the charging method on the basis of an equivalent circuit model or an electrochemical model [37–41]. The model-based charging method with constant heat generation is a typical method. To ensure a constant heat generation rate, a charging current is calculated based on a battery model. During the charging process, the heat is generated and dissipated simultaneously. If the heat generation is concentrated, the battery would suffer from an enormous temperature rise because of the poor heat dissipation. The temperature rise can be reduced by averaging heat generation.

When a single-stage charging method with a constant heat generation is applied, it is necessary to increase the charging current to shorten the charging time. As a consequence, the charging capacity of the battery is reduced when the terminal voltage reaches the upper cutoff voltage. Therefore, a constant heat generation charging method with multi-stage decreasing heat generation rate is needed. When the terminal voltage reaches the upper cutoff voltage caused by the large charging current of the first stage, it shifts to a lower one in the next stage to guarantee the battery charging capacity. A fitness function of charging time and temperature rise is required to determine the heat generation value of each stage.

Take the Thevenin model as an example, charging time and charging temperature rise are set as the optimization targets, and minimized by the genetic algorithm.

Based on Eqs. (7.4)–(7.6), the recurrence relation is [34, 39, 42–44]

$$T_k = \exp\left(-\frac{hS_{\text{sur}}}{mc_{\text{p}}}\right)T_{k-1} + \left[1 - \exp\left(-\frac{hS_{\text{sur}}}{mc_{\text{p}}}\right)\right]\frac{Q_{z,k} + hS_{\text{sur}}T_{\text{amb}}}{hS_{\text{sur}}} \tag{7.23}$$

Then a multi-objective function of charging time and temperature rise is established. The fitness function can be written as [25, 45]

$$F(i_{\mathrm{L}}) = \frac{(1-\beta)t(i_{\mathrm{L}})}{a} + \frac{\beta T_{\mathrm{c}}(i_{\mathrm{L}})}{b} \tag{7.24}$$

where β is the weighting factor of the charging time and temperature rise. t is the charging time, T_{c} is the temperature rise, and a, b are the normalized coefficients.

T_c is calculated as follows:

$$T_{\mathrm{c}} = \max(T_k) - T_{\mathrm{initial}} \tag{7.25}$$

where T_{initial} is the initial temperature of the battery. During the optimization process, in order to ensure the battery safety, the following constraints must be satisfied:

$$\begin{cases} 0 < i_{\mathrm{L}} < I_{\mathrm{c,max}} \\ U_{\mathrm{t,min}} < U_{\mathrm{t}} < U_{\mathrm{t,max}} \\ T_{\mathrm{c}} < T_{\mathrm{max}} \\ \left|SOC_{\mathrm{end}} - SOC_f\right| < \varepsilon_{\mathrm{c}} \end{cases} \tag{7.26}$$

where SOC_f is the charging target SOC, SOC_{end} is the SOC value at the end moment of charging, T_{max} is the upper limit of battery temperature, and $I_{\mathrm{c,\,max}}$ is the maximum allowable current value during the charging process.

A flowchart for optimal charging with a constant heat generation using the genetic algorithm is shown in Fig. 7.24.

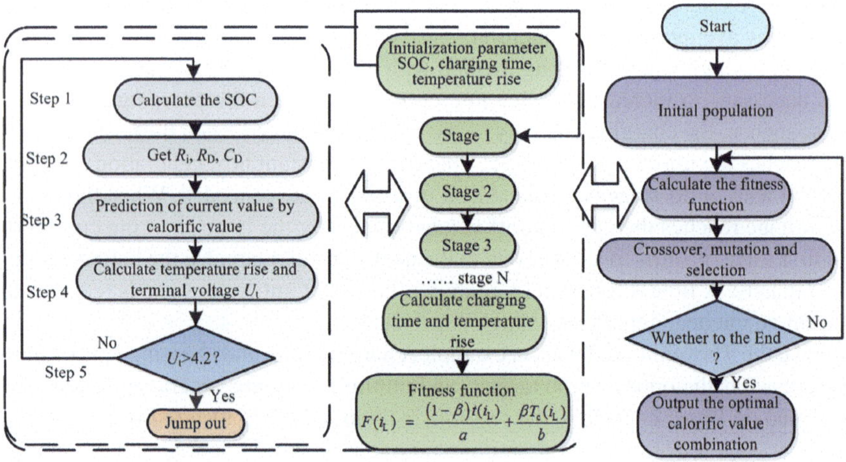

Fig. 7.24 Genetic algorithm optimization process

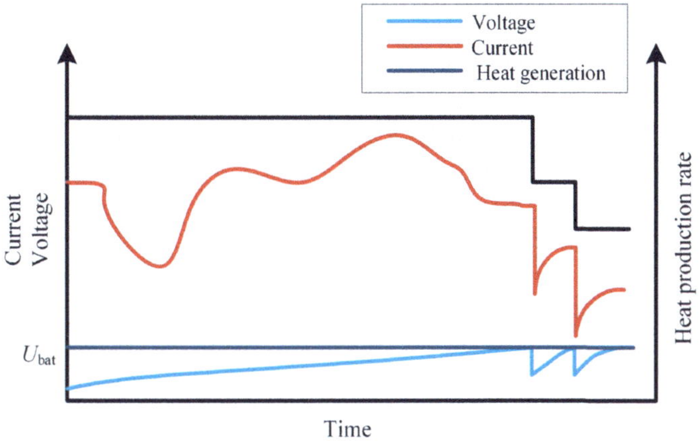

Fig. 7.25 Charging current, voltage, and heat generation

① Initialize the heat generation value of each stage and determine whether the heat generation is gradually decreasing. If so, start calculating the charging time and temperature rise using the model. Else if, eliminate the current generation.
② Perform crossover, mutation, and selection in the generated progeny population, and then carry out parameter optimization.
③ Obtain the multi-stage constant heat generation value corresponding to the charging time and the charging temperature rise. The current at each moment can be obtained by the heat generation, polarization voltage, polarization internal resistance, and ohmic internal resistance at this moment, and can be written as

$$i_L = \sqrt{\frac{J - \frac{U_D^2}{R_D}}{R_i}} \tag{7.27}$$

where J is the constant heat generation value at the moment.

Figure 7.25 shows the schematic diagram of a three-stage charging method with a constant heat generation, and the battery charging current is calculated by a constant heat generation rate. Obtained by the battery circuit model, the charging current curve of each stage is employed to the battery charging.

7.4.5 Case Study

In this section, the three-stage charging method was taken as an example with a constant heat generation. The SOC_f is 0.9, ε_c is 0.05, T_{max} is 50 °C, and $I_{c,max}$ is 2.4 A. The battery 8-cell 06 is taken for analysis.

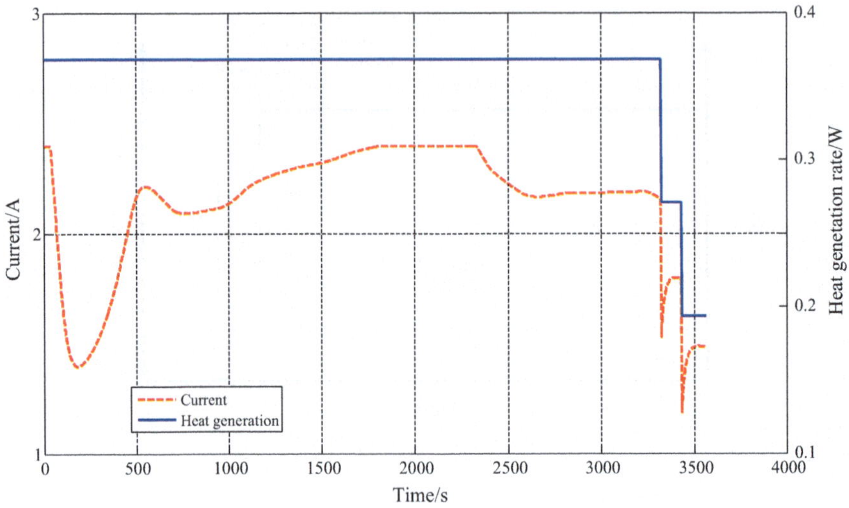

Fig. 7.26 Three-stage constant heat generation rates and charging current

(1) Analysis of simulation results

In this example, β is 0.24, and the three-stage constant heat generation rates calculated by the genetic algorithm are 0.36 W, 0.27 W, and 0.19 W, respectively. Based on the Eq. (7.27), the battery charging current curve was depicted in Fig. 7.26. At this time, the charging time of the battery is 3563 s, and the temperature rise is 1.07 °C.

(2) Analysis of experimental results

In order to evaluate the superiority of the charging method, charging comparison experiments were conducted on the battery 7-cell 01 and cell 02 at 25 °C using the experimental platform of Fig. 7.27. The experimental results are shown in Fig. 7.28.

The experimental results show that the multi-stage charging method with a constant heat generation can reduce the charging time and the temperature rise by 7.89% and 38.8%, respectively, compared with the traditional CCCV method (Table 7.1). It is a feasible optimal charging method.

7.5 Conclusion

This chapter first introduces the common methods and classifications of low-temperature heating strategies for lithium-ion batteries. Then the implementation steps of each method is briefly analyzed and the theoretical basis for AC heating is elaborated. An adaptive enchelon heating method is introduced. It can heat the cell

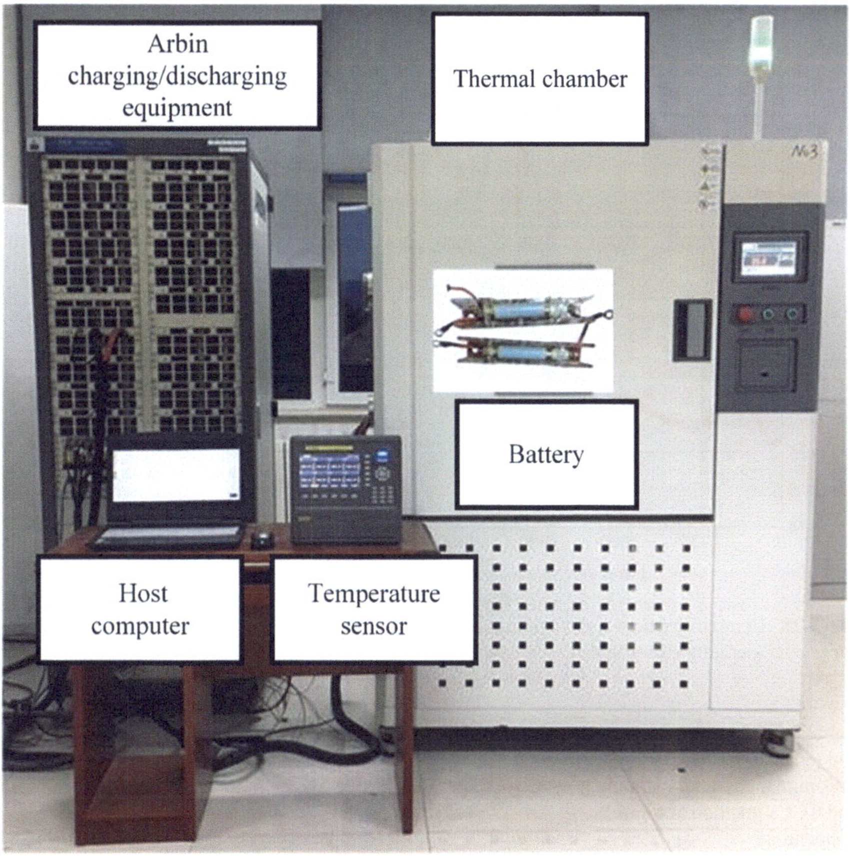

Fig. 7.27 Experimental platform

and the battery pack from −20.3 °C to 10.02 °C in 13.7 min and 12.4 min, respectively, with the respective temperature rise rates of 2.21 °C/min and 2.47 °C/min. There is no significant effect on the battery cycle life and realize the fast and reliable heating of the battery in the low temperature. In terms of optimal charging, several methods especially the multi-stage charging method with a constant heat generation are analyzed. Compared with the traditional CCCV charging method, this method successfully reduces both the charging time and the temperature rise, achieving fast charging.

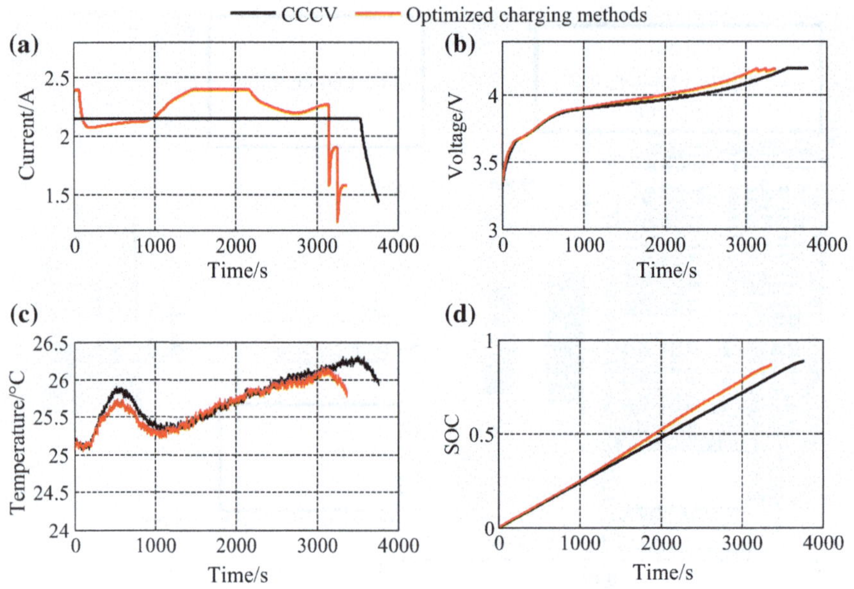

Fig. 7.28 Comparison between the optimal charging method and CCCV: **a** current curve **b** voltage curve **c** temperature rise **d** SOC

Table 7.1 The charging time and temperature rise between the optimal charging method and CCCV method at room temperature

	Charging capacity (A h)	Charging time (s)	Temperature rise (°C)
Optimal charging method	2.07	3376	1.01
CCCV method	2.07	3665	1.65

References

1. Xiong R (2016) A lithium-ion battery pack state of charge and state of energy estimation algorithms using a hardware-in-the-loop validation. IEEE Trans Power Electron 32(6):4421–4431
2. Sun F, Xiong R (2015) A novel dual-scale cell state-of-charge estimation approach for series-connected battery pack used in electric vehicles. J Power Sources 274:582–594
3. Guo S, Xiong R, Wang K, Sun F (2018) A novel echelon internal heating strategy of cold batteries for all-climate electric vehicles application. Appl Energy 219:256–263
4. Guo S, Xiong R, Sun F, Cao J, Wang K (2017) ScienceDirect—An echelon internal heating strategy for lithium-ion battery Assessing the feasibility of using the heat demand-outdoor temperature function district heat demand forecast. Energy Procedia 142:3135–3140
5. Huat L, Mun H, San H, Cai Z, Tong W, Agung N (2018) Novel thermal management system using mist cooling for lithium-ion battery packs. Appl Energy 223:146–158
6. Wang T, Tseng KJ, Zhao J (2018) Development of efficient air-cooling strategies for lithium-ion battery module based on empirical heat source model. Appl Therm Eng 90:521–529

7. Wang T, Tseng KJ, Zhao J, Wei Z (2014) Thermal investigation of lithium-ion battery module with different cell arrangement structures and forced air-cooling strategies. Appl Energy 134:229–238
8. Rahman MM, Rahman HY, Mahlia TMI, Sheng JLY (2016) Liquid cooled plate heat exchanger for battery cooling of an electric vehicle (EV). IOP Conf Ser Earth Environ Sci 32:12053
9. Yang XH, Tan SC, Liu J (2016) Thermal management of Li-ion battery with liquid metal. Energy Convers Manag 117:577–585
10. Zhang Q, Huo Y, Rao Z (2016) Numerical study on solid–liquid phase change in paraffin as phase change material for battery thermal management. Sci Bull 61:391–400
11. Zhang J, Hao G, Li Z, Ding Z (2015) Internal heating of lithium-ion batteries using alternating current based on the heat generation model in frequency domain. J Power Sources 273:1030–1037
12. Hande A, Stuart TA (2002) AC heating for EV/HEV batteries. Power Electron Transp. https://doi.org/10.1109/PET.2002.1185559
13. Stuart TA, Hande A (2004) HEV battery heating using AC currents. J Power Sources 129:368–378
14. Zhu J, Sun Z, Wei X, Dai H, Gu W, Zhu J (2017) Experimental investigations of an AC pulse heating method for vehicular high power lithium-ion batteries at subzero temperatures. J Power Sources 367:145–157
15. Jiang J, Ruan H, Sun B, Wang L, Gao W (2018) A low-temperature internal heating strategy without lifetime reduction for large-size automotive lithium-ion battery pack. Appl Energy 230:257–266
16. Ruan H, Jiang J, Sun B, Zhang W, Gao W, Wang LY (2016) A rapid low-temperature internal heating strategy with optimal frequency based on constant polarization voltage for lithium-ion batteries. Appl Energy 177:771–782
17. Wang CY, Zhang G, Ge S, Xu T, Ji Y, Yang XG (2016) Lithium-ion battery structure that self-heats at low temperatures. Nature 529:515
18. Zhang G, Ge S, Xu T, Yang XG, Tian H, Wang CY (2016) Rapid self-heating and internal temperature sensing of lithium-ion batteries at low temperatures. Electrochim Acta 218:149–155
19. Lee DY, Cho CW, Won JP, Park YC, Lee MY (2013) Performance characteristics of mobile heat pump for a large passenger electric vehicle. Appl Therm Eng 50:660–669
20. Yang L, Tai N, Fan C, Meng Y (2016) Energy regulating and fluctuation stabilizing by air source heat pump and battery energy storage system in microgrid. Renew Energy 95:202–212
21. Dawoud B, Amer E, Gross D (2007) Experimental investigation of an adsorptive thermal energy storage. Int J Energy Res 31:135–147
22. Kim YJ (2015) Heat pipe assembly having heating/cooling functions, battery module for eco-friendly vehicle using the same and method for operating battery module
23. Appen JVN, Braun M (2019) Sizing and improved grid integration of residential PV systems with heat pumps and battery. IEEE Trans Energy Convers 34:562–571
24. Jiang J, Ruan H, Sun B, Zhang W, Gao W, Wang LY (2016) A reduced low-temperature electro-thermal coupled model for lithium-ion batteries. Appl Energy 177:804–816
25. Basu S, Hariharan KS, Kolake SM, Song T, Dong KS, Yeo T (2016) Coupled electrochemical thermal modelling of a novel Li-ion battery pack thermal management system. Appl Energy 181:1–13
26. Greco A, Cao D, Jiang X, Yang H (2014) A theoretical and computational study of lithium-ion battery thermal management for electric vehicles using heat pipes. J Power Sources 257:344–355
27. Li J, Wang L, Lyu C, Pecht M (2017) State of charge estimation based on a simplified electrochemical model for a single $LiCoO_2$ battery and battery pack. Energy 133:572–583
28. Zhao R, Liu J, Gu J (2016) Simulation and experimental study on lithium ion battery short circuit. Appl Energy 173:29–39
29. Zhu J, Sun Z, Wei X, Dai H (2016) Studies on the medium-frequency impedance arc for Lithium-ion batteries considering various alternating current amplitudes. J Appl Electrochem 46:157–167

30. Yang R, Xiong R, He H, Mu H, Wang C (2017) A novel method on estimating the degradation and state of charge of lithium-ion batteries used for electrical vehicles. Appl Energy 207(1):336–345
31. Xiong R, Tian J, Shen W, Sun F (2019) A novel fractional order model for state of charge estimation in lithium ion batteries. IEEE Trans Veh Technol 68(5):4130–4139
32. Fan J, Tan S (2006) Studies on charging lithium-ion cells at low temperatures. J Electrochem Soc 153:A1081
33. Liu K, Li K, Yang Z, Zhang C, Deng J (2016) Battery optimal charging strategy based on a coupled thermoelectric model. IEEE Congr Evol Comput CEC 225:5084–5091
34. Min H, Sun W, Li X, Guo D, Yu Y, Zhu T (2017) Research on the optimal charging strategy for Li-Ion batteries based on multi-objective optimization. Energies 10(5):709
35. Di Yin M, Cho J, Park D (2016) Pulse-based fast battery IoT charger using dynamic frequency and duty control techniques based on multi-sensing of polarization curve. Energies 9(3):209
36. Purushothaman BK, Landau U (2006) Rapid charging of lithium-ion batteries using pulsed currents. J Electrochem Soc 153:A533
37. Tippmann S, Walper D, Balboa L, Spier B, Bessler WG (2014) Low-temperature charging of lithium-ion cells part I: electrochemical modeling and experimental investigation of degradation behavior. J Power Sources 252:305–316
38. Tanim TR, Rahn CD, Wang CY (2015) State of charge estimation of a lithium ion cell based on a temperature dependent and electrolyte enhanced single particle model. Energy 80:731–739
39. Ye M, Guo H, Cao B (2017) A model-based adaptive state of charge estimator for a lithium-ion battery using an improved adaptive particle filter. Appl Energy 190:740–748
40. Zhao XW, Zhang GY, Yang L, Qiang JX, Chen ZQ (2011) A new charging mode of Li-ion batteries with $LiFePO_4$/C composites under low temperature. J Therm Anal Calorim 104:561–567
41. Jiang J, Liu Q, Zhang C, Zhang W (2014) Evaluation of acceptable charging current of power li-ion batteries based on polarization characteristics. IEEE Trans Ind Electron 61:6844–6851
42. Lin C, Mu H, Xiong R, Shen W (2016) A novel multi-model probability battery state of charge estimation approach for electric vehicles using H-infinity algorithm. Appl Energy 166:76–83
43. Chen LR, Hsu RC, Liu CS (2008) A design of a grey-predicted Li-ion battery charge system. IEEE Trans Ind Electron 55:3692–3701
44. Xiong R, Tian J, Mu H, Wang C (2017) A systematic model-based degradation behavior recognition and health monitoring method for lithium-ion batteries. Appl Energy 207(1):372–383
45. Forgez C, Vinh D, Friedrich G, Morcrette M, Delacourt C (2010) Thermal modeling of a cylindrical $LiFePO_4$/graphite lithium-ion battery. J Power Sources 195:2961–2968

Chapter 8
Algorithm Development, Test, and Evaluation

The simplified process of the algorithm in the theoretical design may lead to deviations in practical application. As a result, it is very important to download the algorithm to the real BMS and evaluate it according to the relevant standards and indexes, which helps designers to find and solve some practical problems that are neglected in the theoretical derivation in time and optimize the algorithm. The traditional algorithm development and evaluation methods not only consume a lot of time, manpower cost, but also limited by safety issues. In addition, it is difficult to comprehensively and systematically evaluate some actual controlled objects. Fortunately, the "V" development process based on the rapid prototyping and hardware in the loop (HIL) test can find out the problems in the algorithm and make evaluation efficiently and accurately, which improves the development efficiency. This chapter mainly focuses on the development process of BMS for EVs, and illustrates the evaluation methods of rapid prototyping simulation, HIL test algorithm, and the experiments for vehicles [1].

8.1 Algorithm Development Process

8.1.1 General Process of Algorithm Development

A standardized development process is in demand for the algorithm from the initial design to the final application. The development process of the traditional algorithms in BMS is shown in Fig. 8.1. The designers first need to propose specific quantifiable performance indexes according to the demands of the target controller, and then design the system control strategy and carry out the mathematics simulation. Next, the hardware and software of the controller are designed and the system is integrated. Finally, BMS bench test and vehicle test are carried out.

There are three main issues in the development of traditional algorithms in BMS:

R. Xiong, *Battery Management Algorithm for Electric Vehicles*,
https://doi.org/10.1007/978-981-15-0248-4_8

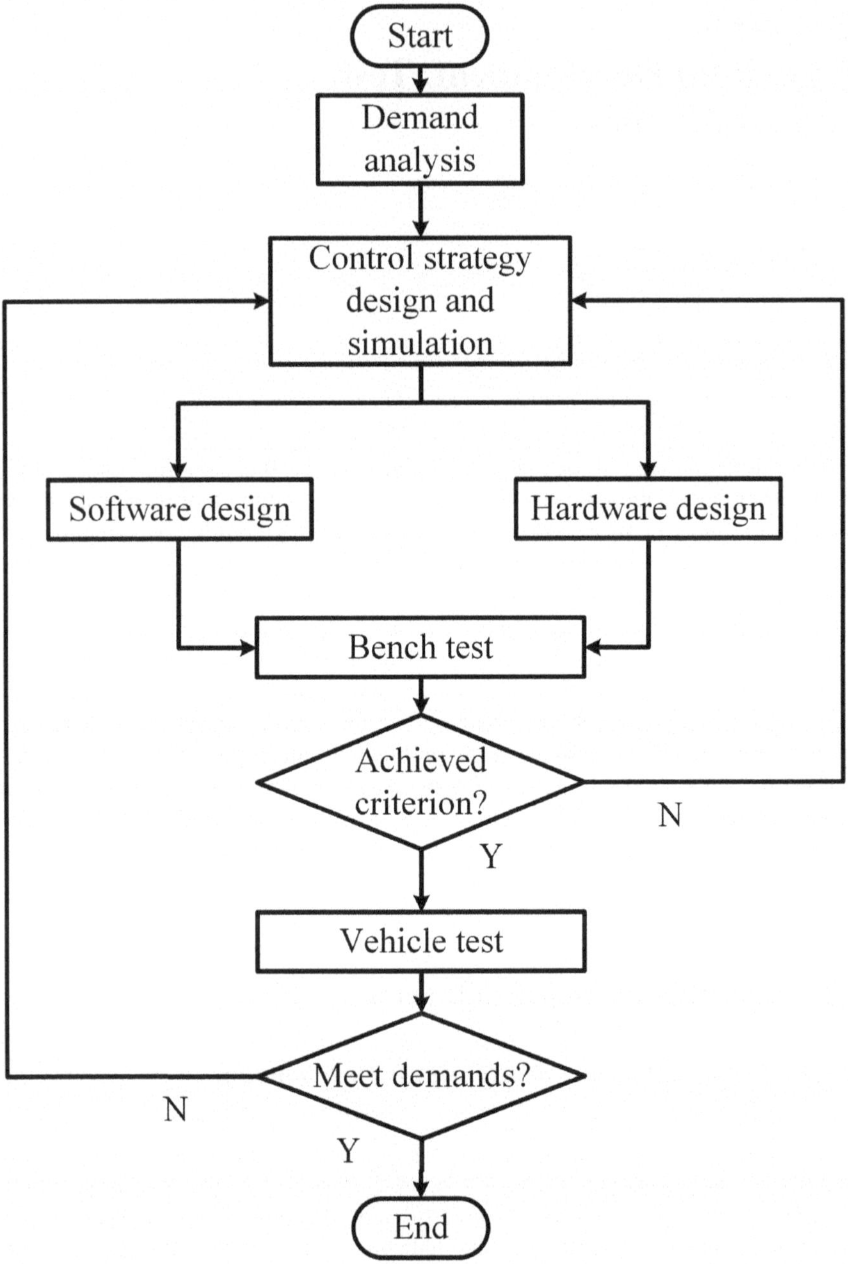

Fig. 8.1 Traditional BMS algorithm development process

(1) Manual programming is inefficient. The reliability of the code cannot be guaranteed by manual programming in the software design. In addition, the manual programming and debugging consume a lot of time and delay the progress of a project.
(2) Control strategy evaluation is not timely. The controller software programming and the hardware circuit designing are directly carried out before the characteristics and effects of the control strategy are determined. The developers need to redesign the software and hardware if the control strategy could not meet the requirements in the test.
(3) Hardware and software problems are difficult to distinguish. Since both the software and hardware of the controller are dependent on the verification of the bench for some design defects, it is difficult to determine the root cause of the problem, reducing the development efficiency.

8.1.2 Model-Based "V" Development Process

Compared with the traditional development process, "V" development process is conducive to detect errors and deficiencies existing in the current algorithm, shortening the development time and saving development costs. Figure 8.2 shows the model-based "V" development process of the BMS core algorithms. Semi-physical simulation is adopted in the process and its characteristic is that some physical objects are directly introduced into the system simulation loop, which makes the simulation results closer to the actual values.

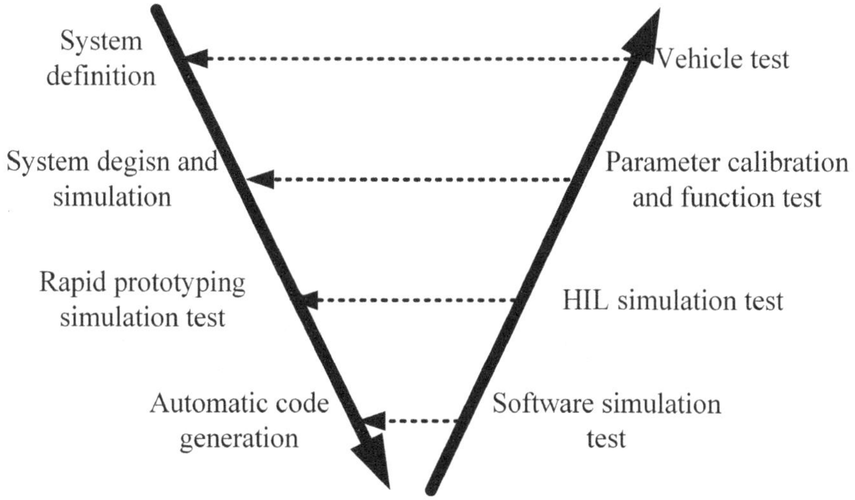

Fig. 8.2 Model-based "V" development process of BMS core algorithms

The semi-physical simulation is usually divided into rapid prototyping simulation and HIL simulation. Compared with numerical simulation, semi-physical simulation can not only verify the real-time performance of the designed algorithm, but also significantly improve the accuracy and facticity of the simulation process. Due to the high confidence coefficient, the semi-physical simulation experiment can reduce the number of road tests, shorten the development time, and reduce the cost and risk in the process of product development. At present, the semi-physical simulation experiment has become a very important part in the development process of BMS, electric motor controller and vehicle controller, which are three core technologies of new energy vehicles.

As the current mainstream development method of the vehicle embedded system, "V" development process is mainly divided into the following parts:

System definition

The specific implementation of system definition is shown in Fig. 8.3, which can be divided into three steps: problem definition, feasibility analysis, and requirements analysis. The significance of problem definition lies in discovering the problems to be solved and making the relevant problem description instructions. The feasibility analysis is to evaluate the rationality and value of the problems in the instructions according to the current industry development status, and formulate the feasibility report. Based on the above two steps and relevant information such as national standards and industry standards, the requirements analysis is committed to determining the functions to be achieved by system, defining the quantifiable performance indexes of each function, and formulating the requirements report.

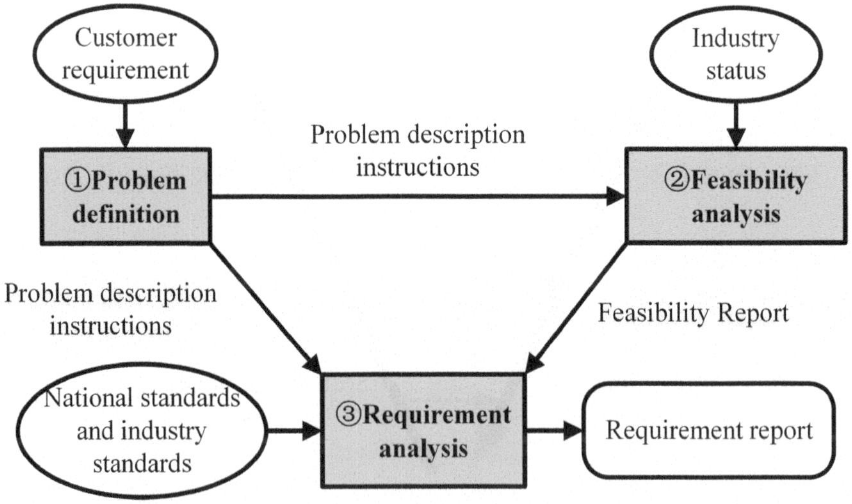

Fig. 8.3 Specific implementation steps of system definition

System design and simulation

According to the demands of the management system, the integrated design of the core algorithms and the control strategies is completed based on the relevant theories, and the whole system is implemented in the computer software environment, that is to realize the modeling and simulation of the virtual controller, the virtual controlled object and the virtual control environment, along with an early and rapid evaluation of the system indexes and errors.

Rapid prototype simulation test

Rapid control prototype (RCP) uses the virtual controller and the actual controlled object to complete the control system simulation test. With the control model downloaded to the standard rapid prototyping simulation platform, the real-time control of the actual controlled object is simulated to verify the functions of the system control strategy. The advantage of rapid prototyping simulation test is avoiding complex and time-consuming work such as hardware design and underlying code writing, which enables designers to focus on the optimal design of control algorithms [2].

Codes automatic generation

The model is built based on the specification, and the code which can be applied to the actual controller is generated automatically after completing the parameter configuration of the standard template. Currently, the code automatic generation technology has become a key technology for the rapid development of embedded system. The MAAB control algorithm modeling specification in MATLAB can be used as a common modeling specification for reference.

Software simulation test

Software in the loop (SIL) simulation test can be used to verify the consistency between the automatically generated code and the design algorithm. Using a large amount of data for software simulation test is conducive to the early detection of the bug in the generated code. It is both required in the second and fifth steps of the "V" development process to complete the construction of virtual controller, virtual controlled object and virtual system environment in the computer software environment, achieving closed-loop simulation and evaluation of the control system. The difference of them is that the virtual controller of the latter is based on the traditional mathematical model for simulation, while the other is based on the generated high-level code for simulation, which is closer to the actual use.

Hardware in the loop simulation test

HIL simulation test is an experimental method which replaces the original virtual controller with the real controller in the simulation loop to make the system simulation closer to the real application environment. As the input and output parameters of the battery system are easy to obtain, it is reasonable to directly take the cell or the battery pack as the controlled object which is usually virtual in the traditional HIL simulation tests. With the advantages of strong real-time performance, easy error correction, short development time, and high efficiency, the HIL simulation system has been widely used in the development of the vehicle and its components.

Parameter calibration and functional test

After completing the HIL test, the debugged controller needs to be connected to the real controlled object and placed in the real environment to conduct the system bench tests. The stability, reliability, and safety of various system functions should be tested under various operating conditions, different ambient temperatures and humidity as well as the various vibration modes. Based on the final results of functional testing, the calibration and design of each system parameter can be completed.

Vehicle test

The vehicle test is to integrate the embedded system containing the core algorithms into the vehicle, test the each function of the control system and judge whether the quantifiable performance indexes of each function meets the standard until the whole "V" design process is finally completed on the basis of the developed requirements specification in the system definition.

8.2 Rapid Prototype Simulation Test

The main idea of RCP technology is designing a product in virtual environment to shorten the development time and reduce the development cost. In the initial stage of system development, the controller model is established quickly, and the whole system is tested offline and online for many times to verify the feasibility of control strategies.

Specially developed for specific application requirements by some companies, the early RCP systems are mostly built with microprocessors such as digital signal process (DSP) and PowerPC, which are technologically advanced but expensive. The most widely used RCP system of this kind used to be the dSPACE in Germany [3], and then the PC-based RCP systems are launched by MathWorks, NI, and other companies. Similar systems have also been developed in China currently such as the RapidECU of Beijing Jiuzhou Huahai technology Ltd., and the cSPACE of Zhengzhou micro-nano technology Ltd., which are much cheaper. The example of the algorithm rapid prototyping test is based on the xPC Target real-time simulation system. Compared with other simulation systems, xPC Target has the advantages of low cost and high computing efficiency under the same simulation accuracy, which has become a high-performance rapid prototyping simulation test platform widely recognized by vehicle design enterprises, universities and research institutes.

8.2.1 System Composition

General composition of the rapid prototype simulation system for the battery

As shown in Fig. 8.4, the general composition of the rapid prototype simulation system for the battery is mainly composed of the host computer, the controller prototype and the actual controlled object. Communication between the host computer and the controller prototype is based on TCP/IP protocol [3]. As the main part of programming and monitoring, the host computer can develop the control strategies and customize the display parameters through graphical programming software, such as MATLAB/Simulink, LabVIEW, to realize the effective compiling and debugging of the program.

The prototype controller includes a real-time system module and a communication module. As the carrier of control strategy, the real-time system module can support the running of the complex and real-time control programs. Combined with the battery parameters transmitted by the communication module, the real-time system module can not only complete the real-time estimation of the implied state variables of the battery and send the control commands, but also fulfill the indirect control of the actuator through the communication module and send the battery status information to the host computer. Through the communication module, different boards can be connected to send the voltage, current, temperature and other parameters of the battery system collected by the sensors to the real-time system, and transmit the control commands of the real-time system module to the actuator [4].

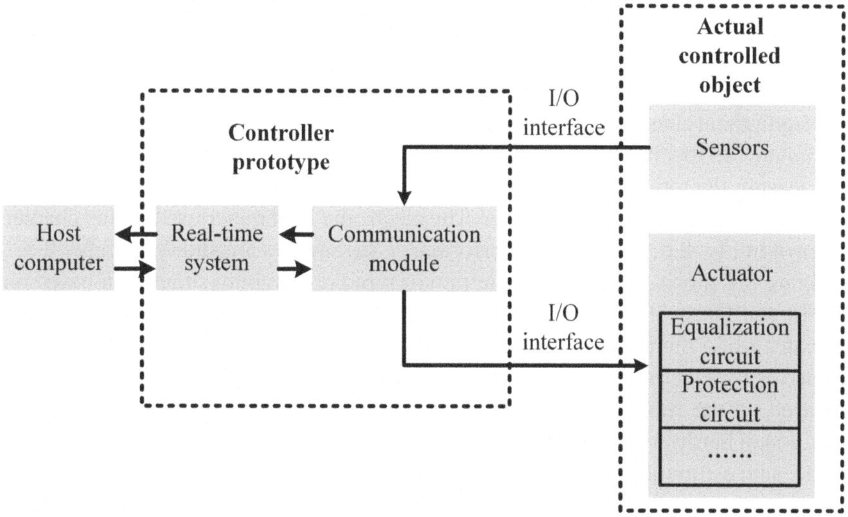

Fig. 8.4 General composition of rapid prototype simulation system of battery

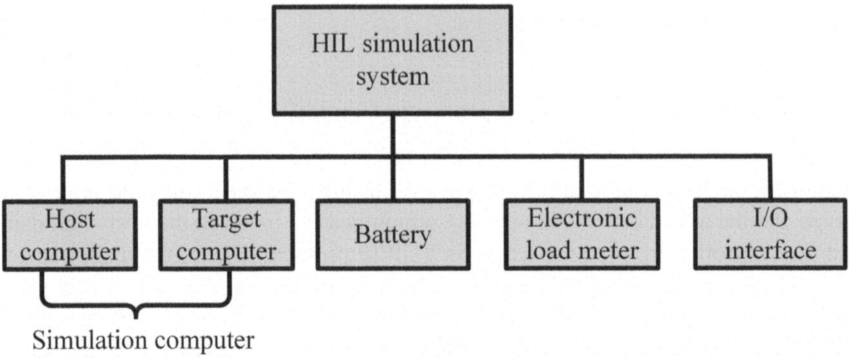

Fig. 8.5 Hardware system composition of rapid prototyping simulation experiment platform based on xPC target

The actual controlled objects are generally divided into two types. One is the sensors including voltage, current and temperature sensors, and the other is the actuators including the protection circuit, equalization circuit, etc.

Rapid prototyping simulation system based on xPC Target

This system consists of both hardware and software, as shown in Fig. 8.5 [5]. The hardware includes simulation computer, high-speed data collected equipment, battery samples, electronic load meter, and corresponding I/O interface. The software includes operating system, simulation software, and simulation model. According to the characteristics of dual PC running mode in xPC Target, Simulink model development and C code compilation are completed by the host computer, and the target computer is used as the actuator of the actual code [5–7].

Batteries are charged or discharged by electronic load meter such as Arbin and other equipments based on the instructions issued by the target computer, which aims to simulate the specified operating conditions. Meanwhile, the electronic load meter can measure the terminal voltage and current of the battery in real time and upload the parameters to the target computer. The electronic load meter used in this chapter is shown in Fig. 8.6, and its main performance parameters are shown in Table 8.1.

Figure 8.7 shows the structure diagram of rapid prototyping simulation based on xPC Target. The host computer provides the development environment of simulation software, while the target computer runs the code generated by the host computer and drives the electronic load meter to charge and discharge the battery.

Based on the framework of the rapid prototyping simulation structure and the definition of hardware device I/O interface, a rapid prototype simulation test platform for the state estimation of lithium-ion battery is built as shown in Fig. 8.8 [8].

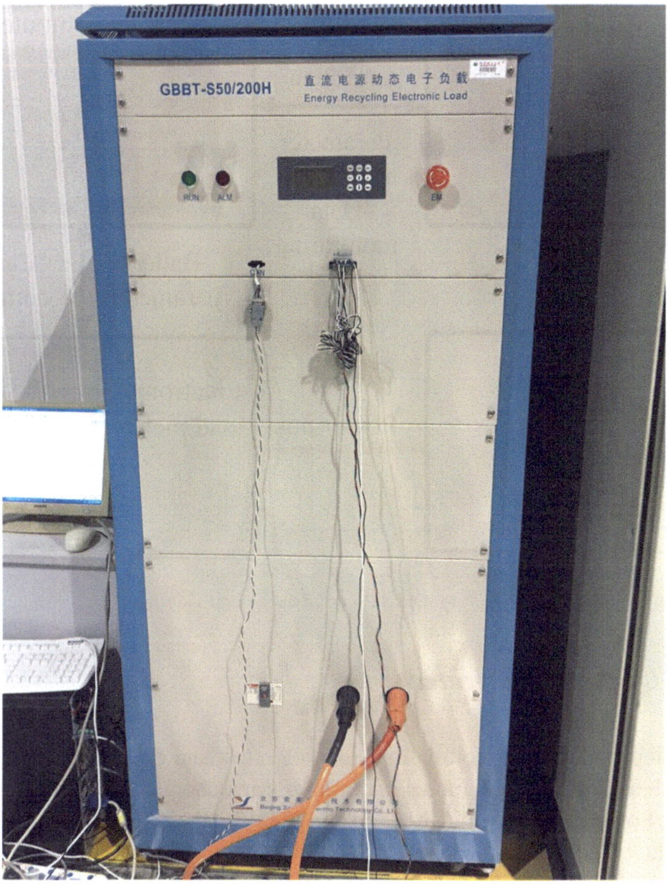

Fig. 8.6 Battery electronic load meter

Table 8.1 Performance parameters of electronic load meter

Indexes	Parameters	Additional information
Rated power	15 kW	–
Charge/discharge DC current	0 A ± 100 A	–
DC voltage range	5–50 V (total voltage)	5–16 V and 18–50 V
Current control accuracy	4%	30 A–100 A
	1.3%	0–30 A
Response time	<5 ms	–

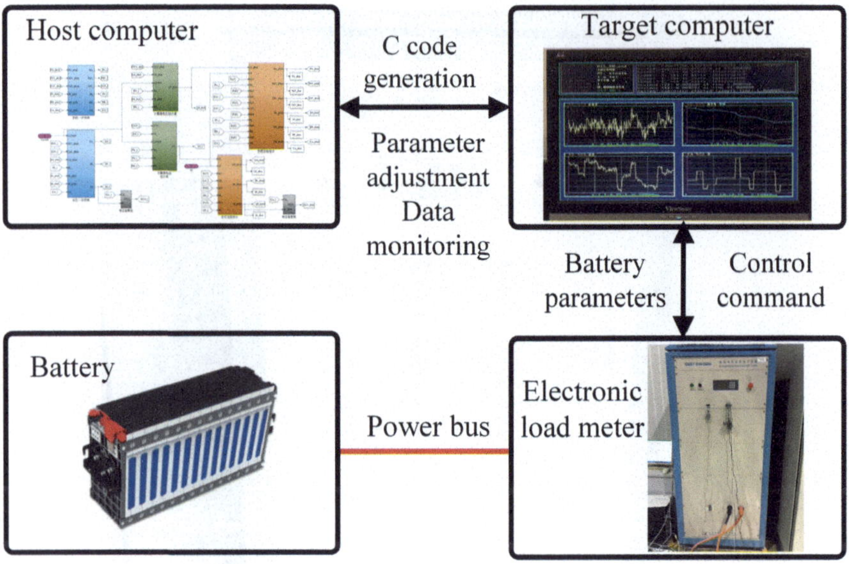

Fig. 8.7 Semi-physical simulation system structure based on xPC target

8.2.2 *Algorithm Integrated*

The core step of the rapid prototyping simulation test and evaluation based on xPC target is to build the Simulink model of the algorithm under the development environment of the host computer, which is mainly divided into the core algorithm simulation model of the battery system and the CAN bus communicationmodel.

The CAN bus communication model is mainly developed to realize the information exchange between the target computer and the electronic load meter, including the control commands from the target computer to the electronic load meter and the current, voltage and other battery parameters collected by the electronic load meter. The CAN bus communication model is shown in Fig. 8.9.

The core algorithm simulation model of the battery system is needed to evaluate. It can feedback current and voltage information in real time supported by the electronic load meter and realize online parameters identification and state estimation. The algorithm is evaluated by comparing the estimated value with the reference value. When the error is large, the algorithm can be directly adjusted and optimized in MATLAB/Simulink. Partial simulation model of MHIF estimation method is shown in Fig. 8.10 [8].

After the establishment of the core algorithm simulation model and CAN bus communication model of the battery system, the established Simulink models need to be transformed into source codes by the code automatic generation technology. Unlike the code automatic generation in the model-based "V" development process,

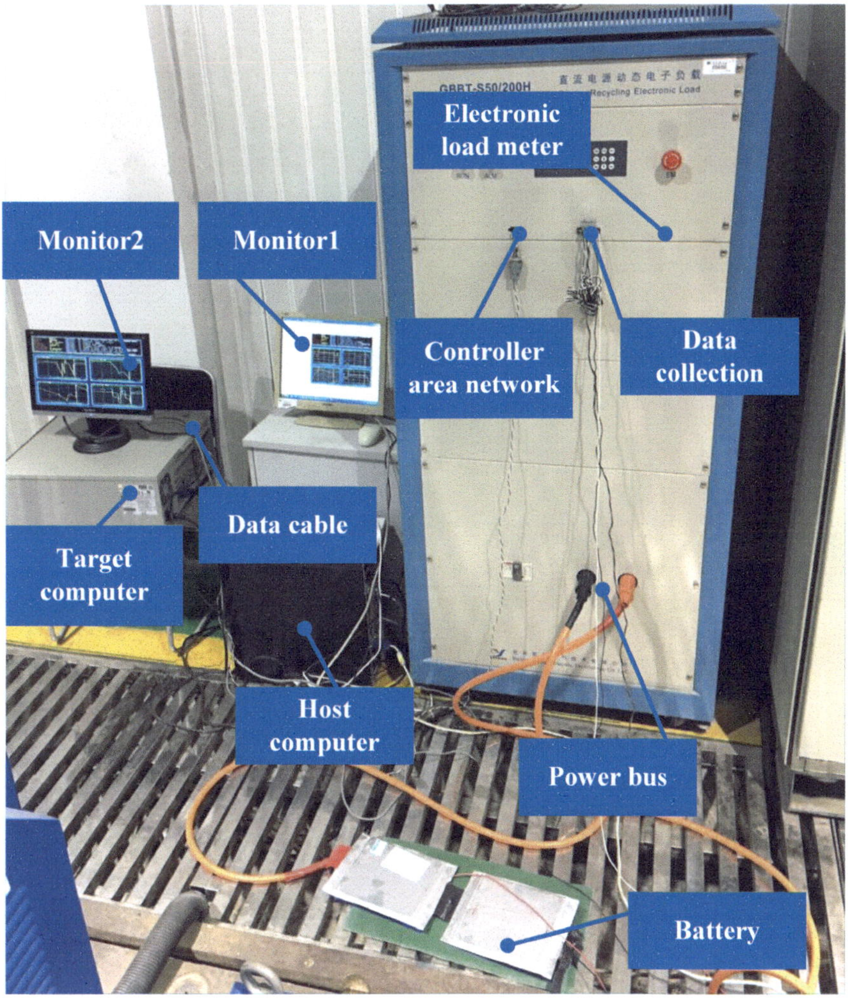

Fig. 8.8 Rapid prototype simulation test platform for state estimation of lithium-ion battery

the source code is based on the virtual controller, which is generally generated through the system target file xpctarget.tlc.

8.2.3 Case Study

The following two cases were conducted with a known static maximum capacity of the battery and an unknown capacity, respectively.

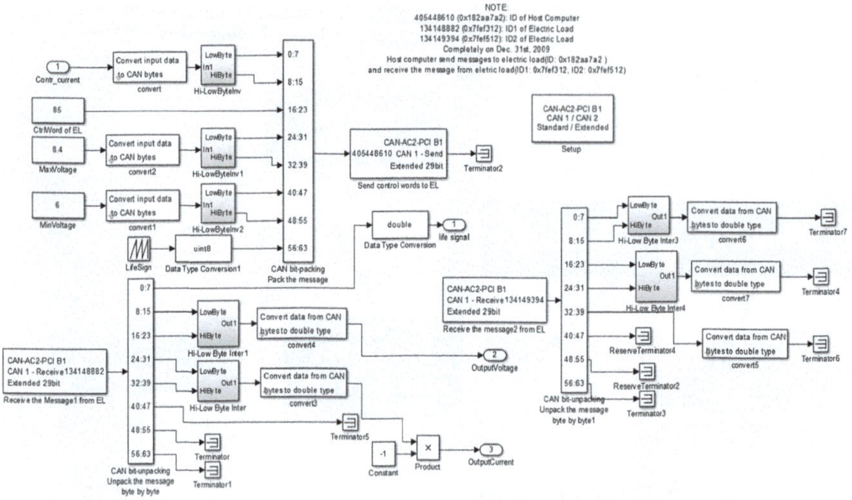

Fig. 8.9 CAN bus communication model

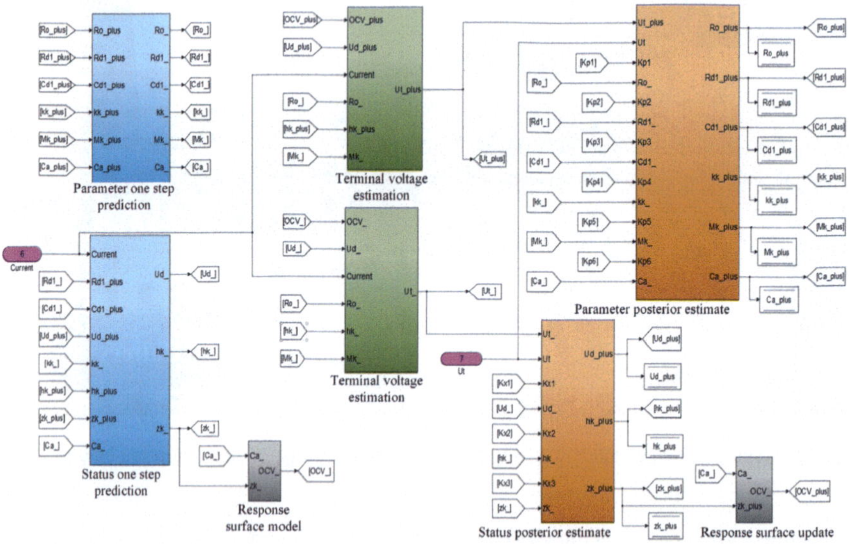

Fig. 8.10 Partial simulation model of MHIF estimation method

SOC estimation based on the known static capacity of battery

Based on the above mentioned rapid prototype simulation test platform for lithium-ion battery state estimation, the evaluation of SOC estimation based on MHIF algorithm was completed. The battery available capacity in MHIF is regarded as a known static value. Since the minimum operating voltage of the electronic load meter is 5 V which is higher than the operating voltage of the NMC cell, two NMC batteries connected in series were used as the controlled physical objects. Under the condition that the series battery pack was fully charged (100% SOC, room temperature), the target computer loads the pack with standard DST profile by electronic load meter, and completes online states and parameters estimation based on the core algorithm simulation model of the battery system to evaluate the accuracy and robustness of the algorithm.

Figure 8.11 shows the interface of the target computer, and Fig. 8.12 shows the estimation results of terminal voltage and SOC. The results indicate that the error of the terminal voltage is within 50 mV, and the SOC estimation results can gradually converge whose error is within 3%, which meet the actual use requirements and prove the feasibility of the algorithm in practical application.

Co-estimation of SOC and available capacity

Based on the above mentioned rapid prototype simulation platform for battery, the battery 3 cell 1 and cell 2 were loaded the DST profile at room temperature (22 ± 3 °C), and the SOC and available capacity online estimation algorithm of the battery based on MHIF were evaluated [8]. The battery available capacity in MHIF is regarded as an unknown quantity. The static available capacity of two cells at room

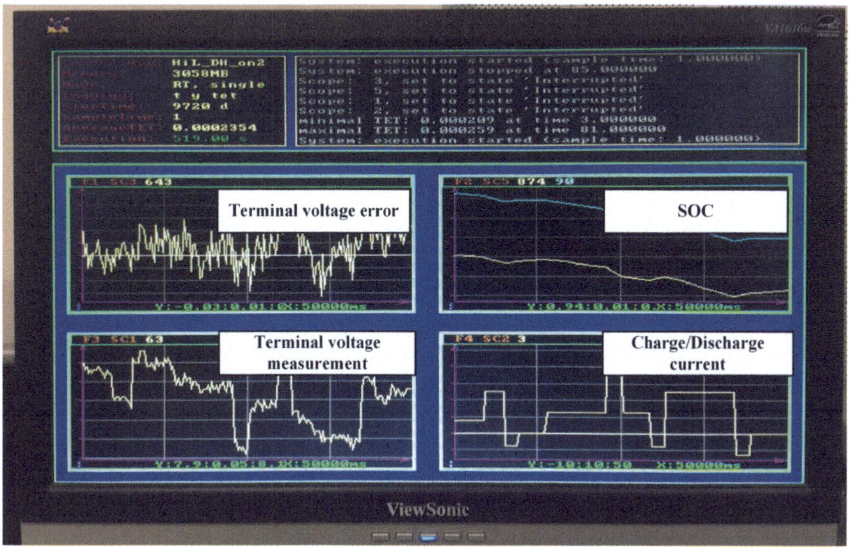

Fig. 8.11 Interface of the target computer

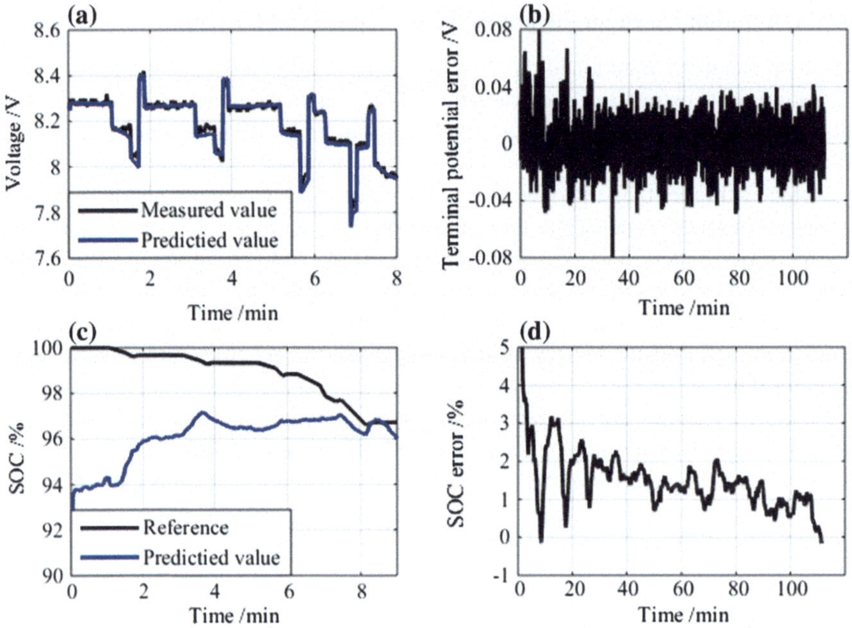

Fig. 8.12 Estimation results of terminal voltage and SOC **a** the comparison between predicted and measured voltages (part); **b** voltage error; **c** the comparison between estimated and reference SOCs (part); **d** SOC error

temperature is 31.54 A h and 31.40 A h, respectively. The simulation results of SOC and available capacity are shown in Fig. 8.13.

The batteries were tested for 300 min with the initial SOC of 80% and the initial capacity of 30 A h. The estimation results of SOC and available capacity are shown in Fig. 8.14. The measured and estimated terminal voltages and the voltage error are shown in Fig. 8.15.

As shown in Fig. 8.14a, the estimated SOC can rapidly converge to the actual value with the maximum estimated SOC error of 0.38%. The available capacity estimation also has excellent convergence characteristics whose overall estimated error is within 2% illustrated in Fig. 8.14b. The maximum estimated error occurs at 9750 s, at which time the estimated capacity is 30.87 A h and the estimated error is 1.78%. As shown in Fig. 8.15, the estimated error of terminal voltage is within 50 mV. According to the above experimental results, the estimated method based on MHIF is feasible and effective, and this method can also converge rapidly under inaccurate initial values, which proves the robustness of the estimated method.

It can be seen that the real-time online estimation of battery dynamic capacity can improve the SOC estimation accuracy, and the SOC and SOH co-estimation of the battery can achieve better application effect.

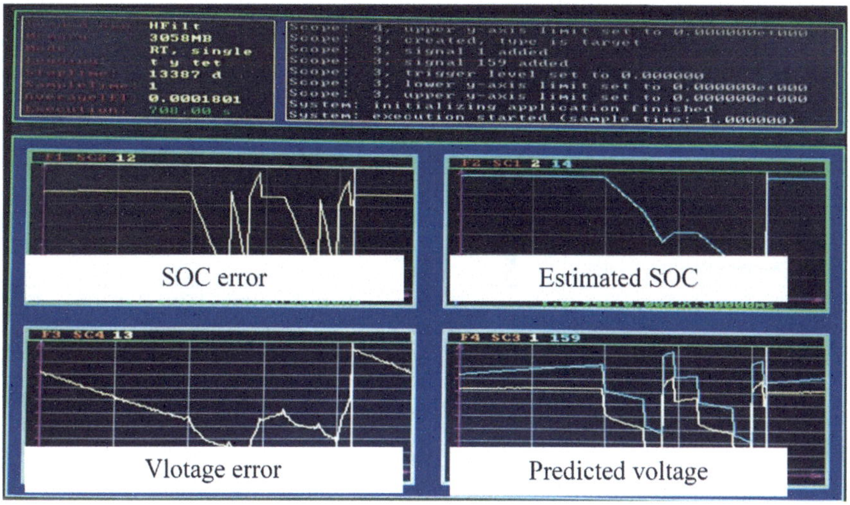

Fig. 8.13 Estimation results of SOC and available capacity for the battery

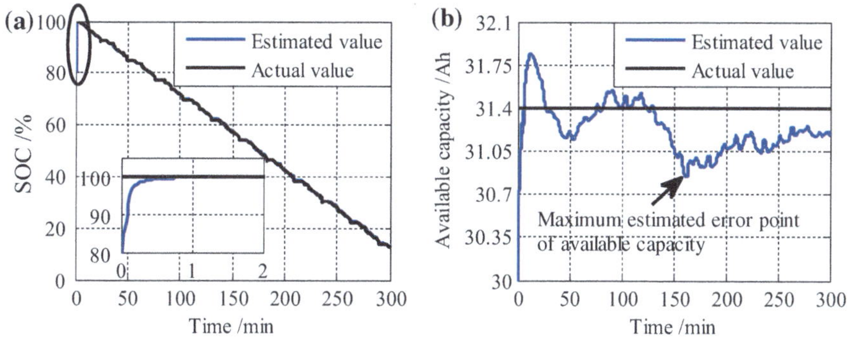

Fig. 8.14 SOC and available capacity estimated results: **a** the comparison between estimated and actual SOCs; **b** the comparison between estimated and actual capacities

8.3 HIL Algorithm Test

The HIL simulation combines both physical and numerical simulation. In the simulation process, the computer is connected to a part of the actual system, and the mathematical model is established in the computer to simulate other parts that are inconvenient for experiment or do not exist. This simulation mode takes advantage of the computer modeling and has the characteristics of simple modeling, low cost, convenient parameter modification and flexibility in practical application. For the parts of the system that are difficult to establish mathematical models, the actual system or physical model can be employed, which can ensure the operation of the

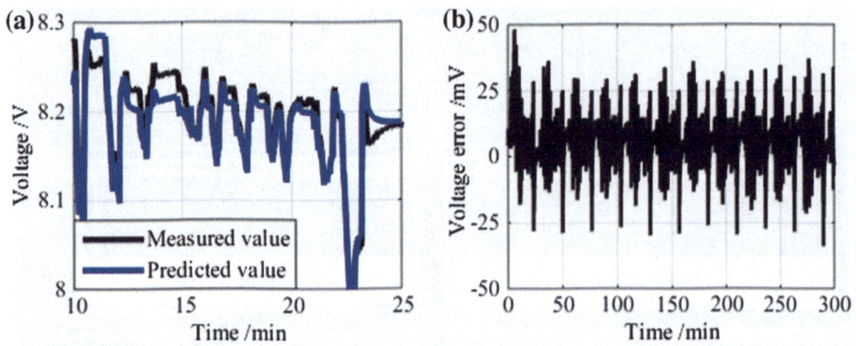

Fig. 8.15 Measured and predicted terminal voltages: **a** the comparison between measured and predicted terminal voltages (part); **b** terminal voltage error

overall system and realize the simulation of the system. Owing to the high authenticity, HIL simulation is generally used to evaluate the feasibility and correctness of the control system scheme. And, it can carry out the fault mode simulation of the product and dynamic simulation of closed loop of the control system, etc. The simulation condition of the HIL is closer to the actual situation, and the product performance can be accurately and objectively reflected in the debugging and test process of the product [9].

In general, the performance of the control strategy can be evaluated most effectively by the joint tests of the actual control system and the actual controlled object (i.e., the bench test and the real vehicle test). However, for some controlled objects with high costs and certain risks, the virtual controlled objects used before the vehicle test can achieve the purpose of reducing development costs and ensuring the safety of the development process.

8.3.1 System Composition

The HIL algorithm test platform shown in Fig. 8.16 [10] is mainly composed of an Arbin BT2000 charge and discharge machine, thermal chamber, DC power supply, host computer, BMS (BCU and BMC), and the tested objects. Because the parameter of the battery is easy to obtain, the platform can use the actual cell as the tested object.

The tested cell that placed in the thermal chamber described in Chap. 2 is charged and discharged by the Arbin BT2000 through the CAN bus. BMC collects the voltage, temperature, and current information of the cell and sends to the BCU for data processing. The algorithm needed to evaluate is downloaded to BCU, and the algorithm results are sent to the host computer through CAN bus and displayed in real time. BMC and BCU are powered by DC power [11–18].

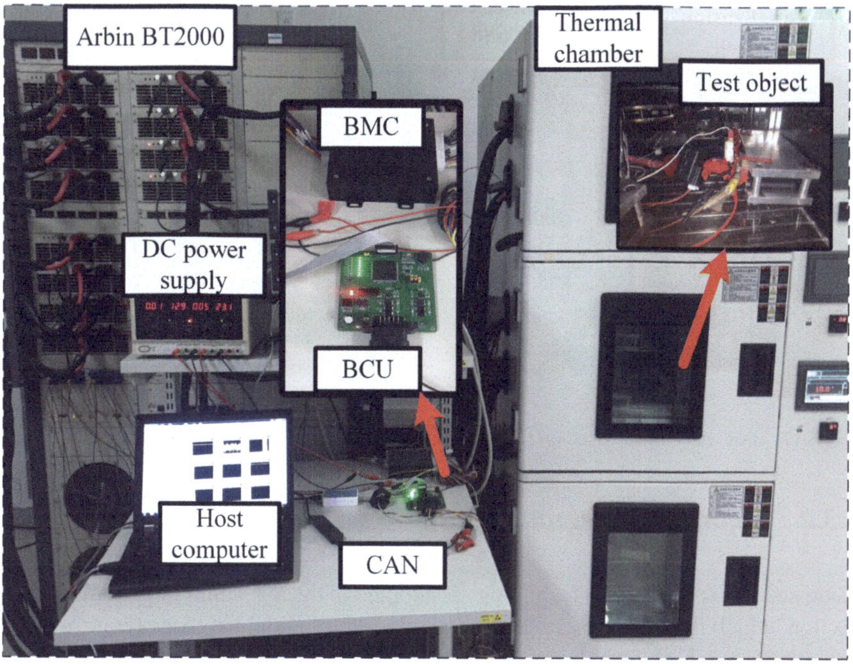

Fig. 8.16 Hardware in the loop algorithm test platform

BCU is developed based on the MPC5644A chip [19], which can realize real-time calculation, system communication, fault diagnosis and so on. The main characteristics of the chip are as follows:

① Power Architecture core with enhanced digital signal processing.
② Up to 150 MHz operating frequency.
③ Up to 4 MB flash memory.
④ 8 KB cache.
⑤ Up to 192 KB static random access memory.
⑥ 3 FlexCAN communication controller modules, up to 64-bit data buffer.

The developed BMC based on LTC6804-1 chip is shown in Fig. 8.17, and its main characteristics are as follows [20]:

① 12 voltage acquisition channels, the error at room temperature is within 2 mV.
② 12 temperature acquisition channels, the error is within 1 °C.
③ 100 ms sampling period.
④ 1 CAN communication channel and 2 SCI communication channels.
⑤ 4-bit coding switch channel for ID coding.
⑥ Power by either external power supply or battery pack.
⑦ Passive equalization and common fault diagnosis.

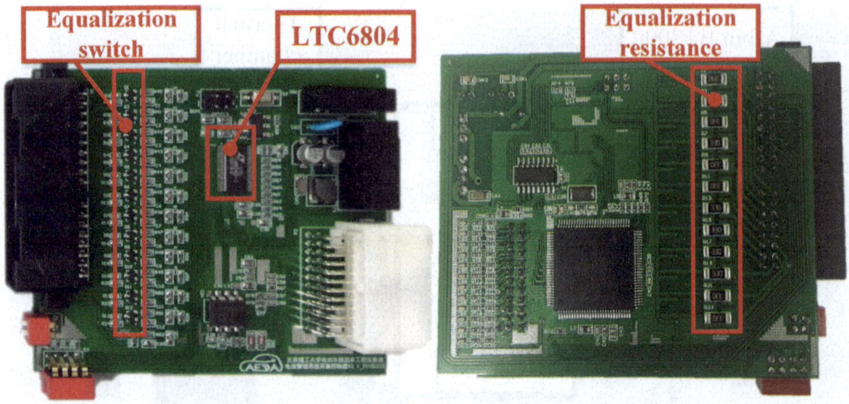

Fig. 8.17 BMC developed based on LTC6804-1 chip

8.3.2 Algorithm Integration

Code automatic generation technology is adopted in the algorithm integration in this section, which is the process of converting algorithm model language (control strategy model) into high-level language (C language, etc.) by developing software. Only an advance transcoding configuration is required to achieve the automatic conversion of the code. In fact, the rapid prototyping simulation process also adopts the code automatic generation method, but the generated code is only applicable to specific virtual controllers, while the automatically generated code in this section can be directly applied to the actual controller.

Compared with traditional manual programming, code automatic generation technology has the following three advantages:

① It frees developers from tedious programming tasks. The developers do not need to master high-level language code programming technology and can focus on the formulation of control strategy, thus reducing the requirements for developers.
② It saves the time of manual coding, greatly improving the development efficiency and shortening the development time of core algorithms.
③ The code automatic generation technology can greatly reduce the error rate in the code generation process, ensure the quality of the code and improve the development efficiency.

The general process of RTW code automatic generation technology is shown in Fig. 8.18. First, the Simulink model is compiled into the rtw file by the RTW compile command, and the rtw file is converted into a series of C files, h files, etc., then the target language compiler (TLC) is used to generate the source code of the model. The generated source code can be applied in two ways. One is using the templates provided by Simulink to automatically generate a Makefile, which can be used to compile the source code generated by the link to obtain the target file. The other is adding the generated source code to the engineering project of the

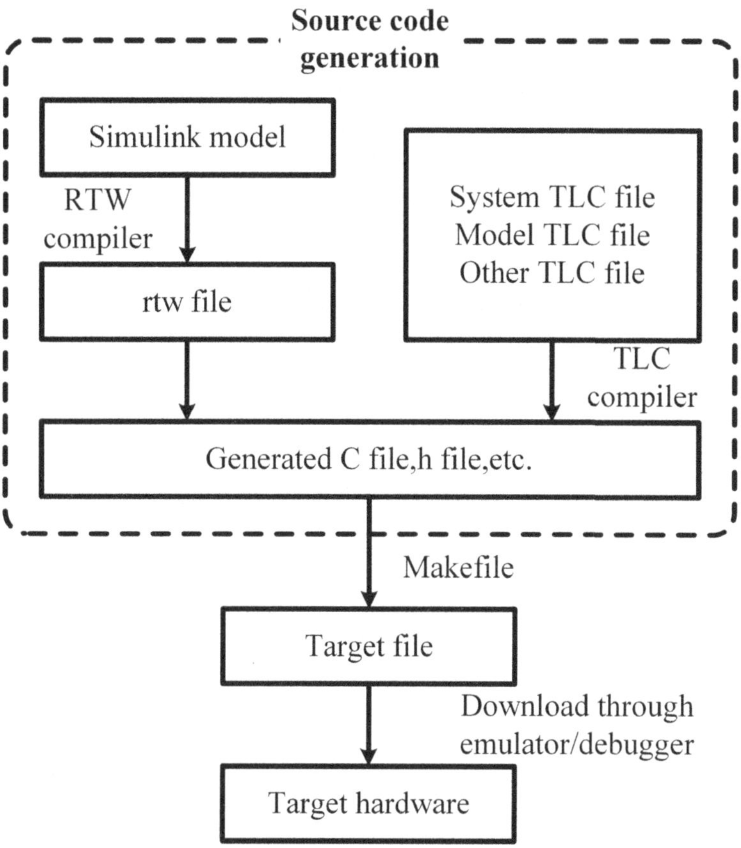

Fig. 8.18 Code generation process

integrated development environment (IDE) used by the target chip. The target file is also obtained by the IDE compilation link. Finally, the target file is downloaded to the target hardware through the emulator or debugger [15, 18].

The following basic setup is required when code is automatically generated by the Simulink model:

① Configure the model solver. Since the RTW can only generate code from a fixed-step model, the solver needs to be configured with a fixed-step size. In the code generation process customized for the target chip, the step size should be equal to the program running period of the actual controller. Since the calculation period of the algorithm in this section is 1 s, the fixed-step size is set to 1 s as shown in Fig. 8.19. For different algorithms and controllers, different solvers and step sizes are set.

② Hardware implementation setup. The hardware manufacturer and model corresponding to the configuration code, and the size of the memory space occupied by the code variables are shown in Fig. 8.20.

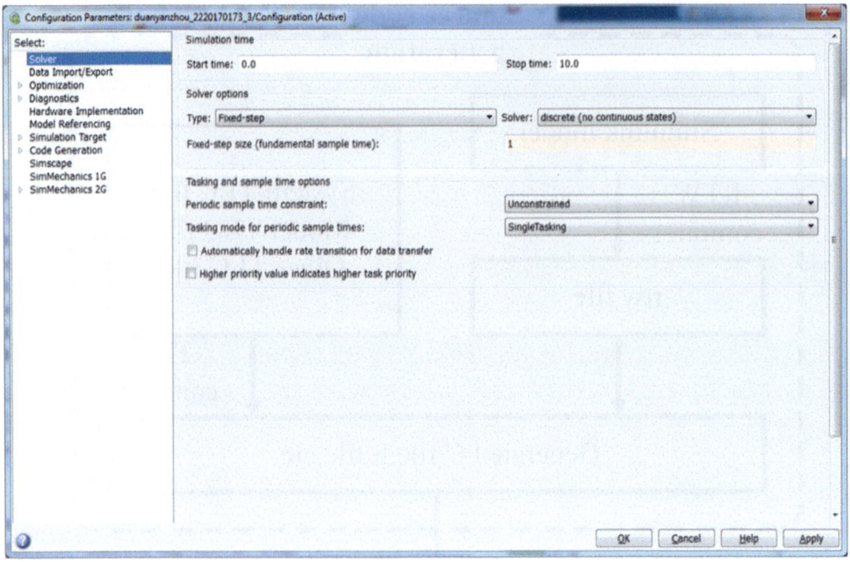

Fig. 8.19 Model solver setup

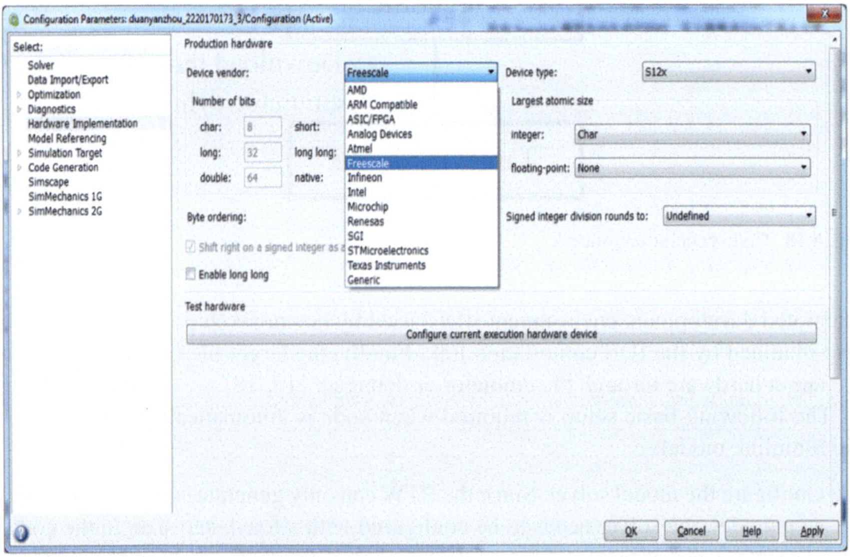

Fig. 8.20 Hardware implementation setup

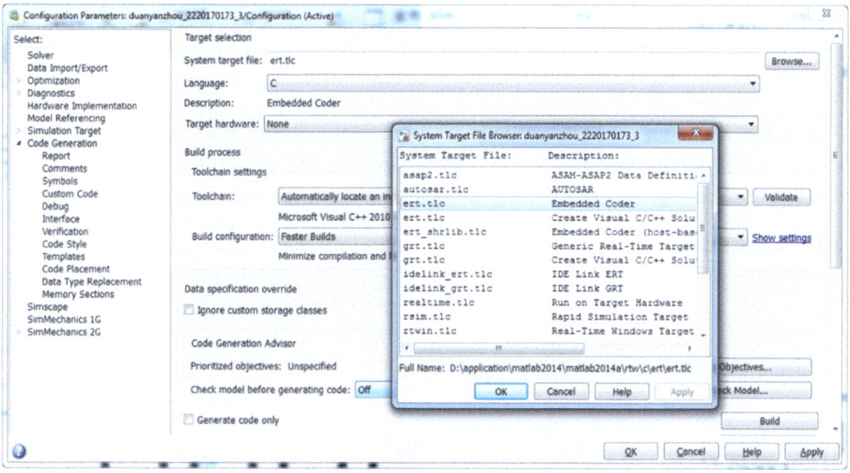

Fig. 8.21 System target file setup

③ System target file setup. As shown in Fig. 8.21, the system target file for the code generation process is configured under the code generation tab. The ert.tlc file is a system target file provided by Embedded Coder that can generate special C code for the embedded system.

Finally, the algorithm source code is generated automatically, which can be modified and embedded appropriately according to the underlying program framework of the controller to realize the running of the automatic generated code.

8.3.3 Testing and Evaluation

The battery model and the estimation and evaluation of SOC/SOH

The battery test samples of this case study are the battery 9 cell 1, cell 2, and cell 3. The experiments include the UDDS-NEDC-CTCDC hybrid stress tests and convergence ability tests of the cell. The UDDS-NEDC-CTCDC hybrid stress tests consist of UDDS,NEDC and CTCDC tests to evaluate the MHIF algorithm with unknown available capacity of the cell. All cells without aging cycle are placed in the thermal chamber (the temperature set at 25 °C). The standard capacity test is carried out before the experiments, and the CCCV is used in the final stage to charge and rest for 1 h. The initial SOC error of the algorithm and the capacity are set to 70% and 2%, respectively. The lower cutoff voltage of the battery is taken as the experimental termination condition of the above mentioned hybrid tests. After the experiment, the estimated results of SOC and capacity are statistically analyzed, and the results based on the ampere hour integral are taken as reference in the corresponding state estimation to calculate the corresponding errors.

The convergence ability test of the algorithm is conducted with uncertain initial SOC values. The cell is placed in 25 °C thermal chamber and evaluated in three cases.

① The reference initial SOC value is 100%, and the algorithm initial value is 70%.
② The reference initial SOC value is 80%, and the algorithm initial value is 50%.
③ The reference initial SOC value is 60%, and the algorithm initial value is 30%.

The algorithm evaluation results are shown in Fig. 8.22. Figure 8.22a, b show that the absolute value of estimation error of SOC based on MHIF algorithm is less than 1%, and the relative errors of capacity and terminal voltage are less than 0.5%.

The convergence time of MHIF is shown in Fig. 8.23. The convergence time of the estimated SOC based on MHIF algorithm is less than 60 s at the reference initial

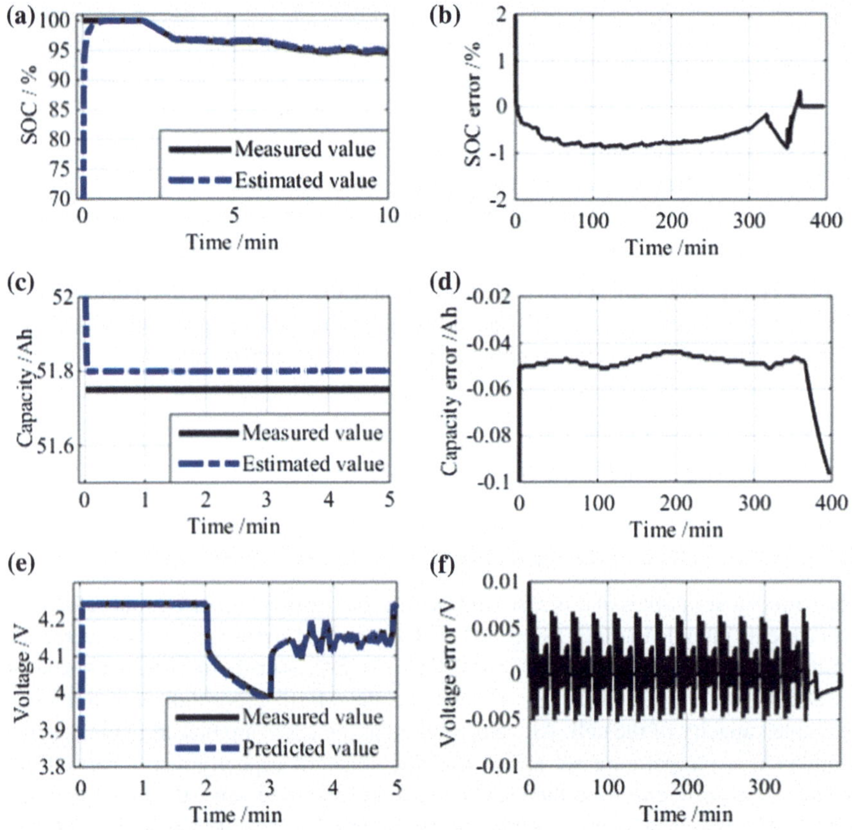

Fig. 8.22 Estimation results and errors of MHIF algorithm: **a** the comparison between measured and estimated SOCs (part); **b** SOC error; **c** the comparison between measured and estimated capacities (part); **d** capacity error; **e** the comparison between predicted and measured voltages (part); **f** voltage error

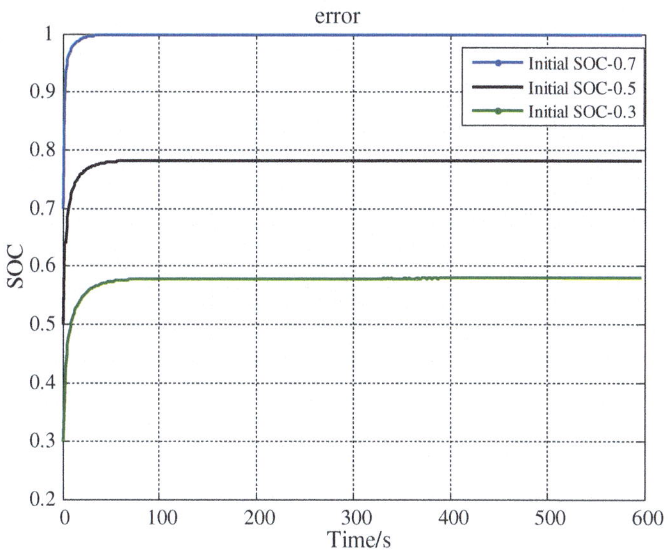

Fig. 8.23 Convergence results of the MHIF algorithm with different initial SOC values

SOCs of 100, 80, and 60%, whereas the algorithm initial SOCs of 70%, 50%, and 30%, respectively.

Evaluation of residual life prediction algorithm for the battery

Based on the actual BMS controller, the battery data is sent to the controller by the host computer for real-time calculation, and the RUL prediction algorithm at different current rates and temperatures is evaluated and analyzed in this case. The HIL evaluation results of the fitting-based battery RUL prediction method (adopting linear model) of partial historical data are shown in Fig. 8.24, where the length of the historical data used in the matching process is 100. It is demonstrated that the results of the actual controller based on the BMS are consistent with the MATLAB simulation results, and the RUL estimation error is less than two cycles [21, 22].

8.4 Vehicle Experiment Verification

The vehicle experiment is to apply the core algorithms in the BMS evaluated by the simulation platform to the vehicle. The purpose of the vehicle experiment is to evaluate the stability and reliability of the core algorithms by fully testing all possible vehicle driving conditions, correspondingly debug the core algorithms according to the test results, and finally make the BMS and its core algorithms meet the specifications. The vehicle experiment is the last step in the model-based "V" development process, which is also the key step before the vehicle products are put into the market.

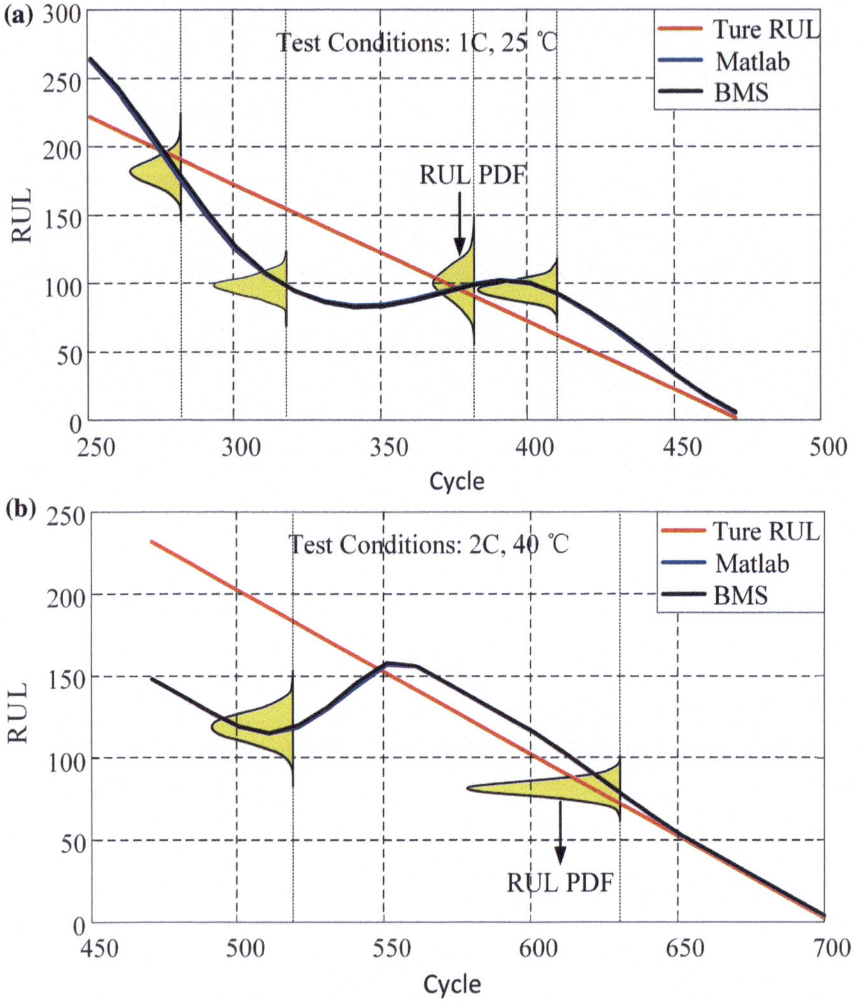

Fig. 8.24 RUL prediction results of HIL evaluation based on moving window method

Generally, the vehicle experiment can be divided into two stages: drum bench test and road test.

8.4.1 Drum Bench Test

In the process of the BMS core algorithm evaluation, a series of experiments can be carried out on the drum test bench in order to fully test the estimation effect of the core algorithms under various typical vehicle driving conditions. The test site of the core

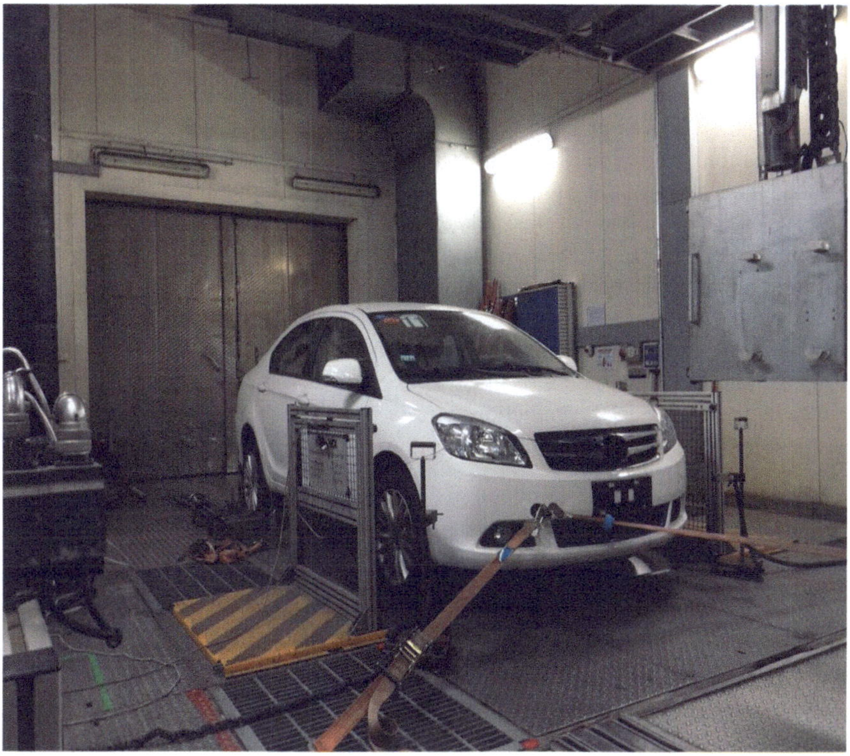

Fig. 8.25 Test bench of the core algorithms

algorithms is shown in Fig. 8.25. The drum test can also be carried out in a special environment. By keeping the ambient temperature within a certain range during the experiment, the testers can evaluate the application of the BMS core algorithms and relevant control strategies at different ambient temperatures.

8.4.2 Road Test

When the vehicle is running on the road, the driving conditions are affected by various factors such as road conditions, traffic conditions and environmental factors, whose randomness is greater than that in the laboratory test environment. It is necessary to further evaluate the performance of the core algorithms of the BMS in road conditions. After the road tests and some adjustments, the entire evaluation process of the algorithms has been completed before putting into practical application.

8.5 Conclusion

This chapter first compares the traditional algorithm development process with the model-based "V" development process. Then, the construction and application methods of the rapid prototyping simulation system based on xPC Target are described in detail. Two cases of the SOC estimation algorithms with known available static capacity and co-estimation of SOC and capacity are given. The results show that the co-estimation of SOC and SOH method can obtain better estimation results. After that, this chapter describes the composition of HIL simulation test platform and specific steps of the HIL simulation test and evaluation. The application effect of MHIF in actual BMS is further verified with the test platform. The results show that the absolute value of SOC estimation error is less than 1%. Finally, the HIL bench test results show that the RUL prediction error is less than two cycles, manifesting an excellent application prospect of the algorithm.

References

1. Sun H, Gu L, Tong X (2014) Development of a hardware-in-loop simulation test bench for electric drives. Drive Syst Tech 28(4):19–26
2. Chin CS, Lum SH (2011) Rapid modeling and control systems prototyping of a marine robotic vehicle with model uncertainties using xPC Target system. Ocean Eng 38(17):2128–2141
3. Zhang Z, Zhou M (2011) Experimental platform design on pure electric vehicle based on dSPACE. Equip Manuf Technol 12:115–117
4. Xiong R, Duan Y (2018) Development and verification of the equilibrium strategy for batteries in electric vehicles. J Beijing Inst Technol 27(1):22–28
5. Mu H (2016) Research on robustness state of charge estimation of lithium-ion power battery for electric vehicles. PhD Dissertation. Beijing Institute of Technology, Beijing
6. Wang C (2018) Modeling and energy management strategy for vehicular hybrid energy storage system. PhD Dissertation. Beijing Institute of Technology, Beijing
7. Cao J (2018) Research on configuration analysis, parameter matching and optimal control for hybrid power system in electric vehicles. MA thesis. Beijing Institute of Technology, Beijing
8. Chen C, Xiong R, Shen W (2018) A lithium-ion battery-in-the-loop approach to test and validate multi-scale dual h infinity filters for state of charge and capacity estimation. IEEE Trans Power Electron 33(1):332–342
9. Wu C, Wei X, Dai H (2014) Battery management system fault diagnosis test based on hardware-in-loop. Meas Control Technol 33(3):25–28
10. Wang J, Xiong R, Li L, Fang Y (2018) A comparative analysis and validation for double-filters-based state of charge estimators using battery-in-the-loop approach. Appl Energy 229:648–659
11. He H, Xiong R, Peng J (2016) Real-time estimation of battery state-of-charge with unscented Kalman filter and RTOS μCOS-II platform. Appl Energy 162:1410–1418
12. Xiong R, Cao J, Yu Q (2018) Reinforcement learning-based real-time power management for hybrid energy storage system in the plug-in hybrid electric vehicle. Appl Energy 211(5):538–548
13. Xiong R, He H, Sun F, Zhao K (2012) Online estimation of peak power capability of Li-Ion batteries in electric vehicles by a hardware-in-loop approach. Energies 5(5):1455–1469
14. Wang C, Xiong R, He H, Shen Zhang YWX (2019) Comparison of decomposition levels for wavelet transform based energy management in a plug-in hybrid electric vehicle. J Clean Prod 210:1085–1097

15. Tian J, Xiong R, Shen W, Wang J (2019) Frequency and time domain modelling and online state of charge monitoring for ultra-capacitors. Energy 176:874–887
16. Wang C, Huang B, Xu WN (2018) An integrated energy management strategy with parameter match method for plug-in hybrid electric vehicles. IEEE Access 6:62204–62214
17. Wang C, Xiong R, He H, Ding X, Shen WX (2016) Efficiency analysis of a bidirectional DC/DC converter in a hybrid energy storage system for plug-in hybrid electric vehicles. Appl Energy 183:612–622
18. Xiong R, Tian J, Shen W, Sun F (2018) A novel fractional order model for state of charge estimation in lithium ion batteries. IEEE Trans Veh Technol 68(5):4130–4139
19. MPC5644A Microcontroller Datasheet, American:Freescale Semiconductor
20. LTC6804-1 datasheet, American: Linear Technology Corporation
21. Zhang Y (2018) State of health identification and remaining useful life prediction of lithium-ion batteries for electric vehicles. Ph.D. Dissertation. Beijing: Beijing Institute of Technology
22. Xiong R, Zhang Y, Wang J (2019) Lithium-ion battery health prognosis based on a real battery management system used in electric vehicles. IEEE Trans Veh Technol 68(5):4110–4121

The manufacturer's authorised representative in the EU is Springer Nature Customer Service Centre GmbH, Europaplatz 3, 69115 Heidelberg, Germany. If you have any concerns regarding our products, please contact ProductSafety@springernature.com

Printed and bound by CPI Group (UK) Ltd, Croydon, CR0 4YY
30/09/2026
02225452-0011